Lecture Notes in Economics and Mathematical Systems

653

For further volumes:
http://www.springer.com/series/300

Philipp Servatius

Network Economics and the Allocation of Savings

A Model of Peering in the Voice-over-IP Telecommunications Market

Philipp Servatius
Université de Fribourg
Department of Quantitative Economics
Bd. de Pérolles 90
1700 Fribourg
Switzerland
philipp.servatius@unifr.ch

ISSN 0075-8442
ISBN 978-3-642-21095-2 e-ISBN 978-3-642-21096-9
DOI 10.1007/978-3-642-21096-9
Springer Heidelberg Dordrecht London New York

Library of Congress Control Number: 2011936793

Cover design: eStudio Calamar S.L.

Printed on acid-free paper

Springer is part of Springer Science+Business Media (www.springer.com)

In loving memory of Poldi Reinl
22.11.1977 – 28.12.2007

Preface

The author would like to express the utmost gratitude to everybody who has contributed to this thesis in one way or another. As there are certainly many more such individuals than I could reasonably accommodate here, I list only the most immediate ones.

First and foremost, I would like to thank my mother, Brigitte Servatius. She has always offered her advice and supported my endeavors, most importantly by letting all final decisions rest with me. Growing up (though some challenge this is indeed accomplished) she gave me all the of freedom I desired and with it the greatest of all goods: The confidence that I would choose to do the right thing, or, as often as this could not be determined a priori, avoid obviously wrong choices. Looking back, I could not have wanted much more.

Next, I want to thank all my of close friends. Countless hours spent in their company pushed my productivity and endowed me with the necessary energy to endure. Names shall not be necessary, you know who you are.

Also, I am very grateful for the support of my colleagues at the various institutions that contributed to my doctoral education. In particular I would like to mention Dr. Kazuhiko "Shiofuki" Kakamu, with whom I not only shared an office at the Institute for Advanced Studies in Vienna but also encountered many challenges there. No less Dr. Barbara von Schnurbein, née Styczynska, my colleague at the Chair of Microeconomics in Fribourg; to her I am indebted for many a constructive criticism and for an atmosphere in our office that was second to none. Also Danielle Martin, then assistant at the Chair of Statistics, who has occasionally let me take advantage of her superb math skills should not go unmentioned. The same is true for Bobo, Suti and Yolanda.

This dissertation was finalized while visiting ECARES at the Université Libre de Bruxelles in Belgium, a stay financed by the Swiss National Science Foundation. To the latter I am most grateful for the generous scholarship. For the convivial atmosphere that welcomed me at ECARES, I would like to thank Prof. Georg Kirchsteiger, who invited me to visit, and all of my colleagues there.

Last but certainly not least, my advisors deserve more than just praise: It is hard to describe what I owe to my first advisor, Prof. Reiner Wolff. Because he did not restrain my research agenda in any way, my curiosity was fueled to a point that lead to something probably neither of us had imagined when I started in Fribourg. This was certainly facilitated by the fact that I had never been overburdened with administrative work (far from it to be honest). I am especially grateful for his painstaking and meticulous revision of this dissertation's first draft. It could hardly have been more detailed. My second advisor, Prof. Hans Wolfgang Brachinger, also deserves my gratitude and a good deal more. He never declined any of my requests and supported me on all academic and extracurricular fronts, especially when it mattered most to me.

But despite all of the helpful input I have received, some errors are bound to remain in this document. Needless to say, the responsibility for these is entirely mine.

Fribourg & Zurich
April 2011

Philipp M. Servatius

Contents

List of Figures

List of Tables

Chapter 1
Motivation and Nontechnical Overview

The telecommunications market is one of the most fascinating witnesses of technological progress. Three decades ago "telecommunications" would refer almost exclusively to fixed-line telephony. And international, let alone intercontinental calls were exotic, mainly due to their high prices.

By now, telecommunications encompasses a wide range of possibilities allowing human beings to communicate over a distance on the basis of many different services other than fixed-line telephony. But telecommunications are not limited to human interaction. Various kinds of computer systems exist whose interaction through a telecommunications network is at least not directly initiated by a human action. Fixed-line telephony is barely identified with the expression telecommunications anymore, and the distances covered are no longer pivotal, if relevant at all in some areas, for the pricing of services.[1]

As such, the telecommunications industry is granted a dual role: First, it naturally constitutes an economic sector of its own. According to Sarrocco and Ypsilanti (2007), communication goods and services are the most rapidly growing item of household consumption in OECD countries. But more importantly, telecommunications support and even give rise to much economic activity in other sectors, with a far greater impact. What The Economist (2010) describes in a special report on information management, titled "Data, data everywhere," crucially hinges on the existence of telecommunications networks. This second role is hard to delineate in its entirety, exactly because of telecommunications' omnipresence in the economy.

Many of the telecommunications services nowadays revolve around the internet, which experienced a massive expansion over the last two decades and by far surpasses the telephone network in data transfer capacity. With the expansion came a tendency to duplicate at least parts of the network structure already present from telephony in order to deliver the new services to the customers. This tendency is now bound to reverse itself, as network operators aim to erect *next generation networks*,

[1] See Fransman (2003) for more on the evolution of the telecommunications industry.

P. Servatius, *Network Economics and the Allocation of Savings*, Lecture Notes in Economics and Mathematical Systems 653, DOI 10.1007/978-3-642-21096-9_1,

which allow to deliver an unspecified menu of services over but a single architecture. The cost advantages are tremendous, as the marginal cost of a service, given the network, is near zero.

Coming back to fixed-line telephony, this is a service that is or will be offered on the next generation networks as well. To the customer, there is little if any difference, but the underlying technology has changed fundamentally: From what used to require a dedicated network, fixed-line telephony is a mere by-product, in terms of both traffic volume and revenue, on those unified network structures. In this context it is called *Voice over IP* telephony (short *VoIP*), *IP* referring to the so-called *Internet protocol* which is used to route data in next generation networks. As the switch to these new networks has only commenced and is far from completed, it is not only the case that both types of phone lines coexist, but the providers of the new kind, we call them VoIP firms, also are not yet comprehensively interconnected among each other. Their networks are, they even connect to the legacy networks, but the services delivering VoIP telephony generally are not. This gives rise to the *re-routing problem*, pivotal to this dissertation: Due to the lack of services interconnection among the VoIP firms, their range of sight is limited to their own customers. When one of them calls a number assigned to a customer of another VoIP firm, the firms are unaware that the call is initiated and terminated on IP-based infrastructure and so it is routed via the legacy telephone network. Had both firms, though, signed a mutual peering agreement, the call would remain entirely on next generation network architecture and costs for transit as well as for termination could be reduced or even eliminated entirely.

Being exactly at the heart of this study, we want to shed some light on those peering relationships, or rather the lack thereof, and on what national regulators could do to create incentives for peering. This is to be achieved by means of economic models from the fields of game and network theory.

The relevance of VoIP telephony is quickly established on the basis of the large growth of its subscribers. In Switzerland alone, the Federal Office of Communication (OFCOM) accounts for 467 874 VoIP subscribers out of a total of 4 704 497 fixed-line telephony contracts for 2009, up from almost none 4 years earlier. The statistics for many other industrialized countries show similar developments. They are even expected to jump significantly, as soon as the formerly state-owned network operators, who still tend to have by far the largest numbers of subscribers, switch entirely to next generation networks. But even without these changes, the growth is nevertheless impressive. Also, a multitude of recent studies focussing on the transition to IP-based telecommunications networks (and some even straight out on VoIP telephony) has been brought forward by national regulators as well as international organisations like the EU, the OECD, or the World Bank. This not only highlights the importance of these inevitable developments, but also shows that there are difficulties to be resolved on the way.

We now provide the reader with a nontechnical overview of what is to come. This work is subdivided into two parts, where the first serves to introduce selected theoretic concepts, and the second contains their application in a telecommunications setting.

Part I is composed of two chapters, one on the theory of games, and one on network theory in the context of economics. The chapter on game theory covers parts of both its major branches, cooperative and noncooperative games, but the emphasis is on the former. The selection of all content is mostly based on what is required for the application in Part II, but some matters are included to give a more complete picture and especially to facilitate a more thorough understanding.

We begin Chap. 2 with an exposition on the theory of noncooperative games. The strategic form game representation is introduced first. It is a concept modelling the strategic interaction between the players of a game so that all participants choose their plan of action, called strategy, simultaneously. Which strategy to select optimally is, thereafter, the subject of selected equilibrium concepts we present. Most notable is the Nash equilibrium and some of its refinements. In such an equilibrium, no player has an incentive to deviate to another strategy, because he could not gain anything by doing so. The refinements extend this notion to concerted deviations by groups of players.

The second part on noncooperative games covers their representation in extensive form, which models games as a sequential process, i.e. one where the players choose their moves one after another. We introduce another variant of the Nash equilibrium, its *subgame perfect* form, which extends the Nash equilibrium concept to so-called subgames of the original game.

Next in Chap. 2 comes the consideration of cooperative games. We first define the notion of a cooperative game, in which players can make binding commitments and achieve a surplus through cooperation within a coalition. A *characteristic function* assigns the corresponding value to each possible coalition of players. Depending on whether there even exist overall gains from cooperation and how these gains specifically arise and change within and over different coalitions, we can classify cooperative games and assign certain properties to them.

Finalizing the section on cooperative games, we draw on the much-cited example of the *Tennessee Valley Authority*: In this example, it is shown how the joint realization of three projects (improved navigation on the Tennessee river, flood control, and the generation of electricity, each through dams) can cut costs significantly as compared to their independent completion.

The rest of Chap. 2 is devoted to some of the most prominent solution concepts for cooperative games. These concepts determine how the gains from cooperation are or can be distributed among the players. We call such distributions allocations as soon as they make no player worse off than without participating in cooperation. The solutions are best distinguished by their size, i.e. whether they provide a set of allocations or only a unique value. We treat them in this order.

Before the most influential solution concept, the *core*, is introduced, the related *von Neumann Morgenstern solution*, the predecessor is being derived. It is based on a notion of domination and compares allocations to ensure that the outcome contains no two allocations violating this domination. The concept does not appear in the subsequent application, but is very instructive.

Next, the solution called *core* contains all allocations which are efficient and cannot be improved upon by any coalition of players. "Efficient" refers to the fact that the sum of the gains up for distribution stems from cooperation of all players and is exhausted. "No improvement" states that no coalition is allocated less than what it could achieve on its own. We show different conditions under which, regarding the underlying game, a solution in terms of the core exists.

The section on single-valued solutions, also called allocation rules, starts with a general definition and some common properties of these rules and how the latter respond to changes to the game they are based on. The first allocation rule we cover is the *Shapley value*, characterized by three straightforward axioms: Again *efficiency*, allowing for no waste of the overall gains in distribution, *symmetry*, stating that players are served on the basis of their characteristics and not by their labels, and finally, *additivity*, which allows Shapley allocations over different games to be added, if these games are merged. The axioms lead to individual allocations that reflect the average marginal values of a given player joining a coalition. Because the Shapley value is computed according to a specific formula, its existence is assured.

In case these marginal values do not reflect the cooperation in a way considered proper in certain situations, the *Weighted Shapley value*, next in our exposition, can be applied. It allows for a shift within the allocation accounting also for a system of weights assigned to the players. Interestingly, the Weighted Shapley value can be calibrated to yield any solution that is also an element of the core.

The chapter concludes with a brief treatment of the so-called bargaining problem and corresponding bargaining solutions of Nash and Kalai–Smorodinsky. With this we include an alternative approach to allocate the gains from cooperation among a number of players.

In Chap. 3 we turn to network theory in the context of economics. We begin with the basic notions from graph theory to describe a network in more detail before we turn to *communication situations*, which in essence are cooperative games that incorporate a network structure. The latter governs how value can be realized through cooperation among the players. We also extend allocation rules to the network context and list some properties they now can exhibit. In conjunction, we present the *Myerson value*, the first allocation rule to incorporate network structures. One can characterize it by means of two concise axioms: *Component additivity*, which does not allow the allocation of value to an isolated part of the network in excess of what this part can create on its own, and *fairness*, according to which any two players' change in payoffs is identical after the formation or deletion of a link between them.

The subject of the last section of Chap. 3 is the formation of networks. It is modelled along the lines of noncooperative games, in both strategic and extensive form, and also relies on the equilibrium notions introduced there. The players correspond to the nodes of the network, and the strategies reflect the willingness of a player to form links with others. Here, we distinguish between bilateral and unilateral link formation, meaning whether the consent of both players or merely of one player is needed in the formation of a link. The weight lies on the former concept.

Last, we include notions of network stability. These are less concerned with the process of network formation, but more with its result. How do changes to given networks influence the payoffs of the players and their consequent incentive to add or delete certain links. Stability notions are closely related to Nash equilibria in that we differentiate between deviations of individual players and concerted deviations of coalitions of players. This concludes Part I of this work.

In Part II, the application of the theory is paramount. The part is subdivided into four chapters, the first of which contains a short introduction to telecommunications. The second chapter, on the basis of a communication situation, offers a model of interaction among VoIP firms, where peering results in possible savings. The third is focussed on the signing of peering agreements between the VoIP firms; its analysis is based on network formation models. The last chapter concludes.

In more detail, Chap. 4 serves as a nontechnical refresher on several topics of telecommunications relevant to us. It is meant to provide a basic understanding. We briefly cover the architecture of both classic fixed-line telephone networks and the internet, as well as their overlap and convergence into next generation networks. We also summarize the structural developments on the telecommunications sector over the last decades, resulting mainly from its liberalization and consequent regulation. There, we concentrate on the interconnection of networks and its pricing. One consequence of the above mentioned liberalization is the emergence of Voice over IP telephony, on which we also dwell, especially for a working definition and in order to classify the providers of such services. With this in mind, we present the re-routing problem, which is at the heart of this work's application. References on related research issues complete the chapter.

In Chap. 5, we are mainly concerned with establishing the *peering game* and the resulting possibilities for allocation. We begin with the environment in which the model is situated and the assumptions imposed. On a market where a finite number of VoIP firms operate, we are not concerned with the maximization of their profits or revenues via prices and possibly related effects on the size of their customer base, but the ability to realize savings from peering with one another. To this respect, we construct a characteristic function which incorporates a network, the links of which represent the peering relationships between the firms. This communication situation is what we baptized the peering game above. Before we go on to the allocation of the realized savings, we inspect the game for desirable properties that facilitate intuitive interpretation or possibly even guarantee the existence of a solution. We highlight the fact that the specific form of the game allows for a further decomposition into player-attributable contributions in addition to merely the common marginal value of cooperation.

In terms of allocations, we emphasize the two components of the peering game: A zero-sum game over the access charges exchanged between the firms, and the savings game regarding the gains from long distance transit fees which cease to apply. We first analyze the (nonempty) core of the game, and its downsized variant, the least-core. We continue with allocation rules, such as the Shapley value, its weighted cousin, and the Myerson value. The final methods of allocation are given by the (weighted) bargaining solutions of Nash and Kalai–Smorodinsky.

The chapter is closed with a comparison of the types of allocations resulting from the different concepts and their applicability in practice.

The subject of Chap. 6 is the network underlying the peering game, or rather its formation. To that end, we elaborate on the functional form of the allocation procedure we apply in this chapter, as well as on its properties. For the first time, we explicitly incorporate a correction for access revenues arising from the zero-sum component of the peering game.

Our first scenario of network formation is bilateral, where firms are free to peer with one another as they deem. In this setting, both the strategic and the extensive form of the network formation game and the corresponding equilibrium notions are applied. Also, network structures are checked for stability.

The second scenario is one of central peering, where members to a peering instance peer unconditionally with one another. Membership is open to all players, but excludes the possibility of selective peering agreements, which adds some flair of onesided or unilateral link formation. We again go through the strategic and extensive form games of network formation and check for stability. The chapter is concluded with implications for national regulators in terms of access fee design; we present an access fee regime leading to an environment that fosters peering among as many VoIP firms as possible.

To summarize, the objective of this study is threefold: The main goal is to establish a model reflecting the interaction among VoIP firms regarding their cooperation through peering and the resulting savings. Based on this model arises the second goal, an analysis of the conditions under which VoIP firms agree to peer with one another in the first place. It aims to serve as a basis for policy advice to national regulators to construct a structure of access fees for the interconnection of VoIP telephony. The latter is still covered by regular fixed-line regulation and hence the different cost structures in IP-based networks are not accounted for. The last goal is more a means to an end: We reproduce the economic theory necessary for the previously mentioned endeavors. In this we try to be as instructive as possible.

As far as the first goal is concerned, we believe to provide a thorough model, taking into account the most relevant and quantifiable factors on the gains of a VoIP firm that engages in peering. Incorporating network theory to model the peering relationships supports intuitive understanding and often simplifies calculations. Given the model, the second goal is also attainable: To analyze the conditions under which peering is deemed favorable by the market participants, even in various settings. As for the third goal, we do not yield any new theoretic results, but reexamine some that were previously established. We prove or define them using more basic analytical tools, hoping to make these concepts accessible to a larger audience.

At this point we also want to make mention of the mathematical appendix, starting from p. 273. It contains selected basics to serve as quick reference and to clarify some concepts which might appear slightly different in some other literature. The readers are encouraged to consult it before beginning with Part I. Also, on pp. 287 ff. we provide an overview of the notation used throughout.

Part I
Selected Theoretical Concepts

Chapter 2
The Theory of Games

2.1 On Game Theory

It is to a certain extent remarkable when two Nobel laureates in economics suggest that *interactive decision theory* might be a more apt name for what has become known as *game theory*.[1] And indeed, the term game theory does not – for the layman – convey much about the theory or its contents at all; it even suggests a recreational context, as the expression "game" predominantly reminds of parlor games.

Generally, situations analyzed by the theory of games can be described as follows: To begin with, they are *strategic*, meaning the agents', or *players*', decisions are *interdependent*. This is where the *interactive* character arises.[2] Also, the players are usually assumed to be *rational* in the sense that they are taking decisions coherent with their objective such as to maximize their own utility or payoff. These decisions are referred to as *strategies* of the players and define the possibilities of their actions. Also, players are assumed to be *intelligent* to the point that they have the comprehensive information about the game situation at hand, or the *rules* of the game, possibly including what they do not know about it. From this, they derive which strategy is in their best interest. Now, with rational players, who want to "win" by choosing a certain strategy, given the rules of the situations, it is probably more clear, why the term *game theory* is not completely off the point.

The field of game theory is subdivided into two main branches, covering *cooperative* and *noncooperative* games. On the cooperative side, players are able to make binding and enforceable commitments, before the game is played. Here, the focus is on the cooperation agreement among some or all players and on the possible distribution of the aggregate gains from cooperation. The emphasis is not

[1] See Myerson (1991, p. 1) and Aumann (2008, p. 529).

[2] In contrast to single-player decisions taken, e.g., for consumer utility maximization, which depend on own preferences as well as the parameters for endowment and prices, but not on other consumers or their specific actions.

P. Servatius, *Network Economics and the Allocation of Savings*, Lecture Notes in Economics and Mathematical Systems 653, DOI 10.1007/978-3-642-21096-9_2,

on the strategy a player should choose in order to maximize his own payoff, as is the case on the noncooperative side. In addition, cooperative games can be subdivided further into games with *transferable* and games with *nontransferable* utility, also known as *TU* or *NTU* games. In the former case, the worth of an agreement, or a cooperation, between players is only specified in the aggregate and separated from possible ways of its distribution among them. In the latter case the worth is assigned individually, as integral part of the agreement. As far as cooperative games are concerned, this work is restricted to TU games. A more detailed consideration of these will follow in Sect. 2.3.

The noncooperative branch of strategic games distinguishes itself by the focus on the individual player's actions. Even if communication among the players is possible before the game is played, i.e. before decisions on actions are taken, no binding agreements can be formed, let alone enforced. This can lead to certain situations in which individual rationality will preclude an outcome that yields strictly higher payoffs to all players involved. The reason for this is the interdependence of the players' strategies.[3] The most popular example of this is known as the *Prisoner's Dilemma* and can be found in almost any economics textbook containing a chapter on game theory.[4]

Noncooperative games are generally represented in two very different yet related variants, i.e. in *strategic* or in *extensive form*. The strategic form, on the one hand, models the game in a way where the players choose their strategies, or plans of action, for all contingencies, once and for all. This, most importantly, happens simultaneous or at least in a way where no player has a priori knowledge about the choice of any other player. The extensive form, on the other hand, models a more sequential process: Different orders of events are possible and players choose an action whenever it is their turn to so, based on what has happened so far. Here, another distinction can arise, namely into games with *perfect* and *imperfect* information. In the latter case, players might not be fully informed about the history of actions taken by their predecessors and, consequently, about the exact outcome of their choice of action.

For both forms of representation there exist multiple *equilibrium* or *solution concepts*. The most influential are the so-called *Nash equilibrium* (see Nash 1951) and the refinements arising from it. Such concepts establish profiles of individual strategies that lead to outcomes, which are robust to payoff-improving deviations in strategies. In other words, no player can find it profitable to deviate from his strategy as assigned by the underlying equilibrium concept. The specific nature of such a deviation then depends on the equilibrium notion at hand and will be discussed for selected concepts below.

Generally speaking, the field of game theory has grown out of the discipline of mathematics. With the seminal work of von Neumann and Morgenstern (1947), it

---

[3]This very problem is ignored in cooperative games, which take a shortcut by assigning a value to a coalition of players. A derivation to do so from noncooperative games is provided on p. 28.

[4]See for example Varian (1999, p. 496 f.) or Mas-Colell et al. (1995, p. 238 f.).

has been established as a scientific discipline of its own. It is commonly applied to analyze situations in social sciences, most notably economics, but also appears in natural sciences, as for example (evolutionary) biology. For a concise, yet detailed overview of the history of the theory of strategic games we refer the reader to Aumann (2008). Also, a somewhat dated but still applicable classification is given by Lucas (1971, p. 494 f.).

We now turn to a more formal analysis of both branches of game theory for the rest of this chapter. We begin with a short treatment of noncooperative games and selected equilibrium concepts around the *Nash equilibrium*. Hereafter, the main emphasis is placed on a more detailed analysis of cooperative games, and especially on various solution concepts and allocation methods. In terms of set-valued solutions, we will investigate von Neumann Morgenstern stable sets and the Core, whereas treated point solutions, or allocation rules, will include introductions to the (Weighted) Shapley value and the Myerson value. The chapter concludes with bargaining solutions.

2.2 Noncooperative Games

2.2.1 *Overview*

We begin our formal treatment of the theory of noncooperative games by defining in turn the two most common forms of representation, the *strategic* and the *extensive form*. While the former is a framework in which all players act simultaneously, the latter is dynamic and players make their moves in a pre-imposed order. Even though both forms of representation appear to be quite distinct, their similarities and relation to one another will be touched upon. Alongside each, we present a number of selected solutions concepts which are either refinements of the basic *Nash equilibrium* or closely related to it. For general orientation we use Ritzberger (2003), from which we mostly adopt the notation, as well as Myerson (1991), Mas-Colell et al. (1995), and Jehle and Reny (2001). More specific references are given below.

2.2.2 *Strategic Form Games*

Having, informally, stated the interactive nature of a game above, we start this section with a more formal definition of a game in strategic form, or simply, a strategic game[5]:

[5]This definition can be found in any textbook on game theory, see for example Myerson (1991, p. 46), Ritzberger (2003, pp. 143 ff.) or Mas-Colell, Whinston, and Green (1995, p. 230).

Definition 2.1. A strategic form game Γ is a triplet

$$\Gamma = \Big(N, \mathsf{S}, \mathbf{u}\Big),$$

where N denotes the finite set of players, S is the set of all pure strategy combinations and $\mathbf{u}$ the payoff function for the n players of the game.

Elaborating on the components of Γ, we begin with S, or rather its most basic elements, the individual strategies of the n players: Given such an n-player game, denote a *pure strategy* of player i by s_i, and his *set of all pure strategies* by S_i.[6] In our context, we assume a finite number of pure strategies, where the cardinality of S_i, denoted $|\mathsf{S}_i|$, is finite for all n players which constitute the (finite) player set $N := \{1, 2, \ldots, n\}$. Consequently, the *set of all pure strategy combinations*, S, is the Cartesian product of the individual pure strategy sets:

$$\mathsf{S} = \times_{i=1}^{n} \mathsf{S}_i = \mathsf{S}_1 \times \mathsf{S}_2 \times \ldots \times \mathsf{S}_n. \tag{2.1}$$

A *pure strategy combination*, or *strategy profile*, given by $s \in \mathsf{S}$, contains for each player exactly one strategy: $s = (s_1, s_2, \ldots, s_n)$ with $s_i \in \mathsf{S}_i$ for all $i \in N$.

Now, let us turn to the payoff function $\mathbf{u}$. The *individual payoff function* of player i for pure strategy combinations, or simply pure strategies, is a mapping $u_i : \mathsf{S} \to \mathbb{R}$. Due to the interactive nature of a game, i's payoff depends not only on his own strategy s_i, but in general also on the strategy choices of the other $-i$ players he faces.[7] This interaction is expressed in the vector function $\mathbf{u} : \mathsf{S} \to \mathbb{R}^n$, where

$$u(s) = \Big(u_1(s), u_2(s), \ldots, u_n(s)\Big) \tag{2.2}$$

assigns each player his payoff under the pure strategy combination s.

Because it fits in well, we now also introduce the notions of mixed strategies and expected payoff functions.[8] A *mixed strategy* of a player i is a probability distribution σ_i over his set of pure strategies S_i. Each such mixed strategy σ_i can be represented by the corresponding probability function defined by

---

[6] A given strategy of a player is a mapping that assigns to each of his moves exactly one of the choices available at this move. It specifies an action for every possible eventuality a player might face during the course of the game. The breakdown in moves and actions will become more clear in the section on extensive form game representation.

[7] The somewhat sloppy notation "$-i$" stands for "all players but player i" and is commonly used in the field of game theory, even though formally unsound. Because the labelling of the players within the Cartesian product of the individual pure strategy spaces is arbitrary, we also often write a strategy profile in the form $s = (s_i, s_{-i})$ when we want to emphasize player i. If written correctly as i and $N \setminus \{i\}$, this notation can naturally be extended to coalitions of players as well.

[8] These are not required for our subsequent treatment, but play an important role when deriving the so-called characteristic function, the foundation of cooperative game theory, in Sect. 2.3.2.

$$\sigma_i : \mathsf{S}_i \to \mathbb{R}_+, \text{ satisfying } \sum_{s_i \in \mathsf{S}_i} \sigma_i(s_i) = 1.$$

The *mixed strategy space* of player i is denoted by Δ_i and can be thought of as the space of all probability distributions σ_i on his set of pure strategies S_i.[9] The *space of all mixed strategy combinations* Θ is the Cartesian product of the mixed strategy spaces:

$$\Theta = \times_{i=1}^{n} \Delta_i = \Delta_1 \times \Delta_2 \times \ldots \times \Delta_n. \tag{2.3}$$

Its elements, $\sigma \in \Theta$, are called *mixed strategy combinations*, where $\sigma = (\sigma_1, \sigma_2, \ldots, \sigma_n)$ lists a probability distribution for each player, over his corresponding set of pure strategies.[10] In this sense, a pure strategy s_i can be interpreted as a degenerate mixed strategy, if and only if all probability mass is concentrated on this single strategy $s_i \in \mathsf{S}_i$ such that $\sigma_i(s_i) = 1$. Given player i's choice of a strategy, his probability distribution is per se unconditional on the other players' strategy choices. Therefore, the probability of any pure strategy combination $s \in \mathsf{S}$ under a given mixed strategy combination $\sigma \in \Theta$ is calculated readily. It is denoted by $\sigma(s)$ and, because of the aforementioned independence of probabilities, is the product of the individual probabilities for pure strategy s_i under mixed strategy σ_i:

$$\sigma(s) = \prod_{i=1}^{n} \sigma_i(s_i). \tag{2.4}$$

This leads to what is called the *expected payoff function*, a mapping from the space of mixed strategy combinations to the n-fold Cartesian product of the real numbers, $U : \Theta \to \mathbb{R}^n$. It is the expected value of the payoff for a mixed strategy combination, according to the probabilities assigned to the pure strategy combinations. For a given $\sigma \in \Theta$, we write for the ith entry of U,

$$U_i(\sigma) = \sum_{s \in \mathsf{S}} \sigma(s)\, u_i(s) = \sum_{s \in \mathsf{S}} \prod_{j=1}^{n} \sigma_j(s_j) u_i(s) \;\text{ for all }\; i \in N. \tag{2.5}$$

Instead of summing over all $s \in \mathsf{S}$, a possible restriction to $s \in \operatorname{supp}(\sigma) \subseteq \mathsf{S}$ would yield the same result, because the support of σ, denoted $\operatorname{supp}(\sigma)$, contains only strategy profiles s where all n constituting individual strategies are played with strictly positive probability. As soon as there exists just one player i, for whom $\sigma_i(s_i) = 0$, the Cartesian product, $\sigma(s)$ assumes the value zero. Also, for

[9] Based on our assumption of a finite number of pure strategies, Δ_i can conveniently be represented by the unit simplex of dimension $|\mathsf{S}_i| - 1$. Then, each mixed strategy is an element of (or point within) this simplex.

[10] As Θ is the product space of n compact and convex simplices, it is itself a compact and convex subset of the Euclidean space with dimension $\sum_{i=1}^{n} |\mathsf{S}_i| - n$.

		Player 2	
		s_2	$\tilde{s}_2$
Player 1	s_1	$u_1(s_1,s_2),u_2(s_1,s_2)$	$u_1(s_1,\tilde{s}_2),u_2(s_1,\tilde{s}_2)$
	$\tilde{s}_1$	$u_1(\tilde{s}_1,s_2),u_2(\tilde{s}_1,s_2)$	$u_1(\tilde{s}_1,\tilde{s}_2),u_2(\tilde{s}_1,\tilde{s}_2)$

Fig. 2.1 Generic normal form (or bimatrix) representation of a game with two players and two strategies each

a degenerate mixed strategy combination σ, where all players assign probability one to a given pure strategy combination $s \in \mathsf{S}$, such that $\sigma_i(s_i) = 1$ for all $i \in N$, the vector given by the expected payoff functions coincides with that given by (2.2) :

$$U_i(\sigma) = u_i(s) \;\text{ for all }\; i \in N.$$

Having defined the ingredients of a game Γ, i.e. the strategies of the players and the resulting payoffs, we can now present an example for a basic 2-player game in strategic form. To facilitate working with strategies and payoffs, the *normal form* representation of a game is often used and so we briefly introduce it here. It provides a concise and intuitive representation of the information contained in games with no more than three players. In the 2-player case, the payoff-pair of each strategy combination is listed in a matrix whose dimensions correspond to the cardinalities of the respective set of pure strategies, e.g. $|\mathsf{S}_1| \times |\mathsf{S}_2|$. For illustrative purposes these matrices are usually given as tables and referred to as *bimatrix form*. An illustration for a 2-player game with $S_i = \{s_i, \tilde{s}_i\}$ for $i = 1,2$ is shown in Fig. 2.1. These tables can include mixed strategies as well as extensions to three (or even more) players, but with an increasing number of either, clarity is bound to suffer. One of the biggest advantages of the normal form representation is the comparable easy with which solutions, or equilibria (if they exist), can be found. The example to follow is taken from Mas-Colell et al. (1995, p. 273 f.) and will serve as a reference point for the treatment of further solution concepts, to which we turn thereafter.

Example 2.1. (Mas-Colell et al. (1995, p. 273 f.)) The following situation, known as the *market entry game*,[11] is reproduced through a simultaneous moves game in strategic form. There are two players, the so-far monopolistic incumbent, player 1, and a new entrant to the market, player 2. Both decide whether to fight (f),

[11]This so-called *market entry game* appears in many different variants throughout the literature. Most generally speaking, it describes a situation of strategic interaction between two players, with two choices of action for each. One player is the incumbent of a given market, the other a potential entrant. The latter considers entering the market or staying out of it, while the former can either accommodate the new entrant or try to deter entry into the market, albeit with costs to himself. Depending on the specific circumstances, the payoffs arising from the (four) possible outcomes usually reflect monopoly- or duopoly-style profits when the incumbent accommodates and zero profits or even losses whenever he fights the entrant.

		Player 2	
		f	a
Player 1	f	$-1, -3$	$-1, -2$
	a	$-2, 1$	$1, 3$

Fig. 2.2 Bimatrix form of the market entry game

for example via an expensive advertising campaign or aggressive pricing, or to accommodate (a) the other market participant. $S_i = \{f, a\}$ for $i = 1, 2$. The payoffs for all possible combinations of behavior, or all strategy combinations, are given by

$$\begin{aligned} u_1(f,f) &= -1 \quad & u_2(f,f) &= -3 \\ u_1(f,a) &= -1 \quad & u_2(f,a) &= -2 \\ u_1(a,f) &= -2 \quad & u_2(a,f) &= 1 \\ u_1(a,a) &= 1 \quad & u_2(a,a) &= 3. \end{aligned}$$

As economic interpretation, fighting is always more expensive (i.e. lowers the individual payoff further) than accommodating the other player. It is even more expensive when the opponent is accommodating (two vs. one unit of payoff reduction). Because it can be tedious to extract this information from such a listing, we employ the bimatrix form of representation. In our case, the dimensions of the resulting matrix are 2×2 and its elements correspond to the payoffs from each strategy combination. The left number in each cell in Fig. 2.2 stands for the payoff of player 1, the right one for that of player 2.

As we claimed above, the normal form representation provides a very intuitive and concise way of delivering all the information contained in the game Γ. In any case, it is commonly used for strategic form games, rather than just listing strategies and resulting payoffs.

We now proceed to introduce selected solution concepts for strategic form games. These make predictions on the final outcome based on the strategies which the players are going to choose in order to maximize their payoffs, given the choices of the other players. Without a lengthy and formal introduction, we jump straight to the most fundamental solution concept in the theory of noncooperative games, the *Nash equilibrium*, bearing the name of its creator, John F. Nash. It was introduced to the literature in Nash (1951) and will serve as starting point for all subsequent refinements. We skip the usual process of deriving this kind of equilibrium via so-called rationalizable strategies and fixed points of best-response correspondences on the set S. We employ as little technical details as possible and focus only on the concepts needed hereafter, instead of giving a comprehensive overview. We also omit conditions and proofs for the existence or uniqueness of an equilibrium.

For those, we refer the reader either to the original literature as indicated, or to a text book on noncooperative game theory, for example Myerson (1991), Ritzberger (2003), or the relevant chapters of Mas-Colell et al. (1995).

Intuitively, an equilibrium is a strategy profile $s \in \mathsf{S}$ leading to an outcome and a corresponding payoff vector, where no deviations can be expected to occur. What exactly qualifies as a deviation depends on the equilibrium notion at hand. The most basic deviation can be achieved by a single player. Depending on the refinement, other deviations by larger subsets of players are also considered. In any case, such equilibrium strategies can be referred to as self-enforcing within the refinement considered. We now state the definition of the original *Nash equilibrium*, which can also be found in Jehle and Reny (2001, p. 273):

Definition 2.2. For a game $\Gamma = (N, \mathsf{S}, \mathbf{u})$ in strategic form, a strategy $s^* \in \mathsf{S}$ constitutes a *Nash equilibrium* in pure strategies, if

$$u_i(s_i^*, s_{-i}^*) \geq u_i(s_i, s_{-i}^*)$$

holds for all players $i \in N$ and all pure strategies $s_i \in \mathsf{S}_i$.

In terms of strategies and depending payoffs, a Nash equilibrium is a strategy profile s^*, where no player i, given the strategy choices of the other players, wants to deviate to any other strategy $s_i \neq s_i^*$, because no such strategy will result in a higher payoff for him. In this sense, the strategy profile s^* is robust to individual deviations, as it holds for all players $i \in N$. A concise discussion of the concept of a Nash equilibrium can be found in Mas-Colell et al. (1995, p. 248 f.).

Returning to Example 2.1, the unique Nash equilibrium in (pure) strategies is readily found: It is the strategy pair (a, a) with payoffs $(1, 3)$. Given this strategy combination, any deviation would cause either player a loss of at least 2.

Prior to the first refinement of the Nash equilibrium, we introduce different notions related to *dominance* regarding the strategies of the players. The first definition covers the case of a *dominated* strategy, i.e. one a utility maximizing player should never choose:

Definition 2.3. In a strategic form game $\Gamma = (N, \mathsf{S}, \mathbf{u})$, a strategy $s_i' \in \mathsf{S}_i$ of player i is *dominated*, if there exists another strategy $s_i \in \mathsf{S}_i$ for which

$$u_i(s_i, s_{-i}) \geq u_i(s_i', s_{-i}) \text{ for all } s_{-i} \in \mathsf{S}_{-i},$$

and strict inequality prevails for at least one $s_{-i} \in \mathsf{S}_{-i}$.

Applying Definition 2.3 reversely on a set of strategies S_i yields the following:

Definition 2.4. In a strategic form game $\Gamma = (N, \mathsf{S}, \mathbf{u})$, if a strategy $s_i \in \mathsf{S}_i$ of player i is not dominated by any other strategy $s_i' \in \mathsf{S}_i$, the strategy s_i is called *undominated*.

Bringing together both previous definitions yields a type of strategy s_i, which maximizes player i's payoff:

Definition 2.5. In a strategic form game $\Gamma = (N, \mathsf{S}, \mathbf{u})$, a strategy $s_i \in \mathsf{S}_i$ of player i is *dominant*, if for all $s_i' \in \mathsf{S}_i$ with $s_i' \neq s_i$

$$u_i(s_i, s_{-i}) \geq u_i(s_i', s_{-i}) \text{ for all } s_{-i} \in \mathsf{S}_{-i},$$

and strict inequality prevails for at least one $s_{-i} \in \mathsf{S}_{-i}$.

A dominant strategy s_i does at least as well as all other strategies player i could choose, when facing s_{-i}, and is strictly better for some s_{-i}. Nevertheless, it should be clear that neither type of strategy introduced in Definitions 2.3–2.5 is guaranteed to occur in a noncooperative game Γ.

We can now state the first refinement of the Nash equilibrium concept, called the Nash equilibrium *in undominated strategies*, or *undominated* Nash equilibrium:

Definition 2.6. For a game $\Gamma = (N, \mathsf{S}, \mathbf{u})$ in strategic form, a strategy $s^* \in \mathsf{S}$ constitutes a *Nash equilibrium in undominated strategies*, if s^* is a Nash equilibrium and is undominated, i.e. not dominated by any other strategy $s \in \mathsf{S}$ for each player $i \in N$.

Now, in addition to having no incentive to deviate, each player must also be playing an undominated strategy in equilibrium.

The next equilibrium concept we cover is the *Strong Nash equilibrium*. It was originally introduced by Aumann (1959, p. 300) as the *strong equilibrium c-point* for cooperative games without side payments.[12] As compared to the Nash equilibrium, it provides strategies that are not only robust against deviations by individual players, but also by any subset of the player set, i.e. by coalitions of players:

Definition 2.7. For a game $\Gamma = (N, \mathsf{S}, \mathbf{u})$ in strategic form, a strategy $s^* \in \mathsf{S}$ constitutes a *Strong Nash equilibrium* in pure strategies, if for all subsets of players $R \subseteq N$ no strategy tuple $s_R \in \mathsf{S}_R = \times_{i \in R} \mathsf{S}_i$ exists, for which

$$u_i\left(s_R, s^*_{\overline{R}}\right) \geq u_i\left(s^*_R, s^*_{\overline{R}}\right) \qquad \text{for all } i \in R\,,$$

and there exists some player $i \in R$ for whom $u_i(s_R, s^*_{\overline{R}}) > u_i(s^*_R, s^*_{\overline{R}})$.

In this equilibrium, no coalition $R \subseteq N$ shall find it profitable to deviate from s^*, given the strategies of $s^*_{\overline{R}}$ of the other players in $\overline{R} := N \setminus R$.[13] In other words, if a coalition R deviates, a gain for one player $i \in R$ must imply a loss for some other member of R, if s^* is to satisfy Definition 2.7.

[12] These games were previously referred to as games with nontransferable utility or NTU games.

[13] In the section on noncooperative games we intentionally use R instead of S to denote coalitions as subsets of the player set N. Even though the latter is generally the norm, we want to avoid confusion with the strategies and strategy sets, which are denoted by the letters s and S, respectively.

A less restrictive (and more commonly used) version of the Strong Nash equilibrium states that a strategy $s^* \in \mathsf{S}$ constitutes such an equilibrium, if for all subsets $R \subseteq N$ no strategy tuple $s_R \in \mathsf{S}_R = \times_{i \in R} \mathsf{S}_i$ exists, for which

$$u_i\left(s_R, s^*_{\overline{R}}\right) > u_i\left(s^*_R, s^*_{\overline{R}}\right) \qquad \text{for all } i \in R,$$

i.e. all players of a deviating coalition R would be strictly better off. In the less restrictive version only deviations where all members i of the deviating coalition R benefit strictly are considered. Hence, deviations from s^* must not allow for a weak Pareto improvement[14] for the members of R. Here, the non-transferability of utility, as referred to in the original literature, comes into play, ruling out the possibility that a strictly better off member of R "bribes" the others.

One concern raised about this notion is its inconsistency regarding the deviation requirements: Based on the whole player set N, no coalition $R \subseteq N$ can deviate in a (weakly) Pareto-efficient manner from the equilibrium strategy s^*. But why is there no subsequent requirement on the deviating coalition R? It is quite plausible that a subset $T \subsetneq R$ itself might find a possibility to alter "its" strategy profile s_T to obtain a (weak) Pareto improvement when compared to the deviation intended by R. As a remedy to this inconsistency, Bernheim et al. (1987) introduce the *Coalition-Proof Nash equilibrium*, which is restricted to strategies $s \in \mathsf{S}$ that are weakly Pareto-efficient for all possible subsets of the player set N, yet self-enforcing at the same time. They build their concept on the basis of the less restrictive version of the Strong Nash equilibrium.

Formally, this concept relies on a recursive definition, starting with single player "games":

Definition 2.8.

1. In a game $\Gamma = (N, \mathsf{S}, \mathbf{u})$ with $n = 1$ player, a pure strategy $s_i^* = s^* \in \mathsf{S}$ is a *Coalition-Proof Nash equilibrium* if and only if

$$s^* \in \arg\max_{s \in \mathsf{S}} u_i(s) \text{ for all } i \in N.$$

2. Suppose that $n > 1$ and that *Coalition-Proof Nash equilibrium* has been defined for games with fewer than n players. Then,

[14] Jehle and Reny (2001, p. 171, pp. 272 ff.) speak of a Pareto improvement whenever someone can be made better off, without making anybody else worse off. A *weak* Pareto improvement is a change in payoffs that will make everybody strictly better off. This is derived from the notions of Pareto efficiency and weak Pareto efficiency. The former describes a state in which no player can be made better off without making somebody else worse off. The latter describes a state, in which not all players can be made strictly better off simultaneously. Note also that the former implies the latter, but not vice versa. For a more formal definition we refer the reader to Myerson (1991, p. 417).

a. For a game $\Gamma = (N, \mathsf{S}, \mathbf{u})$ with n players, a pure strategy $s^* \in \mathsf{S}$ is *self-enforcing*, if, for all coalitions $R \subsetneq N$, the strategy tuple $s_R^* \in \mathsf{S}_R$ is a *Coalition-Proof Nash equilibrium* in the game $\Gamma_{s_{\overline{R}}^*} = (R, \mathsf{S}_R, \overline{\mathbf{u}})$ with $\overline{u}_i(s_R) = u_i(s_R, s_{\overline{R}}^*)$.

b. For a game $\Gamma = (N, \mathsf{S}, \mathbf{u})$ with n players, a pure strategy $s^* \in \mathsf{S}$ is a *Coalition-Proof Nash equilibrium* if it is *self-enforcing* and no other self-enforcing strategy $s \in \mathsf{S}$ exists, for which

$$u_i(s) > u_i(s^*) \qquad \text{for all } i \in N.$$

This refinement solves the inconsistency issue mentioned above, as in any Coalition-Proof Nash equilibrium s^*, all possible subsets of players $R \subsetneq N$ stick to the strategies $s_R^* \in \mathsf{S}_R$, when facing the "reduced" game $\Gamma_{s_{\overline{R}}^*}$ in which the complementary coalition, $\overline{R}$, plays $s_{\overline{R}}^*$ and thereby does not deviate from s^*. Here, we should add that a deviation can also be a degenerate one: The (initially) deviating coalition might contain some players j who have simply agreed to continue playing s_j^* and not to engage in any other deviation. What the notion of Coalition-Proof Nash equilibrium does not capture though is a deviation with players in $\overline{R}$ that could be undertaken by some player(s) in R and might not even be degenerate for either of them. Another, more profound criticism is voiced by van Damme (2002, p. 1586) and applies to both, the Strong, as well as the Coalition-Proof Nash equilibrium notions. In his view, "cooperative refinements" that allow for coordination among subsets of players to yield a profitable deviation are out of place in a framework of noncooperative games. If such possibilities of cooperation are not part of the game itself, they should not be introduced through the "back door", i.e. the solution concept. We agree with his view to a certain extent, but believe that it highly depends on the context, whether such refinements should be used or not.

An alternative, nonrecursive definition of the less restrictive version of the Coalition-Proof Nash equilibrium is presented below. There, the relation to the Strong Nash equilibrium is more apparent.

Definition 2.9. In a game $\Gamma = (N, \mathsf{S}, \mathbf{u})$, a strategy tuple s^* constitutes a Coalition-Proof Nash equilibrium, if no two sets T and R with $T \subseteq R \subseteq N$ and corresponding strategies

$$\widetilde{s}_R = \times_{i \in R} \widetilde{s}_i \text{ and } \widehat{s}_T = \times_{i \in T} \widehat{s}_i$$

exist, for which the following holds true:

$$u_i(\widetilde{s}_R, s_{N \setminus R}^*) > u_i(s_R^*, s_{N \setminus R}^*) \qquad \text{for all } i \in R \text{ and, if } T \neq R,$$

$$u_i(\widehat{s}_T, \widetilde{s}_{R \setminus T}, s_{N \setminus R}^*) > u_i(\widetilde{s}_T, \widetilde{s}_{R \setminus T}, s_{N \setminus R}^*) \qquad \text{for all } i \in T.$$

Whenever $T = R$, Definition 2.9 coincides with the less restrictive version of the Strong Nash equilibrium, excluding only deviations resulting in weak Pareto improvements in the payoffs. We can also make Definition 2.9 coincide with the

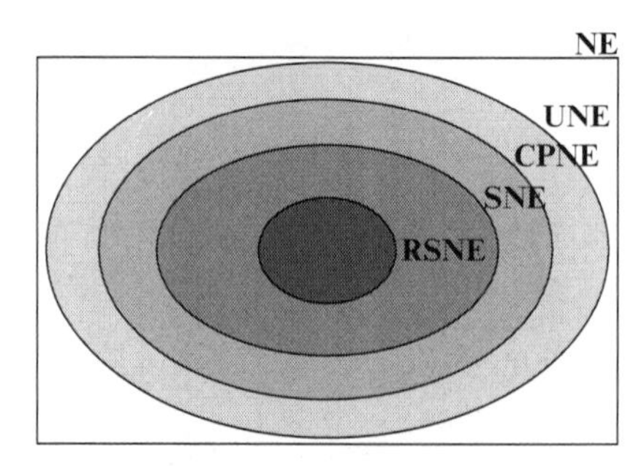

Fig. 2.3 Nash equilibrium refinements in relation to one another

more restrictive version of the Strong Nash equilibrium: Just consider a partition of deviating players R into T and $R \setminus T$. Members of T are those that benefit strictly from the deviation, while members of $R \setminus T$ are the ones with unchanged payoffs.[15]

Because Example 2.1 is a 2-player game, the notion of coalitions is restricted to either singletons or the grand coalition, and so the application of Strong and Coalition-Proof Nash equilibria makes little sense. Also, there is but a single Nash equilibrium, so the concepts would at best coincide. A simple 3-player example, distinguishing the latter two refinements is presented by Bernheim et al. (1987, p. 4). We will do so in Chap. 6, where these concepts are applied to our model.

From Definition 2.9 the different refinements of the Nash equilibrium can be related to one another in a straightforward manner, as we depict in Fig. 2.3 below. Ordered by set inclusion, the set of all Nash equilibrium strategies (**NE**) prepares the stage. Nested within and within each other are in the following order: The set of all Nash Equilibria in undominated strategies (**UNE**), the set of all Coalition-Proof Nash equilibrium strategies (**CPNE**), the set of all Strong Nash equilibrium strategies (**SNE**), and as smallest, the set of all Strong Nash equilibrium strategies in the restrictive sense (**RSNE**). This follows from the definitions of the concepts, as they all contain the "previous" refinements as special cases. **NE** rules out all deviations by single players which might make them strictly better off, given the strategies of the others. **CPNE** excludes weakly Pareto improving deviations by coalitions (possibly singletons), which must furthermore be robust to counter-deviations by subcoalitions. The **SNE** even excludes all weakly Pareto-improving deviations, be they counter-deviation-robust or not. Finally, **RSNE** rules out the set of all Pareto-improving deviations by coalitions of players, thus remaining with the potentially smallest set of solutions, given their existence. And the **UNE**-requirement is fulfilled by the latter three concepts, as this is exactly implied when the deviating coalition consists of the whole players set.

[15]The equivalence of Definitions 2.8 and 2.9 can be shown by contradiction: Suppose an allocation x^* satisfies the former but not the latter definition. Then, some coalitions $T \subseteq R \subseteq N$ must exist that can profitably deviate. For $R = N$, condition 2b of Definition 2.8 is violated. For any $R \subsetneq N$, condition 2a would be violated for any deviation strategies $\widetilde{s}_R$ and $\widehat{s}_T$. The other direction is analogous: If an s^* meeting Definition 2.9 is assumed, no coalition $R \subsetneq N$ will want to deviate and no other self-enforcing strategy can exist for $R = N$.

We now conclude our treatment of noncooperative games in strategic form. In the next section, we turn to the extensive form game representation, which allows for a more detailed description of the game situation at hand.

2.2.3 Extensive Form Games

As mentioned above, games in extensive, or sequential form allow for a more detailed description of the underlying game situation. This is because they consider the choices over actions which a player can make in a stepwise fashion. Rather than condensing them into one strategy, whose application will "instantaneously" lead to a given outcome, games in extensive form represent the players' choices of actions in a sequence: The player whose turn it is, possibly knows all the decisions taken before him and can adapt his choice accordingly. This feature is missing in the strategic form representation, as the sequence of choices is aggregated into one strategy.

Very informally, Jehle and Reny (2001, p. 285) give the following list of elements of an extensive form game:

1. The finite player set
2. *Nature*, or "chance", being a player whose moves are random
3. The rules of the game, which not only specify the order in which players move, but also the available actions and (possibly limited) information about previous moves when it is their turn.
4. The payoffs for the players when the game has reached an end or outcome

For a rigourous formal treatment of the extensive game form, we refer the reader to Ritzberger (2003, pp. 80 ff.) or Mas-Colell et al. (1995, p. 227). We try to remain on a mostly nontechnical level, as the need for a formal approach is very limited in our context. For our purposes, the representation via so-called *game trees*, is sufficient. (See p. 23 in Fig. 2.4, Example 2.2)

The generic game tree is quickly explained: Mainly, it consists of a set of nodes and edges between them, arranged in a cycle-free manner. Both, the nodes and the edges are labelled. The former have players assigned to them, who are supposed to choose one of the available actions at that node; the latter are given the designations of the actions to be taken at a given node by the respective player.

Also the "direction" of the game along the tree is very intuitive. There is an initial node, which represents the first move in the game. This node can be assigned to the chance player, commonly referred to as player 0 or "nature".[16] Depending on which action was selected, some subsequent (and intermediate) node is reached, and, after the last move, the game ends at a terminal node. Each such terminal node,

---

[16]A possible interpretation of this chance player is a probability distribution over different deterministic games that ensue, being then played by the "real" players.

or outcome, is assigned a payoff vector, whose entries correspond to the payoffs the players can expect for this given outcome. In case of *imperfect information*, some players are not comprehensively informed about parts of the previous history of the game and hence which actions were taken by some players who moved before them. Nevertheless, they know what they do not know, that is they are aware of their deficiency and its nature. These situations are identified by a dashed oval containing exactly the nodes a player cannot distinguish. This set of nodes is called an *information set* of the player who is assigned to (all!) these nodes.

We continue using the previous example adapted from Mas-Colell et al. (1995) to illustrate the concept of a game tree:

Example 2.2. (Mas-Colell et al. (1995, p. 273 f.)) This example is taken from Mas-Colell et al. (1995) and describes an extended market entry game, in which first player 2 decides whether or not to enter a monopolistic market. In the next step, given player 2 decided to enter, he and the incumbent, player 1, will decide simultaneously, whether to fight or accommodate each other.[17] The (pure) strategy set of player 1 is given by $S_1 = \{f, a\}$, where "*f*" means to fight player 2 and "*a*" means to accommodate him. Player 2's strategy set is larger, because it accounts for every contingency, even those that should not occur: $S_2 = \{oa, of, ia, if\}$. The strategy "*oa*" instructs not to enter the market and accommodate player 1, just in case. The strategy "*of*" stands for stay out of the market and fight player 1 if necessary. To enter and accommodate is given by "*ia*", and finally "*if*" means to enter and fight player 1. If player 2 plays either "*oa*" or "*of*", the game terminates at the same outcome, regardless of the strategy chosen by player 1. If he plays one of the other two strategies, and hence enters the market, the simultaneous predation game is to follow, where also player 1 decides how to encounter the entrant. Because at the time when taking his decision, player 1 does not know whether player 2 decided to fight or to accommodate him, he cannot distinguish at which node in the game he is. The situation is depicted in Fig. 2.4 below. The decision nodes are labelled with the number of the player whose are supposed to choose an action, the edges represent the available actions and are labelled accordingly. The initial node is the potential entrant's decision, on whether to enter the market or not. The edges are labelled "*i*" and "*o*" for entering the market ("in") or not (stay "out"). Having chosen strategy $s_2 \in \{ia, if\}$, the subsequent node again is assigned to player 2, where he chooses between "*f*" for fighting the incumbent or "*a*" for accommodating him. Finally, player 1 gets to choose between "*f*" and "*a*". As the last two moves are supposed to be simultaneous, player 1's information set contains both nodes, which is indicated by the dashed oval around them. A game like this is called one of *imperfect information*, because player 1 does not (yet) know which action player 1

[17]We deviate from the usual convention to number players in the order that they are called upon to move in the game. We find it more intuitive in this case to label the incumbent as player 1, and the potential newcomer as player 2. Also, this way, we stay in line with Example 2.1.

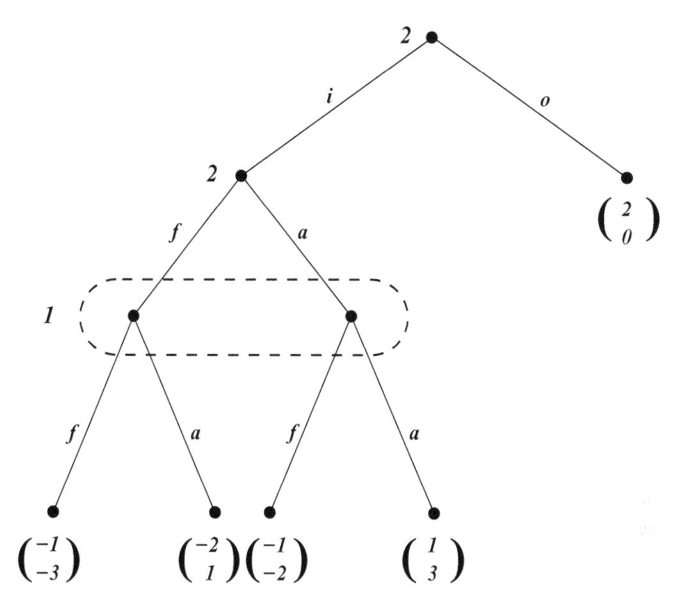

Fig. 2.4 Game tree for the modified market entry game

		Player 2			
		of	*oa*	*if*	*ia*
Player 1	*f*	2,0	2,0	−1,−3	−1,−2
	a	2,0	2,0	−2,1	1,3

Fig. 2.5 Normal form representation of the extended market entry game

has chosen at the previous node.[18] The game then ends at one of the terminal nodes, below which the payoffs are given. The upper entry of each vector corresponds to player 1 and the lower to player 2.

We can also derive a normal form representation of the game described. It can be found in Fig. 2.5 below. In the depicted payoff matrix it is apparent that all four strategy combinations which include player 2 staying out of the market yield identical payoffs.

[18]Please note if one reversed the assignments of the players such that player 1 was to pick first whether to fight or to accommodate, the payoffs would be identical. Only, in this case player 2 would have to "suffer" from imperfect information, not knowing the decision of player 1. The simultaneous character is unaffected by this change. Such interchanging of simultaneous decisions is called an *I-transform*. These operations are treated extensively (no pun intended) in Ritzberger (2003, p. 155 ff.) and Thompson (1997, p. 43 f.).

As we have done with the game in Example 2.2, a strategic form representation can be derived for all games in extensive form.[19] According to its definition (see Footnote 6), it is possible to construct a pure strategy space from all the possible decisions that players have to take at their respective nodes. The payoff functions are then simply the payoffs associated with the outcomes, which result from a specific choice of strategy. Even though such a derivation of a strategic form might not accurately reflect the richness of the extensive form, it is convenient due to its striking simplicity. We can also see that the Nash equilibrium notions as defined above are still applicable in extensive form games. Nevertheless, some equilibrium concepts have been devised specifically with the extensive form in mind. We introduce the two most basic, *backward induction strategies* for perfect information scenarios and *subgame perfect Nash equilibria* for those with imperfect information.

As the name suggests, backward induction strategies are found by rolling up the game tree in the other direction, i.e. from the terminal nodes to the initial node. The following process is called the *backward induction algorithm*: We begin with the penultimate nodes, for which all successive nodes are terminal nodes. At such a node, we look which is the payoff-maximizing action for the respective player. We then replace all edges and terminal nodes with the payoff vector from the chosen terminal node. What was initially a penultimate node, is now a terminal node itself, and the initial game has been reduced. This procedure is repeated until we reach the initial node. Consequently, payoff-maximizing actions have been chosen at every decision node, giving rise to a pure strategy $s \in \mathsf{S}$. This strategy is called a *pure backward induction strategy*. Kuhn (1953) shows in what has become known as *Kuhn's Lemma* that every pure backward induction strategy also constitutes a Nash equilibrium in pure strategies for the corresponding game in strategic form.[20]

For games with imperfect information, we turn to the *subgame perfect Nash equilibrium* as solution concept of choice. In this scenario, the backwards induction algorithm is bound to fail in certain situations: When we try to find a payoff-maximizing action at a penultimate node which belongs to a information set with more than one element, the payoff-maximizing action might not be the same for all nodes. Because the player does not have the ability to distinguish between nodes, he might not be able to choose an action maximizing his payoffs.[21] Hence, we can not replace the terminal nodes and edges with the payoff corresponding to the action chosen by the player. The subgame perfect equilibrium concept will provide a way to overcome this obstacle. Before we come to the concept itself, we introduce the notion of a subgame:

[19]See Ritzberger (2003, pp. 143 ff.).

[20]The converse does not necessarily hold.

[21]This is the case in Example 2.2, where player 1, depending on which penultimate node he is at, should either choose action "*a*" or "*f*".

Definition 2.10. A *subgame* of an extensive form game fulfills the following three properties:

1. It has a unique initial node, i.e. no information set with multiple elements.
2. It contains all the subsequent nodes that can be reached via the game tree's edges, but no other nodes.
3. Each information set contained in the subgame must be identical to its counterpart in the original game.

Relating these properties to the graphical representation in a game tree, the following can be said: A subgame always begins at a single node in the tree which does not belong to some information set including other nodes in the original game. The subgame contains everything "below" its initial node in tree, but nothing else; it does neither cut through edges nor through information sets containing more than one element. Hence, subgames cannot overlap one another, unless through inclusion.

Of course, for some games, no proper subgame might even exist. In Example 2.2, Fig. 2.4, the only proper subgame begins with the second decision node of player 2 and includes everything below it. Its strategic form representation is given by Fig. 2.2 of Example 2.1.

The subgame perfect equilibrium concept, originally introduced by Selten (1965) as *perfect equilibrium point*, is defined as follows:

Definition 2.11. For a game Γ in extensive form, a strategy $s^* \in \mathrm{S}$ constitutes a *subgame perfect Nash equilibrium* (*SPNE*), if it induces a Nash equilibrium in every subgame of Γ.

To derive SPNE, we use the *generalized backward induction algorithm*, similar to the one introduced above: We begin with the smallest possible subgames, i.e. subgames that cannot contain proper subgames themselves; replace these in the game tree with the payoffs corresponding to Nash equilibrium strategies for the respective subgame. Repeat this process until the initial node is reached and every move has been determined. The deduced strategy $s \in \mathrm{S}$ constitutes an SPNE. However, as there could exist multiple Nash equilibria for any subgame, SPNE does not necessarily yield a unique outcome.

We have seen that an extensive form game may consist of more than just strategies and a payoff function. Its foundation are the rules of the game, i.e. the order in which players take turns and how much they know about the history of the game. From this, strategy sets for each players can be derived. In strategic form games, it is more the strategies that are emphasized while the underlying rules are kept more or less implicit. A player i once has the choice to pick a strategy $s_i \in \mathrm{S}_i$, and then his payoff is determined given the one-off strategy choices of the other players.

In the extensive form, the strategies are compiled from the possibly multiple choices each player has to make over the course of the game. Here they are merely the product, but not a necessary ingredient for the form of representation. This detailed treatment allows for a much richer description of the game situation.

Now that the reader has learned the basics about the two most common forms of representing a noncooperative game and was introduced to some selected equilibrium notions, it should be possible to see the close relation between the two that is not apparent at all on first sight. Indeed, it is possible to transform every strategic form game into one in extensive form, and vice versa, albeit not on a one-to-one basis. We have shown this exemplarily above, but omit all further details on this matter. The interested reader can find more on this in Chap. 2 of Myerson (1991), which also contains a much more elaborate consideration of equilibrium notions for both forms of representation in Chaps. 3 and 4.

We now conclude our treatment of noncooperative game theory. In the next section we turn to the theory of cooperative games, on which the focus of this work lies.

2.3 Cooperative Games

2.3.1 Overview

The rest of this chapter is dedicated to the theory of cooperative games and possible solutions to them.[22] In this section, we begin with the games themselves, defining them, especially in contrast to noncooperative games. After deriving the heart of each cooperative game, its characteristic function, we continue to classify such games. Mainly, we distinguish between games that are zero-sum and therefore without any aggregate gains, and general-sum games. Also, it is shown how unanimity games, roughly describable as "all-or-nothing" games, can be used to construct any cooperative game. Similarly, under the heading of strategic equivalence, we show how apparently different games can be considered identical in their treatment. Then, via duality, we relate different uses of the same underlying information on coalitions' values to one another and provide the instructive example of the Tennessee Valley Authority, which is a standard reference for applied cooperative game theory. The section concludes with a listing of common properties cooperative games can possibly exhibit.

On all accounts, we do not claim to be exhaustive in our choice of topics. It has been made according to the requirements for subsequent parts of this work as well as the attempt of the author to produce a somewhat coherent and concise introduction to the theory of cooperative games. Therefore, some concepts have been omitted deliberately, while others might not appear later on but are very instructive or simply needed to build upon.

[22] As we noted earlier, we are only concerned with cooperative games where utility is transferable among the players (TU) and ignore even the existence of games with nontransferable utility (NTU).

2.3.2 The Cooperative Game

The distinguishing feature of cooperative game theory is the assumption that all (coalitions of) players can make waterproof contractual agreements that are unconditionally binding, before the game is played. Bilbao (2000, p. 2) defines a cooperative game as "*a game in which the players can make binding commitments, as opposed to a noncooperative game, in which they cannot.*" Whereas the noncooperative branch focusses on actions and/or strategies of players, cooperative game theory basically skips this step. It emphasizes merely the payoff or value a certain coalition of players, ranging from individuals to the whole player set, can achieve as a whole. How this payoff is reached in detail, i.e. what underlying strategies the players might have chosen is not part of the game. The decision here is merely whether to cooperate and with whom. This is the essence of games with transferable utility. A very concise description of transferable utility, to which we have nothing to add, has been brought forward in Harsanyi (1959, p. 326):

> *Transferable utility means that side payments between players are allowed and have no effect on the sum of all players' utilities, the payee's utility gain being always exactly equal to the payer's loss. This requires that money transfers should not be subject to costs or subsidies, and that the marginal utility of money be constant to all players over the relevant range.*

Another classification into cooperative and noncooperative game theory can be made on the basis of the objectives of the ($n \geq 2$) players: Whether they have a common one (cooperation) or multiple (strategies).

In cooperative game theory the most precise representation of a game is the *characteristic function form*:

Definition 2.12. A characteristic function form game Γ is a pair

$$\Gamma = (N, v),$$

where N is a finite set of players and v is the characteristic function assigning a value to each possible subset of players $S \subseteq N$.

The finite player set $N = \{1, 2, \ldots, n\}$ needs no further elaboration, but certainly the *characteristic function*, also referred to as the *value function* of the game: It is a mapping $v : 2^N \to \mathbb{R}$ from the power set of the set of all players, $2^N = \{S | S \subseteq N\}$, into the real numbers, $\mathbb{R}$. Hence, it assigns every *coalition* of players, $S \subseteq N$, a *value* or *worth*. By convention $v(\emptyset) = 0$, because a cooperation of no players cannot achieve a nonzero payoff. In case all $n := |N|$ players cooperate, we speak of N as the *grand coalition*. Coalitions of size one, where $|S| =: s = 1$, are called *singletons* or *trivial* coalitions.

The set of players N defines and is part of its power set, $N \in 2^N$, the latter of which is the domain of the function v. Therefore, the game is also commonly referred to only by its characteristic function v. This function v is also denoted often as vector $\mathbf{v}$, where $\mathbf{v} \in \mathbb{R}^{2^n - 1}$ and each entry of $\mathbf{v}$ corresponds to the value $v(S)$ of an element

$\emptyset \neq S \subseteq N$ in its domain. We alternate between these expressions depending on the context. The idea of a characteristic function was first introduced to the literature in von Neumann (1928, pp. 316 ff.), as part of a formal treatment of parlor games. The concept then had its breakthrough with von Neumann and Morgenstern (1947, pp. 238 ff.). A more advanced treatment can also be found in Shapley and Shubik (1973).[23]

Denote by $\mathbf{V}_N$ the space of cooperative games with player set N. Its elements are the vectors $\mathbf{v}$, and, as we will see below, $\mathbf{V}_N$ can be considered a vector space. As the set of players is fixed at N throughout our work, we usually try to simplify notation by omitting super- or subscripts N or n that refer to the player set or its cardinality. This of course will only occur, when no confusion is bound to arise.

Generally, for a coalition $S \subseteq N$, the value $v(S)$ can be interpreted as the maximum payoff achievable by S without the cooperation of the remaining players $N \setminus S$. How this is brought about in detail is given in the next step, where the characteristic function v of a cooperative game with transferable utility is derived from a noncooperative game; more specifically, merely from the components of a noncooperative game in strategic form, i.e. the set of all strategy profiles S and the payoff function $\mathbf{u}$, which we introduced in Sect. 2.2.2. Our approach is based on von Neumann and Morgenstern (1947, pp. 46 ff., pp. 238 ff.) and Weber (1994).

Define for any coalition $S \subseteq N$ its space of mixed strategy combinations to be

$$\Delta_S := \times_{i \in S} \Delta_i, \tag{2.6}$$

and, analogously for the complement of S in N, $\overline{S} := N \setminus S$:

$$\Delta_{\overline{S}} = \times_{i \in \overline{S}} \Delta_i. \tag{2.7}$$

The corresponding mixed strategy tuples are denoted σ_S and $\sigma_{\overline{S}}$. As the order of the subspaces within the product spaces Δ_S and $\Delta_{\overline{S}}$ can be chosen at will, we have $\sigma_S \times \sigma_{\overline{S}} = \sigma$ and $\Delta_S \times \Delta_{\overline{S}} = \Theta$.

Expected payoff functions for coalitions $S \subseteq N$ are merely aggregated over the members i of S:

$$U_S(\sigma) = U_S(\sigma_S, \sigma_{\overline{S}}) = \sum_{i \in S} U_i(\sigma), \text{ for all } S \subseteq N, \ \ \sigma \in \Theta. \tag{2.8}$$

Now, even under the most adverse circumstances no coalition $S \subseteq N$ can be made worse off than

$$\max_{\sigma_S \in \Delta_S} \min_{\sigma_{\overline{S}} \in \Delta_{\overline{S}}} U_S(\sigma_S, \sigma_{\overline{S}}) =: v(S). \tag{2.9}$$

[23]Luce and Raiffa (1957, pp. 180 ff.) provide a very interesting introduction on characteristic functions, especially on p. 189 f., where they draw up striking parallels and differences between characteristic functions and probability measures.

This is the amount the players in S can secure for themselves even when the players in $\overline{S}$ act with the goal to minimize S's payoff. The worst-case-scenario given in (2.9) is the conceptual basis for all characteristic functions from hereon. This approach leads to what is known as the α-characteristic function, sometimes denoted v_α.[24]

Having introduced the characteristic or value function of a cooperative game, we can now proceed to some general classifications of certain types of games, as well as to some features of and relations between games.

2.3.3 Important Classes and Types of Games

In this section we take a look at some possibilities to classify the space of coalitional games $\mathbf{V}$ and to introduce some features that are somehow related to these classes of games. First, we consider the notion of *strategic equivalence* which allows us to group games with certain similarities in their characteristic functions. A normalization procedure is presented, too. We continue with probably the most important distinction regarding coalitional games, that between so-called *constant-sum* and *general-sum* games. In the former, the aggregate gain achievable is, loosely speaking, a constant, and coalition formation only determines how much a particular coalition and its complement do or do not get. The latter also include those constant-games, but are usually referred to as games where cooperation increases the size of the pie that can be distributed among the players. We conclude this section with the class of *unanimity games*, which we will see are the cornerstones of the space of coalitional games.

2.3.3.1 Strategic Equivalence

Definition 2.13. A game (N,v) is *strategically equivalent* to the game (N,w), if there exist constants $a \in \mathbb{R}_{++}$, $b_1, b_2, \ldots, b_n \in \mathbb{R}$, such that for all coalitions $S \subseteq N$,

$$v(S) = a \cdot \left(w(S) - \sum_{i \in S} b_i \right). \tag{2.11}$$

[24]It is the type commonly used in the literature, though there are rarely any remarks concerning its origin. Another concept is the β-characteristic function v_β: Here, the order of actions is reversed, i.e. first coalition S maximizes its payoff and then the remaining players in $\overline{S}$ try to hold it down to a minimum. The functional is given by

$$v_\beta(S) = \min_{\sigma_{\overline{S}} \in \Delta_{\overline{S}}} \max_{\sigma_S \in \Delta_S} U_S(\sigma_S, \sigma_{\overline{S}}). \tag{2.10}$$

Unlike stated in von Neumann and Morgenstern (1947, p. 240) for zero-sum games, functions v_α and v_β need not assume identical values in general-sum games. These distinctions are treated more detailed in Friedman (1991, pp. 240 ff.).

Intuitively, strategic equivalence of two different games (N,w) and (N,v) can have two reasons, possibly simultaneous: First, the one game might be some multiple of the other, lying on some straight line from the origin of $\mathbf{V}$ through the latter. This is represented by the scale factor a, which affects all values of v likewise. The second cause is rooted in the individual factors b_i, which are an individual scaling of the players' (linear) utilities underlying the coalitional game.[25] It can be shown that strategic equivalence is an equivalence relation, i.e. one that is reflexive, symmetric, and transitive. Denoting the class of strategically equivalent games of (N,v) by $[v]$, it follows that two equivalence classes, say $[u]$ and $[v]$, corresponding to coalitional games (N,u) and (N,v), are either disjoint or equal. They are equal if and only if (N,u) and (N,v) are strategically equivalent.

We now introduce a normalization procedure for coalitional games which can be considered a special case of strategic equivalence.

2.3.3.2 The (0,1)-Normalization

A special case of strategic equivalence is the $(0,1)$-normalization of a game (N,v), correspondingly denoted by $(N,\overline{v})$, or simply $\overline{v}$:

$$\overline{v}(S) = \frac{v(S) - \sum_{i \in S} v(\{i\})}{v(N) - \sum_{i \in N} v(\{i\})} = \begin{cases} 0 & \text{if } |S| = 1 \\ \in [0,1] & \text{if } 1 < |S| < |N| \\ 1 & \text{if } |S| = |N| \end{cases} , \text{for all } S \subseteq N. \quad (2.13)$$

Comparing this expression to (2.11), the following ensues:

$$a = \frac{1}{v(N) - \sum_{i \in N} v(\{i\})} \quad \text{and } b_i = v(\{i\}) \text{ for all } i \in N. \quad (2.14)$$

In this form, by construction, $\overline{v}(\{i\}) = 0$ for all $i \in N$ and $\overline{v}(N) = 1$, and hence the name $(0,1)$-normalization. This normalization is not defined for inessential games, in which the denominator $v(N) - \sum_{i \in N} v(\{i\})$ is equal to zero. This normalized form is referred to as a special case of strategic equivalence, because for each class

[25] Under equal restrictions on the constants, strategic equivalence is also often equally defined by the expression

$$v(S) = a \cdot w(S) - \sum_{i \in S} c_i, \quad (2.12)$$

where for all $i \in N$ we have $c_i = a \cdot b_i$. In the literature, slightly different notions of equivalence have surfaced: von Neumann and Morgenstern (1947, pp. 245 ff.) define *strategic equivalence* on zero-sum games, consequently dropping the multiplicative constant a. On the other hand, *S-equivalence*, as introduced by McKinsey (1950, p. 120) exhibits this multiplicative constant but is equal to (2.12) otherwise. In the literature, both terms usually refer to the concept defined McKinsey (1950), which we also adopted.

of strategic equivalent games $[v]$, there exists a unique $(0,1)$-normalized game[26]: By normalizing the expression of strategic equivalent games (as given by (2.11)) according to expression (2.13), it is immediate that the constants a and b_i cancel out, regardless of their value. This leaves but a unique $(0,1)$-normalized game to represent each class of strategically equivalent games. A technical elaboration can be found in Jianhua (1988, p. 99), as well as in Friedman (1991, pp. 251 ff.). The $(0,1)$-normalization is itself a special case of so-called games in *reduced* form, where the restriction $v(\{i\}) = 0$ for all $i \in N$ still applies but the grand coalition's values are given by $v(N) \in \{-1,0,1\}$. (See Gillies 1959, p. 67.)

Next, we turn to a fundamental distinction regarding coalitional games. It is per se independent of notions of strategic equivalence and groups games according to the profitability of cooperation.

2.3.3.3 Constant-Sum and Zero-Sum Games

Definition 2.14. A coalitional game (N,v) is a *constant-sum* game, if the elements of any two-element partition of the player set N yield the same aggregate value:

$$v(S) + v(N \setminus S) = v(N) \qquad \text{for all } S \subseteq N. \tag{2.15}$$

Often, some constant $a \in \mathbb{R}$ is used instead of $v(N)$ on the right-hand side of (2.15). But since this equation holds for all possible $S \subseteq N$, hence also for N, and by convention $v(\emptyset) = 0$, it automatically follows that $a = v(N)$.

In the special case, where we have $v(N) = 0$, we refer to the game as a *zero-sum* game. Then, from (2.15), we have $v(S) = -v(N \setminus S)$ for all possible coalitions $S \subseteq N$. In both notions, the size of aggregate gains is fixed, and the choice of coalition only determines which subset of players S gets how much, when compared to its complement in N. It is worthwhile to note that for every constant-sum game, it is possible to find a zero-sum game which is strategically equivalent, most easily by choosing the constants b_i in (2.11) such that $\sum_{i \in N} b_i = v(N)$ and $a = 1$. A more detailed account of the relation between zero-sum and constant-sum games can be found in von Neumann and Morgenstern (1947, pp. 346 ff.). We now relax the restriction on the partition of N and allow for finer partitions, up to the finest possible partition into singletons.

2.3.3.4 Rational, Essential, and Inessential Games

Definition 2.15. A game (N,v) is called *rational* if forming the grand coalition is at least as good as no cooperation at all:

---

[26]There exists a bijective mapping from the space of games' equivalence classes to the space of $(0,1)$-normalized games.

$$v(N) \geq \sum_{i \in N} v(\{i\}). \tag{2.16}$$

Both cases arising from a rational game are now considered in turn:

Definition 2.16. A game (N, v) is called an *inessential* game, if expression (2.16) holds with equality:

$$v(N) = \sum_{i \in N} v(\{i\}). \tag{2.17}$$

In this case, there are no gains to be made by forming the grand coalition as opposed to no cooperation. But this does not exclude the possibility that some proper subsets of N reveal strict gains from cooperation. Also note that an inessential game does not contradict the definition of a constant-sum game.[27]

Definition 2.17. A coalitional game (N, v) is *essential*, if a strict inequality prevails in expression (2.16):

$$v(N) > \sum_{i \in N} v(\{i\}). \tag{2.18}$$

In this case, the grand coalition N can achieve strictly positive gains as compared to the aggregate of the individual players. Yet again, without further assumptions on (N, v), nothing is said about the value that non-singleton coalitions can achieve. When we refer to *general-sum* games, we usually mean games that are not necessarily constant-sum, but rather unspecified in this respect a priori. Before we advance to the class of unanimity games, we want to introduce two features of which unanimity games make ample use.

2.3.3.5 Dummy Players

Definition 2.18. A dummy player $i \in N$ is a player who contributes always his stand-alone value to any coalition he joins:

$$v(S \cup \{i\}) - v(S) = v(\{i\}) \text{ for all } S \subseteq N \setminus \{i\}. \tag{2.19}$$

Following this definition a dummy player does not create a marginal value exceeding his stand-alone value, whenever he joins a coalition $S \subseteq N \setminus \{i\}$.[28] Applying this idea reversely provides more insight on the dummy player. In other

[27] There is also the much stronger notion of *inessential* coalitions: A coalition S is inessential, if as a whole it can not achieve more than the sum of any conceivable partition of S. This concept is much stricter than plain inessentiality, as it compares not only the set with the sum of its singletons, but with all possible partitions. With respect to a notion defined below, one could say on an inessential coalition S, the value function v is decomposable to all possible partitions of S.

[28] In general, the *marginal value* of a player i when joining a coalition $S \subseteq N \setminus \{i\}$ is given by the left-hand side of (2.19). It is defined in more detail in (2.32) below.

words, the presence of a dummy player in a coalition has no influence on the marginal values of other, non-dummy players. If we remove the dummy player from any coalition S, where $i \in S$, the marginal loss is identical to the marginal value $v(\{i\})$ when joining the coalition T with $S \subseteq T \setminus \{i\}$, i.e. regardless of how many other players might have joined in the meantime.

Often referred to as dummy player, even though a special case of one, is the so-called *null-player*: His marginal contributions are identically zero for all coalitions $S \subseteq N \setminus \{i\}$, and so $v(\{i\}) = 0$ as well.

2.3.3.6 Carriers

Definition 2.19. In a coalitional game (N, v), a coalition T is a *carrier*, if

$$v(T \cap S) = v(S) \;\; \text{for all} \;\; S \subseteq N. \tag{2.20}$$

According to this requirement, the value a coalition S can create originates only from its members which are also members of T. Therefore, for all members of S *and not* T, the marginal value to any given coalition must be equal to zero. From this definition we can conclude that every superset of a carrier again fulfills the requirements of a carrier. Therefore, when ordered by set inclusion, there is only one minimal and one maximal carrier, and no two distinct carriers $T_1, T_2 \subseteq N$ can exist. Either $T_1 \subseteq T_2$ or $T_2 \subseteq T_1$ must hold and no two carriers with empty intersection or "partial overlap" are allowed. Also, given a carrier T,

$$v(T) = v(R) \qquad \text{for all} \; T \subseteq R \subseteq N. \tag{2.21}$$

If this was not the case and for some superset $R \supseteq T$ the inequality $v(R) > v(T)$ would prevail and the definition of T as a carrier would be violated. The inequality implies the existence of a set S and a player i for which $i \in S$, $i \notin T$ and $v(T \cap S) < v(S)$, contradicting (2.20).

Clearly, it is a necessary condition to belong to a carrier, in order not to be a null-player. But it is sufficient only, if that carrier happens to be the minimal carrier in terms of set inclusion. Carriers are not only essential to the concept of a unanimity game, its definition incorporates them directly, as they coincide exactly with the set of players required for nonzero values.

2.3.3.7 Unanimity Games

Based on the notions of carriers and null-players, we now introduce *unanimity games*. They are a fundamental class of games, which, as we will see, form a possible basis for the space of all coalitional games. More precisely, the set of unanimity games (N, u_S) with $\emptyset \neq \subseteq N$ is a possible basis for the linear space $\mathbf{V}_N$.

Definition 2.20. A game (N, u_S) with $\emptyset \neq S \subseteq N$ is said to be a *unanimity game*, if

$$u_S(T) = \begin{cases} 1 & \text{if } S \subseteq T \\ 0 & \text{else} \end{cases}, \quad \text{for all } T \subseteq N. \tag{2.22}$$

A coalition T can realize a value, normalized to 1, if all the members of the crucial coalition S are also members of T. Otherwise, T is not able to realize any value other than zero. In reference to the concept of a carrier introduced above, S is called the *carrying coalition* for the unanimity game u_S, being its minimal carrier. Any player $i \in N \setminus S$ is consequently a null-player.[29]

We now show that any general game $\mathbf{v}$ can be written as a linear combination of unanimity games $\mathbf{u}_S$:

Proposition 2.1. *If (N, v) is a coalitional game, there exist $2^n - 1$ real numbers γ_S for $\emptyset \neq S \subseteq N$ such that*

$$\mathbf{v} = \sum_{S \subseteq N} \gamma_S \mathbf{u}_S, \tag{2.23}$$

where $\mathbf{v}$ and $\mathbf{u}_S$ are vectors of dimension $2^n - 1$ each.

Before proving this result, we expand (2.23):

$$\begin{pmatrix} v(S_1) \\ v(S_2) \\ v(S_3) \\ \vdots \\ v(S_{2^n-1}) \end{pmatrix} = \begin{pmatrix} \gamma_{S_1} u_{S_1}(S_1) & + & \gamma_{S_2} u_{S_2}(S_1) & + \ldots + & \gamma_{S_{2^n-1}} u_{S_{2^n-1}}(S_1) \\ \gamma_{S_1} u_{S_1}(S_2) & + & \gamma_{S_2} u_{S_2}(S_2) & + \ldots + & \gamma_{S_{2^n-1}} u_{S_{2^n-1}}(S_2) \\ \gamma_{S_1} u_{S_1}(S_3) & + & \gamma_{S_2} u_{S_2}(S_3) & + \ldots + & \gamma_{S_{2^n-1}} u_{S_{2^n-1}}(S_3) \\ \vdots & & \vdots & \ddots & \vdots \\ \gamma_{S_1} u_{S_1}(S_{2^n-1}) & + & \gamma_{S_2} u_{S_2}(S_{2^n-1}) & + \ldots + & \gamma_{S_{2^n-1}} u_{S_{2^n-1}}(S_{2^n-1}) \end{pmatrix}.$$

Now it is more apparent that every one of the $2^n - 1$ rows on the right-hand side corresponds to a sum of $2^n - 1$ elements, adding up to the value $v(S)$, of some nonempty coalition $S \subseteq N$. While the rows vary over the arguments of the unanimity games, the columns vary over the constants γ_S and the carrying coalitions of the unanimity games. The individual summands of this system of equations assume either the value 0 or γ_S, as the unanimity games take values of either 0 or 1.

Proof. As expression (2.23) is a system of linear equations in the coefficients $\gamma_{S_1}, \gamma_{S_2}, \ldots, \gamma_{S_{2^n-1}}$ and the vectors $\mathbf{u}_{S_1}, \mathbf{u}_{S_2}, \ldots, \mathbf{u}_{S_{2^n-1}}$ representing the unanimity games are linearly independent, a solution must exist. If we denote by t, s, and r the cardinality of the sets T, S, and R, respectively, the solution for nonempty $S \subseteq$ is given by

$$\gamma_S := \sum_{T \subseteq S} (-1)^{s-t} v(T), \tag{2.24}$$

[29] It is standard in the literature to use the letter u for characteristic functions of unanimity games, as well as for utility functions in noncooperative games. We hope to separate both domains sufficiently to avoid any ambiguity.

as we now proceed to show. Picking an arbitrary, nonempty, coalition R, we choose a particular row, the one assigned to R, from expression (2.23). By substituting (2.24) into this row, we can rewrite the corresponding line to

$$\sum_{S\subseteq N} \gamma_S u_S(R) = \sum_{S\subseteq R} \gamma_S = \sum_{S\subseteq R}\left[\sum_{T\subseteq S}(-1)^{s-t}v(T)\right]$$
$$= \sum_{T\subseteq R}\left[\sum_{S \mid T\subseteq S\subseteq R}(-1)^{s-t}\right]v(T). \tag{2.25}$$

Regarding the second expression in square brackets, we know there exist $\binom{r-t}{r-s}$ sets S with cardinality s that satisfy $T \subseteq S \subseteq R$.[30] Hence, we can replace it with

$$\sum_{s=t}^{r}\binom{r-t}{r-s}(-1)^{s-t},$$

which can also be written as

$$\sum_{s=t}^{r}\binom{r-t}{r-s}(-1)^{s-t}(1)^{r-s} = (-1+1)^{r-t}.$$

The last expression corresponds to the standard binomial coefficient

$$\sum_{k=0}^{n}\binom{n}{k}(a)^{n-k}(b)^{k} = (a+b)^{n},$$

where $n = r-t$, $k = r-s$, $b = 1$, and $a = -1$. Its value is either 0 for all $t < r$ or 1 for $t = r$. By coincidence of the cardinality of the sets T, S, and R we obtain

$$\sum_{S\subseteq N} \gamma_S u_S(R) = v(R) \quad \text{for all } R \subseteq N.$$

This is exactly the row of (2.23) corresponding to coalition R. ■

We showed that all unanimity games on N can indeed be used to span the respective (linear) space of general coalitional games on the same player set N.

[30] For a given set T from the outer summation of (2.25), there are $r-t$ elements from which we can draw. The set S consists of the union of T with these elements. Depending on how many elements we add to T, $r-s$ denotes this number in decreasing order as s increases. This follows from the symmetry of the binomial expression.

This property will be useful in later applications, where concepts of linearity are attributed to functions whose domain is the space of games $\mathbf{V}_N$. Next, we introduce a generalized notion of unanimity games, called partnerships.

2.3.3.8 Partnerships

A *partnership* is often considered a generalization of unanimity games on general games, because certain restrictions are imposed on the creation of value within a coalition referred to as the partnership. Like in a carrying coalition of a unanimity game, no proper subset of the partnership can create any value, neither on its own, nor by "defecting" and cooperating with players outside the partnership. But, unlike in unanimity games, there is no restriction on the possible gains which coalitions in the complement of the partnership in N can realize.[31] In short, members of a partnership can only create value as a whole, but the remaining players are not directly restricted in terms of creating value through the partnership. Formally, this is summed up in the following definition:

Definition 2.21. In a coalitional game (N,v), a coalition $S \subseteq N$ is called a *partnership*, if for all proper subsets $T \subsetneq S$ and all coalitions $R \subseteq N \setminus S$ the following holds:

$$v(R \cup T) = v(R) \tag{2.26}$$

The restriction to proper subsets is imposed on purpose, because nothing prevents the partnership as a whole from cooperation with some other coalition to create an even higher value. Consequently, the value created by the partnership $v(S)$ need not equal that of the grand coalition $v(N)$, as is the case for carrying coalitions in unanimity games.

2.3.4 Properties of a Game

In this subsection we will list some basic properties that coalitional games may or may not exhibit. All of them are very common in the literature and can be found in any textbook on cooperative game theory, for example Forgó et al. (1999) or Bilbao (2000). Our list is by no means exhaustive. We only try to gather those properties used in subsequent parts of this work or with a direct relationship to it. The order, when logically possible, is intended to move in what we believe from the strongest assumption (symmetry) to the weakest one (monotonicity), if possible arranged to imply one another along the way. This of course always depends on the context

[31] In a unanimity game u_S, not only the value of all coalitions $T \subsetneq S$ but also that of any $R \subseteq N \setminus S$ is zero.

and could be interpreted vastly different in another situation. We finish with some examples showing that certain properties, even though intuition would say so, need not imply one another.

2.3.4.1 Symmetry

Definition 2.22. A coalitional game (N,v) is said to be *symmetric*, if its characteristic function $v(S)$ depends only on the cardinality of its argument $S \subseteq N$:

$$v(S) = v(S') \text{ for all } S, S' \subseteq N \text{ such that } |S| = |S'| \tag{2.27}$$

In symmetric games, the identity of the individual players is negligible for the creation of value, only their number in any cooperation agreement counts. This is often assumed for games of voting or whenever a majority is crucial. Also in so-called k-games the size of a coalition is very important, albeit the identity of the players will not be lost:

2.3.4.2 k-Games

Definition 2.23. A coalitional game (N,v) is a *k-game*, if its characteristic function v takes the following form:

$$v(S) = \sum_{T \subseteq S: |T| = k} v(T) \text{ for all } S \subseteq N \tag{2.28}$$

Here, the value of a given coalition $S \subseteq N$ depends not on its cardinality per se, but on the number of its possible subsets with cardinality k and their respective value: Only within these the actual gains are accrued. Note though that no symmetry assumption rests on the value of the k-sized subsets, so their values do depend on the individual players. In this context, a 1-game is simply an additive game with $v(S) = \sum_{i \in S} v(\{i\})$ for all $S \subseteq N$, where each member can potentially realize some gain, but coalition building does not lead to actual gains from cooperation. Since this also holds for $S = N$, the 1-game is also inessential. (See (2.17).)

With increasing cardinality of a set S the number of possible subsets $T \subseteq S$ increases as well, which can be concluded from rearranging the binomial coefficient:

$$\binom{s}{k} = \frac{s!}{k! \cdot (s-k)!} = \frac{(s-1)!}{k! \cdot (s-k-1)!} \cdot \frac{s}{s-k}. \tag{2.29}$$

Unless $s = k$, for which $\binom{s}{k} = 1$ by definition, the factor $\frac{s}{s-k}$ in (2.29) is strictly larger than one, which implies that the number of possible subsets of S with cardinality k rises when we increase the cardinality of S. From this it can be shown that all nonnegative k-games, where $v(S) \geq 0$ for all $S \subseteq N$, are *convex*, a concept introduced below.

2.3.4.3 Supermodularity and Convexity

Informally, the notions of supermodularity and convexity impose on the characteristic function of a game (N,v) are nondecreasing returns of joining ever-larger coalitions for cooperation. The first is not so transparent in its definition, but the second, based on marginal values of cooperation, is more so.

Definition 2.24. A coalitional game (N,v) is supermodular, if

$$v(S \cap T) + v(S \cup T) \geq v(S) + v(T) \ \text{ for all } \ S,T \subseteq N. \tag{2.30}$$

This not very intuitive expression states that a merger of two coalitions S and T must create no less gains than the sum of S and T. In case the two coalitions overlap, the value of the overlap must be considered as well.

More intuitive, because in terms of marginal values of individual players, the same is realized through the notion of convexity, as we will show below.

Definition 2.25. A game (N,v) is *convex*, if the value of a coalition increases no slower when these coalitions grow in size:

$$v(S \cup \{i\}) - v(S) \leq v(T \cup \{i\}) - v(T) \ \text{ for all } \ S \subseteq T \subseteq N,\ i \in N \setminus T. \tag{2.31}$$

When adding some player i to the smaller coalition S, his marginal value cannot exceed his marginal value for joining the larger coalition T. It is important to note that this criterion does not, like symmetry, rely only on the size of the coalitions, but also on the fact that one is nested within the other. It is worthwhile to note that the existence of dummy, and even null-players does not contradict convexity. They are a borderline case, as their marginal values are constant.

The idea of the marginal value of a player i to any coalition can be formalized through a function $d_i : 2^N \to \mathbb{R}$. It is defined for all $S \subseteq$ as

$$d_i(S) := \begin{cases} v(S \cup \{i\}) - v(S) & \text{if } i \notin S \\ v(S) - v(S \setminus \{i\}) & \text{if } i \in S \end{cases}. \tag{2.32}$$

This also allows us to rewrite expression (2.31) in terms of d_i, to state convexity more concise:

$$d_i(S) \leq d_i(T) \ \text{ for all } \ S \subseteq T \subseteq N,\ i \in N. \tag{2.33}$$

As we will show in Example 2.4, the nondecreasing marginal values which define convexity, can also appear as nonincreasing marginal burdens. Both will ensure increasing returns of a growing coalition.

2.3.4.4 How Supermodularity Implies Convexity and Vice Versa

We show now that supermodularity and convexity indeed imply one another, even though from expressions (2.30) and (2.31) this is not readily apparent.[32]

Proposition 2.2. *A coalitional game (N,v) is supermodular if and only if it is convex.*

Proof. To show that supermodularity implies convexity of v, we use in the set $S\cup\{i\}$ instead of S in (2.30):

$$v((S\cup\{i\})\cap T)+v((S\cup\{i\})\cup T)\geq v(S\cup\{i\})+v(T) \text{ for all } S,T\subseteq N. \quad (2.34)$$

The definition of convexity relies on the assumption that $S\subseteq T$, and so (2.34) simplifies to

$$v(S)+v(T\cup\{i\})\geq v(S\cup\{i\})+v(T) \text{ for all } S,T\subseteq N\,,\ S\subseteq T. \quad (2.35)$$

Rearranging the terms yields (2.31), which is the definition of convexity.

The other direction is more complicated: Because for convexity of a game, expression (2.31) holds for all $i\in N$ and for all $S,T\subseteq N\setminus\{i\}$, we can expand it in a particular fashion. For this, let us index from 1 to r all "remaining" players from the complementary set $N\setminus T=\{i_1,i_2,\ldots,i_r\}$ in an arbitrary order. By successively adding those players to S and T, a system of inequalities arises:

$$\begin{gathered}
v(S\cup\{i_1\})-v(S)\leq v(T\cup\{i_1\})-v(T)\\
v(S\cup\{i_1\}\cup\{i_2\})-v(S\cup\{i_1\})\leq v(T\cup\{i_1\}\cup\{i_2\})-v(T\cup\{i_1\})\\
\vdots\\
v(S\cup\{i_1\}\cup\cdots\cup\{i_r\})-v(S\cup\{i_1\}\cup\cdots\cup\{i_{r-1}\})\\
\leq v(T\cup\{i_1\}\cup\cdots\cup\{i_r\})-v(T\cup\{i_1\}\cup\cdots\cup\{i_{r-1}\}).
\end{gathered}$$

Regardless of the order chosen, at each step of the above summation, i.e. for all $R\subseteq N\setminus T$, the inequalities add up to

$$v(S\cup R)-v(S)\leq v(T\cup R)-v(T) \text{ for all } S\subseteq T\subseteq N,\ R\subseteq N\setminus T. \quad (2.36)$$

Rename the sets in (2.36) such that $S=:S_0\cap T_0$, $T=:T_0$, and $R=:S_0\setminus T_0$. Note, this does note violate the requirements $S\subseteq T\subseteq N$ and $R\subseteq N\setminus T$. Then,

$$v((S_0\cap T_0)\cup S_0\setminus T_0)-v(S_0\cap T_0)\leq v(T_0\cup S_0\setminus T_0)-v(T_0), \quad (2.37)$$

[32]Our approach follows closely that of Moulin (1991, pp. 112f.).

which can be simplified to

$$v(S_0) - v(S_0 \cap T_0) \leq v(T_0 \cup S_0) - v(T_0) \quad \text{for all} \quad S_0, T_0 \subseteq N. \tag{2.38}$$

Note that S_0 and T_0 are arbitrary coalitions from 2^N, and so (2.38) is easily rearranged to the definition of supermodularity, as given by (2.30). ■

As a special case of supermodularity, one where the sets S and T in expression (2.30) have empty intersection, we introduce superadditivity.

2.3.4.5 Superadditivity

Definition 2.26. A coalitional game (N, v) is said to be *superadditive*, if for any pair of coalitions $S, T \subseteq N$ with empty intersection, $S \cap T = \emptyset$, the following holds:

$$v(S \cup T) \geq v(S) + v(T). \tag{2.39}$$

It is assumed that two disjoint coalitions when merged will do at lest as well as they did in the aggregate before the merger. This notion can be generalized to an arbitrary number of disjoint coalitions:

$$v\left(\bigcup_k S_k\right) \geq \sum_k v(S_k), \quad \text{with } S_k \subseteq N \quad \text{for all} \quad k, \; S_k \cap S_l = \emptyset \text{ for } k \neq l. \tag{2.40}$$

If we take apart (2.39) player by player, we arrive at the following chain of weak inequalities:

$$\begin{aligned} v(S \cup T) &\geq v(S) + v(T) \\ &\geq v(S \setminus \{i\}) + v(\{i\}) + v(T \setminus \{j\}) + v(\{j\}) \\ &\vdots \\ &\geq \sum_{i \in S \cup T} v(\{i\}). \end{aligned} \tag{2.41}$$

Simplified, we obtain

$$v(S \cup T) \geq \sum_{i \in S \cup T} v(\{i\}) \quad \text{for all } S, T \subseteq N, S \cap T = \emptyset. \tag{2.42}$$

Choosing coalitions S and T so that they partition N, yields (2.16), and hence a superadditive game is also rational.

Corollary 2.1. *If a coalitional game $(N, v,)$ is superadditive, it is also rational.*

Proof. Follows directly from the definition of superadditivity and expressions (2.41) and (2.42) when choosing sets S and T such that $S \cup T = N$ and $S \cap T = \emptyset$. ■

The other direction is generally not fulfilled. Related to this is the notion of weak superadditivity:

Definition 2.27. A coalitional game (N, v) is *weakly superadditive*, if

$$v(N) \geq v(S) + \sum_{i \in N \setminus S} v(\{i\}) \quad \text{for all } S \subseteq N. \tag{2.43}$$

We can quickly show that superadditivity of a coalitional game also implies that it is weakly so:

Proposition 2.3. *If a coalitional game $(N, v,)$ is superadditive, it is also weakly superadditive.*

Proof. Take a partition P of the player set $N = \{1, 2, \ldots, k, \ldots, n\}$ such that $\mathsf{P} = \{S, S_k, S_{k+1}, \ldots, S_n\}$ with $S = \{1, \ldots, k-1\}$ and $S_l = \{l\}$ for $l = k, \ldots, n$. Substituting this partition step by step for N under the assumption of superadditivity of v yields a chain of (in)equalities,

$$\begin{aligned} v(N) = v(S \cup S_k \cup \ldots \cup S_n) &\geq v(S \cup S_k \cup \ldots \cup S_{n-1}) + v(S_n) \\ &\geq v(S \cup S_k \cup \ldots \cup S_{n-2}) + v(S_{n-1}) + v(S_n) \\ &\;\vdots \\ &\geq v(S) + \sum_{l=k}^{n} v(S_l) = v(S) + \sum_{i \in N \setminus S} v(\{i\}), \end{aligned}$$

whose first and last element make up (2.43) from Definition 2.27. ■

The next notion we want to analyze is that of monotonicity. It is not particularly strong but has some far-reaching implications.

2.3.4.6 Monotonicity

Definition 2.28. A coalitional game (N, v) is *monotonic*, if the addition of members to a coalition cannot lower its value under v:

$$v(T) \geq v(S) \quad \text{for all } S, T \subseteq N,\ S \subseteq T. \tag{2.44}$$

In other words, the marginal value of a player or a coalition of players is always nonnegative. Because of the convention that $v(\emptyset) = 0$, it follows that any monotonic game is also nonnegative: The empty set, as the smallest possible coalition, yields a value of 0 and because no non-degenerate coalition can result in a strictly smaller value, $v(S) \geq 0$ for all $S \subseteq N$. Again, the other direction does not generally hold true, i.e. a nonnegative games need not be monotonic as well. Reversely, if we assume nonnegativity of v to begin with, we can show that superadditivity implies monotonicity:

Proposition 2.4. *If a coalitional game (N,v) is superadditive and nonnegative, it is also monotonic.*

Proof. Pick two coalitions S and T with $S \subseteq T$. Because $T = S \cup (T \setminus S)$, we can establish the following chain:

$$\begin{aligned} v(T) &= v(S \cup (T \setminus S)) \\ &\geq v(S) + v(T \setminus S) \\ &\geq v(S). \end{aligned} \tag{2.45}$$

The first equality is the result of an elementary set theoretic operation. The second line results from superadditivity, and the third from nonnegativity. Then, the first and last element of (2.45) constitute (2.44) of Definition 2.28. ■

Two derivatives of monotonicity appear in the literature on occasion, that of *zero-monotonicity* and that of *N-monotonicity*.

Definition 2.29. A coalitional game (N,v) is *zero-monotonic*, if its $(0,1)$-normalized form $(N,\overline{v})$ is monotonic.

That monotonicity does not imply zero-monotonicity is quickly established: When we try to transform (2.44) according to the normalization in (2.13), we already fail in the first step. After subtracting the sum of the stand-alone values of the members of S and T on the left- and right-hand side, respectively, (2.44) could be violated. Because they are strategic equivalents, the reverse case holds: Any $(0,1)$-normalized game which is monotonic can be retransferred into its original game with monotonicity still intact. The inequality of (2.44) is never violated when retransformed back from (2.13).

Definition 2.30. A coalitional game (N,v) is *N-monotonic*, if it is nonnegative and no coalition can attain a higher value than the grand coalition:

$$0 \leq v(S) \leq v(N) \quad \text{for all} \quad S \subseteq N. \tag{2.46}$$

This allows for constellations where players or sets of players might have a negative marginal value, albeit never when the resulting set is the grand coalition, and within the boundary of nonnegativity.

The following property, decomposability, is again related to convexity and superadditivity.

2.3.4.7 Decomposability

Definition 2.31. A coalitional game (N, v) is decomposable with respect to partition P, if for all $S \subseteq N$ it holds that

$$v(S) = v(S \cap P_1) + v(S \cap P_2) + \cdots + v(S \cap P_m). \tag{2.47}$$

The restrictions $v(S \cap P_i)$, $i = 1, \ldots, m$, of the characteristic function v on the m elements of P are called components of the game (N, v). They are denoted $(P_i, v|_{P_i})$, where $v|_{P_i}$ is the function v with reduced domain 2^{P_i}.

According to this definition the characteristic function v is additive over the elements of the partition P.[33] Its values are the sum of the values attained in each component game $(P_i, v|_{P_i})$. The finest decomposition theoretically possible is a partition of N into singleton coalitions, which would be based on an inessential game as presented in (2.17). How fine a decomposition can be found always depends on the underlying value function v. Take, for example, strictly convex games, where (2.31) holds with strict inequality. Then, a decomposition is not possible at all. Shapley (1971, p. 15) provides more on the relation between decomposability and convexity of a coalitional game. For us, the notion of decomposability will become especially useful in the context of games on network structures.

We conclude this section with two examples to show that the properties of *convexity* and *monotonicity* of game, or rather the underlying value function v, need not be prerequisites of one another, even though the former property often includes the latter.

The first, Example 2.3, offers a game, that is monotonic but not convex. The second, Example 2.4, treats just the opposite case. In both examples we have constructed the value functions such that the highlighted properties are somewhat systematic and do not break down due to an arbitrary change in one instance, i.e. as given by an exception.

Example 2.3. Consider the game (N, v), where $N = \{1, \ldots, 5\}$. The characteristic function takes the following form:

$$v(S) = \begin{cases} |S|^2 - 4 & \text{if } |S| = 4 \text{ and } i = 3 \in S \\ |S|^2 & \text{otherwise} \end{cases}, \text{ for all } S \subseteq N. \tag{2.48}$$

Player 3 has been picked arbitrarily, the lower value he imposes as a member of coalitions with $|S| = 4$ could just as well have been assigned to any other player. The game is clearly monotone, as $v(S) \leq v(T)$ whenever $S \subseteq T \subseteq N$. But when we look at convexity, we can see that this is not fulfilled in the following case: Pick any coalition S such that $|S| = 2$ and another coalition T, where $|T| = 3$, $S \subsetneq T$, and

---

[33] Let $\mathsf{P} = \{P_1, P_2, \ldots, P_m\}$ be a partition of N, where $P_i \cap P_j = \emptyset$ for all $i \neq j$ and $\cup_{i=1}^m P_i = N$.

$i = 3 \notin T$. Now, (2.31) is violated for $i = 3$, because adding player 3 to the smaller coalition S yields a higher increase in value than adding him to coalition T. So the game (N, v) is monotone but not convex.

The next example provides the converse case, a game which is convex but not monotonic:

Example 2.4. Again, consider a game (N, v), where $N = \{1, \dots, 5\}$. This time, the characteristic function is given by

$$v(\{i\}) = 2 \quad \text{for all } i \in N, \tag{2.49}$$

and,

$$v(S) = \begin{cases} (|S|-1) \cdot \sum\limits_{i \in S \setminus \{3\}} v(\{i\}) - \dfrac{1}{|S|} & \text{if } i = 3 \in S \\ |S| \cdot \sum\limits_{i \in S} v(\{i\}) & \text{if } i = 3 \notin S \end{cases}, \quad \text{for all } S \subseteq N,\ |S| > 1. \tag{2.50}$$

Once again, the choice of player 3 is arbitrary, any other player would serve the purpose equally well. This game is not monotonic, as the addition of player 3 to any coalition will always result in an overall value that is strictly lower than without player 3. Convexity, on the other hand is not violated, which can be checked in three steps: In one case player 3 is not involved at all. Here, the marginal value of a player $i \neq 3$ to any coalition $S \subseteq N \setminus \{3\}$ is always $4 \cdot |S| + 2$. This expression is increasing with the size of the coalition to which the player is added, and so (2.31) must hold.

Given any coalition $S \subseteq N$ for which $3 \in S$, adding a player yields the positive margin

$$4 \cdot |S| + \frac{1}{|S| \cdot (1 + |S|)} - 2.$$

This margin is increasing with the size of the coalition S, and again, (2.31) is fulfilled.

Finally, for any coalition $S \subseteq N$ with $3 \notin S$, adding player 3 yields the negative margin $-1/(|S|+1)$. Because this negative margin decreases with the size of coalition S, (2.31) still holds.

This example involves a somewhat peculiar case of convexity with respect to player 3: Instead of positive margins that increase with the size of the coalition to which player 3 is added, we observe negative margins that shrink. This is sufficient to construct the desired violation of monotonicity, but does not clash with the definition of convexity as given in (2.31).

We now turn out attention to dual games and an appropriate application.

2.3.5 Dual Games and the Tennessee Valley Authority

So far, we have only referred to the characteristic function of a coalitional game as a *value* function, one that measures the value, or gain, players might realize through cooperation. The main reason for this is the fact that subsequent parts of this work only treat value-based characteristic functions focussing on the gains from cooperation. Nevertheless, we do not want to deprive the reader of other uses and applications of a characteristic function in the realm of coalitional games. One prominent use, at least in practice, is the notion of a cost function, whereby costs are assigned to various projects represented by the players. Then, coalitions are joint undertakings of these projects and ideally the cost of such a joint venture is less than the sum of their individual costs. From these considerations arise so-called cost allocation problems, e.g. as in accounting. The aim is to resolve which of the cooperating projects should benefit to what extend from the reduction in costs, or savings, realized through cooperation. One of the most famous applications, which we illustrate below, is known as the *Tennessee Valley Authority problem*. Before we approach this example, we want to introduce the notion of a *dual game*. It is used to relate the two most common applications of coalitional games: Cost- and surplus-sharing games, which are duals of one another. In addition, we will also introduce the concept of a savings game, based on the cost-sharing game. These concepts are then illustrated along the aforementioned application.

First, let us define what a dual game is, according to Bilbao (2000, p. 4):

Definition 2.32. Given a coalitional game (N,v), its *dual game*, denoted by (N,v^*) is defined as:

$$v^*(S) = v(N) - v(N \setminus S) \quad \text{for all } S \subseteq N. \tag{2.51}$$

The reverse relationship is quickly checked to ensure that (N,v) is the dual game of (N,v^*) as well, and $v(S) = v^*(N) - v^*(N \setminus S)$ for all $S \subseteq N$. We simply substitute:

$$\begin{aligned} v(S) &= v^*(N) - v^*(N \setminus S) \\ &= v(N) - v(\emptyset) - v(N) + v(N \setminus (N \setminus S)) \\ &= v(N) - v(N) + v(S) \\ &= v(S). \end{aligned}$$

For singleton coalitions $S = \{i\}$, the value of v^* can be interpreted as the marginal contribution of player i to the grand coalition; in case of a cost game, it is known as the *separable cost* of player $\{i\}$. For larger coalitions $|S| > 1$ this intuitive approach breaks down, as the marginal value or cost of a coalition is not as meaningful, and especially less useful when it comes to splitting up the costs among the players. Note also that $v(N) = v^*(N)$.

The definition of a savings game is equally straightforward and has even more intuitive appeal:

Definition 2.33. Given a cost game (N,c), its corresponding *savings game*, denoted by (N,s) is defined by:

$$s(S)=\sum_{i\in S}c(\{i\})-c(S)\ \text{ for all }\ S\subseteq N. \tag{2.52}$$

The value of $s(S)$ equals the savings that the members of S can achieve by cooperation, as opposed to individual undertakings. It cannot, though, be applied to considerations regarding marginal savings, when some player joins an already existing coalition. Such marginal savings are rather calculated through subtracting the savings created by coalition S from those created by $S\cup\{i\}$, when $i\notin S$. Next, let us demonstrate by means of an example how cost- and surplus-sharing games are duals, and how the savings game relates to either one.

2.3.5.1 Duality Applied: The Tennessee Valley Authority

Possibly the most famous example on cost allocation is referred to as *Tennessee Valley Authority*, or simply TVA. It is a corporation owned by the federal government of the United States, set up by Congress in 1933. Back then, its goal was to realize three different but overlapping purposes to be undertaken in the Tennessee River basin: Improving conditions for navigation on the river, better flood control through dams along the banks of the river, and the generation and provision of electricity. All three projects involved the construction of certain dams along the Tennessee River. Each undertaking could be realized on its own, but at a higher cost than when realized in cooperation. The resulting cost allocation problem was originally considered by Ransmeier (1942), before any of the now standard allocation or solution concepts for coalitional games had been introduced to the literature. The subject was revisited with a more game theoretic approach by Straffin and Heaney (1981). We limit ourselves to a comparison of the types of characteristic functions introduced above.

Let us denote the standard game at hand by (N,c), where the player set $N=\{a,f,e\}$ includes the three purposes n**a**vigation, **f**lood, and **e**lectricity. The characteristic function, or more precisely, the cost function $c:2^{\{a,f,e\}}\to\mathbb{R}_+$ assigns costs to the possible combinations of the three purposes. From it, we derive the dual function $c^*:2^{\{a,f,e\}}\to\mathbb{R}_+$, as well as the savings function $s:2^{\{a,f,e\}}\to\mathbb{R}_+$, and, again, its dual function $s^*:2^{\{a,f,e\}}\to\mathbb{R}_+$. All values are given in Table 2.1 below.

The column corresponding to $c(S)$ is the basic information given by Straffin and Heaney (1981, p. 41), listing the cost for each individual undertaking, as well as for all their possible combinations. Clearly, cooperation is beneficial in all (nontrivial) coalitions, because the resulting cost is always lower than the sum of the cost of its parts. This is depicted in the column corresponding to $s(S)$. The column in between contains the dual game of the original cost function and is harder to interpret. For the

Table 2.1 Costs given in units of *USD* 1000 at historical values

$S \subseteq N$	$c(S)$	$c^*(S)$	$s(S)$	$s^*(S)$
$\{a\}$	163 520	45 214	0	118 306
$\{f\}$	140 826	33 763	0	107 063
$\{e\}$	250 096	110 977	0	139 119
$\{a,f\}$	301 607	162 488	2 739	141 858
$\{a,e\}$	378 821	271 758	34 795	141 858
$\{f,e\}$	367 370	249 064	23 552	141 858
$\{a,f,e\}$	412 584	412 584	141 858	141 858

first three entries, corresponding to the individual purposes, the value equals their marginal cost of achieving the grand coalition. The latter three are equally marginal costs, but rather those of coalitions than those of individual undertakings.

Interesting is also the rightmost column, corresponding to $s^*(S)$. It lists the marginal savings of a purpose, or a coalition of purposes, with respect to the grand coalition. But only the marginal savings of the single undertakings are really instructive. As, by definition, no single purpose can have nonzero savings under s, for coalitions with cardinality $n-1$ the marginal savings equal the total savings. And because $n=3$, the four last entries must be identical in this column.

2.3.5.2 Relationship

Intuitively, a dual game always provides the marginal values of a coalition S joining its complement in N, whether it be costs or savings. As we have seen above, the dual relationship holds in both directions. This is not the case for the savings function: Take for example the cost function

$$\widetilde{c}(S) := c(S) + |S| \cdot \kappa \text{ for all } S \subseteq N,$$

where $\kappa \in \mathbb{R}$. Plugging this in Definition 2.33, and rearranging, yields

$$\begin{aligned} s(S) &= \sum_{i \in S} \widetilde{c}(\{i\}) - \widetilde{c}(S) \\ &= \sum_{i \in S} (c(\{i\}) + |\{i\}| \cdot \kappa) - c(S) - |S| \cdot \kappa \\ &= \sum_{i \in S} c(\{i\}) + |S| \cdot \kappa - c(S) - |S| \cdot \kappa \\ &= \sum_{i \in S} c(\{i\}) - c(S), \end{aligned}$$

which does not allow for a bijection between cost and savings functions. It is always possible to deduce a unique savings function, but the reverse is not true.

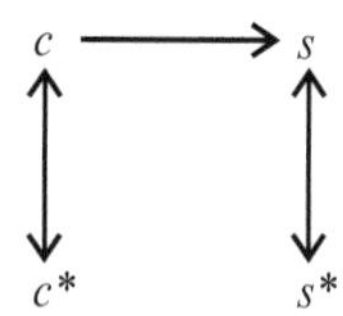

Fig. 2.6 Relations between cost and savings functions and their duals

When we compare dual cost and dual savings functions, generally no direct relation can be established: Applying Definition 2.33 to the dual of the cost function c^* and expressing this in terms of the original cost function c leads to the following:

$$s^*(S) = \sum_{i \in S} c^*(\{i\}) - c^*(S) = \sum_{i \in S} (c(N) - c(N \setminus \{i\})) - c(N) + c(N \setminus S).$$

Likewise, solving the dual of the savings function s^* in terms of the underlying cost function results in

$$s^*(S) = \sum_{i \in S} c(\{i\}) - c(N) + c(N \setminus S).$$

Both expressions coincide only if

$$\sum_{i \in S} (c(N) - c(N \setminus \{i\})) = \sum_{i \in S} c^*(\{i\}),$$

i.e. if the players' marginal values to the grand coalition are equal to their stand-alone values. This is for example the case in additive games where all players are dummies, but does not have to hold for general games. The relationship between characteristic functions for costs, savings and their duals is depicted in Fig. 2.6. With this, we conclude the section on coalitional games, their specific classes and properties. The next section is devoted entirely to finding solutions for value-based games, i.e. on how to distribute the gains from cooperation among the players. As we will see, a wide variety of solution concepts has been brought forward in the literature. They range from less specific ones that provide a breath of possible outcomes or boundaries within which outcomes are deemed to be acceptable, to solutions that only provide a single answer to the question of distribution.

2.4 Solutions Concepts for Cooperative Games

2.4.1 Overview

So far, in our study of cooperative game theory, we have only considered gains from cooperation which are represented by a single number assigned to the respective coalition. No thought has been given as to how these gains should be distributed among the cooperating players. This exactly is the main motivation of this remaining chapter, where we analyze multiple solution concepts for coalitional games. Such solution concepts, or short, solutions, are procedures specifying how the gains in a given game should be distributed. Depending on one's aim, there are many solutions to a game, in principle as many as there are possibilities to allocate the gains, i.e. infinitely many. We will of course restrict ourselves to the solutions that suit our endeavor best. Coarsely, one can distinguish between two classes of solutions: On the one hand, those delivering potentially more than one possibility of distribution, referred to as set-valued solutions. On the other hand, those resulting in a unique distribution of the gains to the players, called point-valued solutions. We begin our treatment with the former type, narrowing in on the set of all possible distributions. The first concept we look at are *von Neumann Morgenstern solutions*, or *stable sets*, which originated in the seminal work von Neumann and Morgenstern (1947). We then turn to a related solution, the *core*, formally introduced in Gillies (1953). Broadly, both methods designate as outcome such distributions to which no coalition of players would object, because they could not do better on their own. For the core we also consider some conditions for existence, especially in connection with convex games. We end our reflections on set-valued solutions by relating core and stable sets, and by introducing a method with which to possibly narrow down the set of solutions provided by the core.

Before we turn to specific point-valued solution concepts, we introduce them in all generality as *allocation rules*, along with selected properties they can exhibit. The first specific such rule we consider is the *Shapley value*, introduced to the literature in Shapley (1953b). It distributes to the players their average marginal values upon joining a coalition. We also consider a variant of the Shapley value, the *Weighted Shapley value*, which allows to skew the resulting allocation within certain boundaries by assigning weights to the players. Surprisingly, despite polar approaches in construction, these boundaries are closely related to the core.

The last concept we introduce are so-called *bargaining solutions*. These solutions are not directly based on cooperative games, but as we will show, a transformation to create what is called a *bargaining problem* is fairly straightforward. The bargaining solutions we are particularly interested in are the *Nash solution* and the *Kalai–Smorodinsky solution* as well as their weighted variants. The original concepts appeared first in the literature in Nash (1950) and Kalai and Smorodinsky (1975). To conclude this overview, we introduce some basic notation related to the possible distributions of the gains of a coalitional game. Our approach is, at least conceptually, based on Forgó et al. (1999, pp. 222 ff.) and Shapley and Shubik (1973, pp. 29 ff.).

To begin with, what was clumsily referred to so far as "possible distribution of gains among players" is called simply an *allocation*. This expression is used for an arbitrary allotment of payments to the n players of the game (N, v). It is represented by a vector $\mathbf{x}$, which is n-dimensional and real-valued, $\mathbf{x} \in \mathbb{R}^n$. Its n entries are denoted by x_i for all $i \in N$. To simplify notation, especially within passages of text, we use an additive function $x : 2^N \to \mathbb{R}$, defined for a specific allocation $\mathbf{x}$:

$$x(S) = \sum_{i \in S} x_i \quad \text{for all } S \subseteq N. \tag{2.53}$$

For allocations **y** or **z**, function (2.53) is denoted $y(\cdot)$ or $z(\cdot)$, respectively, and so on.

We can now start to impose certain restrictions on the broad set of allocations $\mathbb{R}^n$ in order to narrow down the space in which we operate, possibly even before considering a certain solution concept. The first cut we make, is to restrict ourselves to feasible payoff vectors:

Definition 2.34. An allocation $\mathbf{x} \in \mathbb{R}^n$ in a coalitional game (N, v) is *feasible*, if can be achieved by the grand coalition:

$$\sum_{i \in N} x_i = x(N) \leq v(N). \tag{2.54}$$

Feasibility limits the sum (in balance) of what is to be distributed, $x(N)$ for an allocation **x**, to what is available, in our case the proceeds from forming the grand coalition $v(N)$. This requirement is a boundary from above and sets the limit to what can be allocated. We have restricted the set of all allocations, $\mathbb{R}^n$ to a half-space $H_-(\mathbf{1}, v(N))$, bounded above by the hyperplane $H(\mathbf{1}, v(N))$. The set of such feasible payoff vectors for a game (N, v) is defined as

$$F(N,v) := \{\mathbf{x} \mid \mathbf{x} \in \mathbb{R}^n,\, x(N) \leq v(N)\} = \{\mathbf{x} \mid \mathbf{x} \in H_-(\mathbf{1}, v(N))\}. \tag{2.55}$$

Next, we restrict ourselves to efficient allocations:

Definition 2.35. An allocation $\mathbf{x} \in \mathbb{R}^n$ in a coalitional game (N, v) is *efficient*, if it exhausts the gains from forming the grand coalition:

$$\sum_{i \in N} x_i = x(N) = v(N). \tag{2.56}$$

This refines feasibility, as (2.54) now must hold with equality, and ensures that none of the proceeds from the grand coalition are wasted. We now are confined to the hyperplane $H(\mathbf{1}, v(N))$, which bounds $F(N, v)$. The set of efficient payoff vectors of a game (N, v) is in turn defined as

$$E(N,v) := \{\mathbf{x} \mid \mathbf{x} \in \mathbb{R}^n,\, x(N) = v(N)\} = \{\mathbf{x} \mid \mathbf{x} \in H(\mathbf{1}, v(N))\}. \tag{2.57}$$

Because $H(\mathbf{1}, v(N)) \subsetneq H_-(\mathbf{1}, v(N))$, also $E(N, v)$ is a proper subset of $F(N, v)$, $E(N, v) \subsetneq F(N, v)$. In addition, $E(N, v)$ happens to coincide with the set of *Pareto-optimal* allocations within $F(N, v)$. As long as $x(N) = v(N)$, no player can be allocated more without lowering what other players receive. Unfortunately, as there are no other restrictions yet, an arbitrarily high payoff to player i can always be offset by its negative counterpart to some other player j, which still leaves an $(n-1)$-dimensional real-valued space. $E(N, v)$ is also called the set of *preimputations*.

Finally, individual rationality is imposed:

Definition 2.36. An allocation $\mathbf{x} \in \mathbb{R}^n$ in a coalitional game (N,v) is *individually rational*, if no player is allocated less than his stand-alone value:

$$\sum_{i \in N} x_i \geq v(\{i\}) \qquad \text{for all } i \in N. \tag{2.58}$$

Now, players must receive at least as much under allocation $\mathbf{x}$ as if they would not cooperate at all. This requirement can be seen as an individual boundary from below, so players cannot be exploited by allocations that fall short of their singleton values. By itself, individual rationality is not so much a restriction, but imposing this condition on the set of preimputations $E(N,v)$ will curtail the hyperplane $H(\mathbf{1},v(N))$ such that it becomes a bounded set. We obtain the set of *imputations*[34]:

$$I(N,v) := \{\mathbf{x} \mid \mathbf{x} \in \mathbb{R}^n,\ x(N) = v(N),\ x_i \geq v(\{i\}) \text{ for all } i \in N\}. \tag{2.59}$$

The resulting allocations are then not only efficient but also ensure that no excessive payoffs for some players on the account of others can be realized. In short, a payoff vector $\mathbf{x} \in \mathbb{R}^n$ that is individually rational, Pareto-optimal or efficient, and feasible, qualifies as an imputation. Interestingly enough, the highest achievable gains are always attributed to the grand coalition N. This reminds of N-monotonicity (see Definition 2.30), implying that no player i can have a strictly negative marginal value upon joining $N \setminus \{i\}$. This is closely related to both superadditivity and monotonicity of v. Regardless, from now on, we restrict the playing field of possible allocations to the set of imputations $I(N,v)$, unless otherwise specified.

When comparing allocations to one another, the prevalent means of doing so is the binary relation of *dominance*. It comes in two variants:

Definition 2.37. An allocation $\mathbf{x}$ *dominates* another allocation $\mathbf{y}$ *through* coalition S, denoted $\mathbf{x} \succ^S \mathbf{y}$, if

$$x_i > y_i \text{ for all } i \in S, \text{ and} \tag{2.60}$$

$$x(S) \leq v(S). \tag{2.61}$$

Coalition S is called the *effective set*.

According to (2.60), coalition S can obtain strictly more under allocation $\mathbf{x}$ than under $\mathbf{y}$. Also, following (2.61), S can obtain this higher payoff on its own, without cooperation as part of the grand coalition.

Definition 2.38. An allocation $\mathbf{x}$ *dominates* another allocation $\mathbf{y}$, denoted $\mathbf{x} \succ \mathbf{y}$, if there exists some coalition $S \subseteq N$ for which $\mathbf{x} \succ^S \mathbf{y}$ is true.

---

[34] The expression *imputation* was introduced by von Neumann and Morgenstern (1947, p. 263), albeit for a zero-sum game with slightly different conditions.

This definition does no longer rely on a specific effective set, it is merely required that one exists. Also, there might be more than one for a given pair of vectors.

Both definitions imply that the underlying binary relations are irreflexive, because the strict inequalities in (2.60) guarantee that $\mathbf{x} \neq \mathbf{y}$. For the same reason, they are also transitive. While Definition 2.37 is an asymmetrical relation, this is not per se the case for Definition 2.38: For two given allocations $\mathbf{x}$ and $\mathbf{y}$, $\mathbf{x} \succ \mathbf{y}$ and $\mathbf{y} \succ \mathbf{x}$ can hold true, albeit for two disjoint underlying effective sets S and $\tilde{S}$.

Unless otherwise noted, we subsequently use the relation of domination from Definition 2.37, as originally introduced by von Neumann and Morgenstern (1947). A more recent account can be found in Owen (1999, p. 259) and on the domination-through relation, Definition 2.38, in Lucas (1992, pp. 550f.). Three more pieces of notation are required before we turn to the solution concepts themselves:

Definition 2.39. An allocation $\mathbf{x}^{\pi}$ is a *marginal vector*, if

$$x_i^{\pi} = v(S_{\pi,\pi(i)}) - v(S_{\pi,\pi(i)} \setminus \{i\}) \text{ for all } i \in N, \tag{2.62}$$

where $S_{\pi,i} = \{j \in N \,|\, \pi(j) \leq i\}$.

For each permutation π, the marginal vector $\mathbf{x}^{\pi}$ contains as elements the marginal values of the players when added one by one to an ever-growing coalition in the order given by π. Each game has $n!$ marginal vectors (which need not be distinct), and we speak of *neighboring* marginal vectors $\mathbf{x}^{\pi}$ and $\mathbf{x}^{\pi'}$, if the orderings π and π' differ only in one pair of players. This is also called a transposition.

Definition 2.40. The *characteristic vector* of a coalition $S \subseteq N$, denoted $\mathsf{X}_S \in \{0,1\}^n$, is composed of n binary components assuming value

$$\mathsf{X}_S^i = \begin{cases} 1 & \text{if } i \in S \\ 0 & \text{otherwise} \end{cases}, \quad \text{for all } \; i \in N, S \subseteq N. \tag{2.63}$$

Definition 2.41. An *indicator function* is a mapping denoted $w_S : N \to \{0,1\}$. It assumes binary values, depending on i being an element of S or not:

$$w_S(i) = \begin{cases} 1 & \text{if } i \in S \\ 0 & \text{otherwise} \end{cases}. \tag{2.64}$$

Having set the stage, we can now proceed to out first solution concept, the *von Neumann Morgenstern solution*.

2.4.2 Stable Sets: The von Neumann Morgenstern Solution

This section is devoted to a solution concept introduced in the seminal contribution von Neumann and Morgenstern (1947). It can be taken as a run-up to the subsequent treatment of the core, which we cover most thoroughly in terms of set-valued solutions within this work. The latter can also be considered to have superseded the von Neumann Morgenstern solution in the literature, being more intuitive and plausible.

In their treatise of game theory, von Neumann and Morgenstern (1947, pp. 33 ff.) begin their considerations with a more general approach of what a solution should be. They decide to relinquish the totalitarian concept of a set of rules that give instructions to all players in any given situation in favor of something less complex: A number assigned to each player, representing the minimal payoff obtainable to him under "rational" (i.e payoff-maximizing) behavior, and instructions on how to obtain it. But instead of a solitary imputation as solution, von Neumann and Morgenstern (1947, p. 36) focus on more general sets of imputations, possessing the vague – and at first undefined – characteristics of "balance and stability." Each such set (if it exists at all) is made up by imputations that do not "dominate" one another.[35] At this point it must be noted that von Neumann and Morgenstern (1947) base most of their theory either directly or implicitly on games that are zero- or constant-sum. Where necessary, we amend their approach to conform with our treatment of general-sum coalitional games. We state the definition of their solution straight away:

Definition 2.42. A set of imputations, denoted by $\mathfrak{S} \subseteq I(N,v)$, is a *solution*, if no two of its elements dominate one another, and if for each element outside the *solution*, there exists one within, which dominates the former. More formally, these two requirements are given by:

$$\nexists\, \mathbf{x}, \mathbf{y} \in \mathfrak{S} \text{ such that } \mathbf{x} \succ \mathbf{y}, \text{ and} \tag{2.65}$$

$$\text{for all } \mathbf{z} \notin \mathfrak{S},\ \exists\, \mathbf{x} \in \mathfrak{S} \text{ such that } \mathbf{x} \succ \mathbf{z}. \tag{2.66}$$

The first requirement, (2.65), rids us of any inner contradictions: A *solution*[36] cannot contain any two imputations that dominate each other. The second requirement, (2.66), guarantees that each imputation which is not part of the *solution* is dominated by at least one element of the *solution*.

These two requirements, as innocuous as they might appear at first, have a lot more to them: It is, for example, possible that an imputation not in the *solution*, say $\mathbf{z}' \notin \mathfrak{S}$, dominates some element of the *solution* $\mathbf{y}' \in \mathfrak{S}$, i.e. $\mathbf{z}' \succ \mathbf{y}'$. But according to (2.66), there must exist at least one element of the *solution* $\mathbf{x}' \in \mathfrak{S}$, for which $\mathbf{x}' \succ \mathbf{z}'$. This allows the construction of cycles of dominance that do include elements of the *solution*.

---

[35] So far, "dominate" is to be understood as a relation comparing two imputations at a time. But the criterion for "domination" is rather vague, as no formal definition is presented and only a certain number of players are required to prefer the one imputation over the other. The formal approach is presented much later in von Neumann and Morgenstern (1947, pp. 263 ff.)

[36] For now, we *emphasize* the word "solution" whenever it appears not in its general meaning, but in the sense of von Neumann and Morgenstern (1947).

Also, it is possible, that for a given element of the solution, $\mathbf{x}'' \in \mathfrak{S}$, no element outside the solution exists, which is dominated by $\mathbf{x}''$. In such a case the imputation $\mathbf{x}''$ does not dominate *any* other imputation, not only those in the *solution*. But nevertheless, as established above, $\mathbf{x}''$ can itself be dominated by some imputation outside the *solution*. It seems counterintuitive that an imputation $\mathbf{x}'' \in \mathfrak{S}$, which does itself not dominate any other imputation can be dominated by some imputation $\mathbf{z}'' \notin \mathfrak{S}$ and still be an element of the solution. But if $\mathbf{x}''$ was not a member of the solution, (2.66) would require the existence of some other imputation $\mathbf{y}'' \in \mathfrak{S}$ for which $\mathbf{y}'' \succ \mathbf{x}''$ would have to hold, clearly contradicting (2.65). Hence, an imputation like $\mathbf{x}''$ can be an element of the *solution*.

Shapley and Shubik (1973, p. 57) refer to condition (2.65) as *internal stability* and to condition (2.66) as *external stability* of the *solution*. Therefore, as the *solution* is "*any set of imputations that is both externally and internally stable*", it is also called *stable set*. We subsequently use both designations.

The *solution* is a very precise concept, as removing or adding even one element will already contradict either (2.65) or (2.66): Suppose we add some imputation $\mathbf{z} \notin \mathfrak{S}$ to the *solution*. Immediately, we have a contradiction to (2.65), because according to (2.66) there must be some $\mathbf{x} \in \mathfrak{S}$ with $\mathbf{x} \succ \mathbf{z}$. Conversely, removing some imputation $\mathbf{y} \in \mathfrak{S}$ from the solution yields a contradiction to (2.66), because by (2.65), $\mathbf{y}$ is not dominated by any of the remaining elements in the *solution*.

From these considerations we can not, though, rule out the possibility that a given coalitional game (N, v) has more than one *solution*. But, by their definition, these multiple *solutions* would have to be mutually exclusive: Each one would constitute a *solution* in its own right but be in conflict with the others. For a more detailed look on the non-uniqueness of stable sets, see Example 2.5 below. One fact limiting the multiplicity of solutions can be stated right away: No *solution*, or stable set, is a proper subset of another stable set. This is concisely claimed by Shapley and Shubik (1973, p. 58): "*A stable set is, simultaneously, a minimal externally stable set and a maximal internally stable set.*" We sum up this statement in a proposition:

Proposition 2.5. *Any stable set, as defined by (2.65) and (2.66), is both a minimal externally stable set and a maximal internally stable set.*

Proof. We begin by showing that any stable set $\mathfrak{S}$ is a maximal internally stable set, and no imputations can be added. Suppose this was not the case, and there exists some imputation $\mathbf{z} \notin \mathfrak{S}$ fulfilling requirement (2.65) for internal stability: For all $\mathbf{y} \in \mathfrak{S}$ we have $\mathbf{z} \not\succ \mathbf{y}$ and $\mathbf{y} \not\succ \mathbf{z}$.[37] This directly contradicts the definition of a *solution*, because according to requirement (2.66), for all $\mathbf{z} \notin \mathfrak{S}$ there must exist some $\mathbf{y} \in \mathfrak{S}$ for which $\mathbf{y} \succ \mathbf{z}$ is true.

Analogously, any stable set $\mathfrak{S}$ is a minimal externally stable set, from which no imputation can be removed. Suppose this was not so, and pick an arbitrary $\mathbf{y} \in \mathfrak{S}$ as candidate for removal. This again leads to an immediate contradiction, because

---

[37]Here, "$\not\succ$" stands for "not $\succ$" or "is not dominated by".

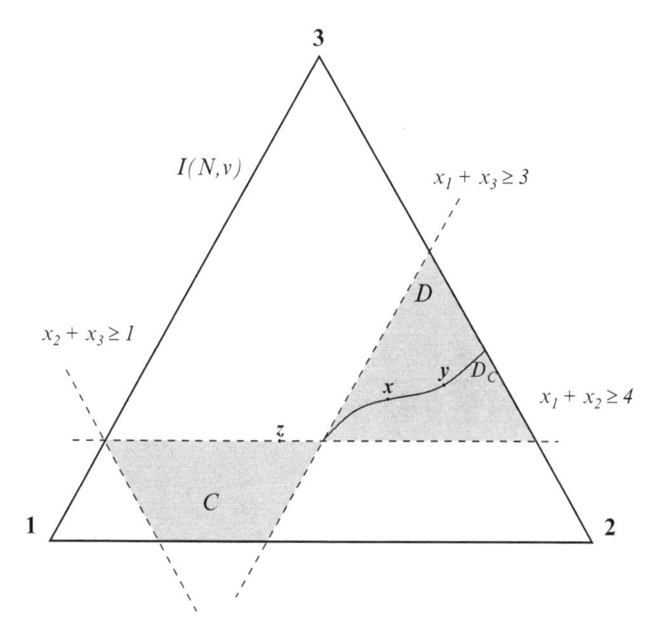

Fig. 2.7 A 3-player game with multiple stable sets

(2.66) requires that there exists some $\mathbf{x} \in \mathfrak{S} \setminus \{\mathbf{y}\}$, for which $\mathbf{x} \succ \mathbf{y}$ holds true. But since we did not change $\mathfrak{S}$ other than remove $\mathbf{y}$ from it, according to requirement (2.65) there exists no such $\mathbf{x}$. ∎

Especially the first part of the preceding proof, among other things, is also shown by von Neumann and Morgenstern (1947, pp. 266 ff.). They take a much broader approach, focussing in more detail on the effects of possible symmetries of the employed binary relations.

Corollary 2.2. *No stable set, as defined by (2.65) and (2.66), is a proper subset of another stable set.*

Proof. Follows directly from Proposition 2.5. ∎

Corollary 2.2 must be understood with care. It does not allow for a stable set being contained in another one. Nevertheless, it remains silent about the multiplicity of stable set solutions as such. The following example, taken from Lucas (1992), illustrates this potential multiplicity.

Example 2.5. (Lucas (1992, pp. 556 f.)) Consider the coalitional game (N,v) with the player set $N = \{1,2,3\}$. The characteristic function v takes the following values: $v(\emptyset) = 0$ and $v(\{i\}) = 0$ for all $i \in N$, $v(\{1,2\}) = 4$, $v(\{1,3\}) = 3$, $v(\{2,3\}) = 1$, and $v(N) = 5$. An illustration is given in Fig. 2.7 below, where the set of imputations

$I(N,v)$ is depicted in form of a simplex.[38] For a formal proof, we refer the reader to the comprehensive treatment of this topic in von Neumann and Morgenstern (1947, pp. 403 ff.). We proceed instead with a constructive argument to show why there can and must be more than one solution for this example. We begin by drawing an almost arbitrary curve,[39] denoted D_C, through the set D. It stretches from the corner of C to the border D shares with the simplex. We claim that $C \cup D_C$ is a stable set for the game (N,v).

To substantiate this claim, we first demonstrate that the requirement of internal stability, (2.65), holds for all $\mathbf{x} \in C \cup D_C$: According to (2.61), no two elements in the interior of C, $intC$, can dominate each other, because they are not affordable to any of the 2-player coalitions. This also applies to any two imputations on different edges of the boundary of C, bdC, because there exists no effective set. The comparison of imputations on the same side of bdC fails for the same reason as it violates the strict inequalities in (2.60).

For any two elements in D_C, we can restrict ourselves to the possibly effective sets $S_a := \{1,2\}$ and $S_r := \{1,3\}$, because for $S_l := \{2,3\}$ no allocation on D_C (even the shaded region D) is affordable and (2.61) does not hold. Now, for any two imputations $\mathbf{x}, \mathbf{y} \in D_C$, (2.60) is violated, because a move from $\mathbf{x}$ to $\mathbf{y}$ (or vice versa) will always make one member of the sets S_a and S_r better off, while the other one will be worse off. So neither set can be effective.

Regarding possible domination in between elements of C and D_C, we have two cases, depending on the "direction". Any element of C, say $\mathbf{z}$, cannot be dominated by, say $\mathbf{x}$, because no effective set can be found as (2.60) is going to be violated: In S_a, it is player 2 who benefits, as player 1 loses, while in S_r player 3 benefits, again at the cost of player 1. When we consider the other direction, it suffices to look at elements in the boundary of C, more precisely in the edges given by S_a and S_r, as all other elements violate affordability via (2.61). And there again, one player benefits on behalf of another (inversely to the above described) when moving from, say, $\mathbf{x}$ to $\mathbf{z}$.

Next, we establish the validity of external stability, (2.66). For the white areas above, to the right, and to the left of C, the sets S_a, S_r, and S_l, respectively, are the effective sets. For each imputation in one of these white areas, we can find some imputation on the corresponding boundary of C that dominates the former. For any imputation in $\mathbf{k} \in D \setminus D_C$ there exists some $\mathbf{y} \in D_C$ such that $\mathbf{y} \succ \mathbf{k}$ holds: If $\mathbf{k}$ lies above D_C, the effective set is S_a, if $\mathbf{k}$ lies below D_C, it is S_r. Hence, the set $C \cup D_C$ is a stable set as defined by (2.65) and (2.66).

We now explain, why this stable set is an "almost arbitrary" selection from uncountably many: As far as the nature of the curve D_C is concerned, we can

[38] The use of dashed lines in Fig. 2.7 has exclusively illustrative purposes. No indication of openness or closedness is supposed to be given by these lines. Such properties are solely described by weak or strict inequalities.

[39] What "almost arbitrary" means will become more clear during the course of the argument and is also elaborated at the end.

impose some restrictions, abusing vocabulary from functional analysis in doing so. The starting point of D_C is the nonempty intersection of C and the closure of D, clD. The endpoint of D_C is on the part of the boundary of $I(N,v)$ that intersects with D.[40] In between these points, the curve must be continuous (albeit not differentiable) and cannot exhibit any "thick" parts. Horizontal segments are possible, and even vertical ones as long as the curve satisfies what would be weakly increasing monotonicity in a function. Any deviation from these properties would result in a possible violation of (2.65) or (2.66). Other than that, the curve can be picked arbitrarily, even including the dashed lines bounding D from the left and below.

Concluding, we can see why *solutions* in this example are not only not unique, but also uncountably many stable sets exist.

Another fundamental question to the relevancy of the solution concept is that of existence of stable sets in general coalitional games (N,v). Even though von Neumann and Morgenstern (1947, p. 42) consider existence "*most desirable*", they do not provide a proof (not even within their framework of constant-sum games), and rely on the fact that "*in all cases considered so far solutions were found.*" After considerable contributions to the literature[41] concerning often special classes of games, Lucas (1968) is the first to come up with an example for a general-sum coalitional game (in his case with $n = 10$) not to have a *solution*. This is backed up by a proof in Lucas (1969). But, for a certain class, namely that of convex games, Shapley (1971) has not only proved the existence of stable sets but also their uniqueness and coincidence with another solution concept, the core. The close relation of the two concepts is already striking from their definition alone, as we shall see. Nevertheless, we dedicate a more detailed section to this topic in Sect. 2.4.3.9. For now, we conclude our considerations on stable sets and continue with the above mentioned core, a very popular solution concept.

2.4.3 *The Core*

2.4.3.1 Overview

This section introduces the most widely used (set-valued) solution concept within the theory of cooperative games, the *core*. At least conceptually it dates back to Edgeworth (1881, pp. 48 ff.). It is attributed to his considerations on the indeterminateness of contracts when "*co-operative associations*" form and how to resolve this indeterminateness with a "*principle of arbitration*". The core was first

---

[40]The curve must extend all the way to the boundary of $I(N,v)$, because we have shown that such a set is a stable set and according to Corollary 2.2, no proper subsets of stable sets are stable sets themselves.

[41]Shapley and Shubik (1973, pp. 76 f.) give an extensive and well structured overview of the literature on stable sets. So does Shubik (1982) in the appendix.

formalized in game theory and referred to under its name by Gillies (1953). The allocations resulting from the core all satisfy one characteristic: No coalition of players, from singletons to the grand coalition, can, by denying cooperation with the rest of the players, achieve more on its own. On these grounds, the allocations proposed should satisfy all players and possible coalitions and keep them from deviating into other cooperations.

We begin with different definitions of the core's concept, and continue then to conditions for its existence in coalitional games. The properties of the core in convex games is of special interest for us. Before we conclude with a method to possibly reduce the size of the set of allocations proposed by the core, we point out the latter's relation to stable sets in more detail.

2.4.3.2 Definition and Characterization of the Core

As we mentioned above, the first formalization of the core is due to Gillies (1953). Since then, the concept has appeared in countless places in the literature. Owen (1995), for example, defines the core along the lines of the domination relation, as given in Definition 2.38:

Definition 2.43. (Owen (1995, p. 218)) The set of all undominated imputations for a game (N, v) is called the core. It is denoted by $C(N, v)$.

Another similar definition is provided by Shapley (1971):

Definition 2.44. (Shapley (1971, p. 11)) The core of an n-person game (N, v) is the set of feasible outcomes that cannot be improved upon by any coalition of players.

The expression "improved upon by" can be interpreted roughly as the binary relation "dominate through" from Definition 2.37. It states that some coalition $S \subseteq N$ can improve upon an allocation $\mathbf{x}$ (which is feasible, i.e. $x(S) \leq v(S)$), if there exists another feasible allocation $\mathbf{y}$ for which $y_i \geq x_i$ holds for all $i \in S$. This coalition is what we called the effective set and the difference are the weak inequalities employed here.[42] Feasible allocations that cannot be improved upon possess the *core property* and are therefore elements of the core.

With definitions such as the above, the core is often characterized to be an n-person generalisation of Edgeworth's contract curve.[43] This usually occurs in the context of exchange economies, where no productive operations are considered. For this, we point to the introductory chapter of Hildenbrand and Kirman (1988).

More formal, but otherwise equivalent definitions of the core are available in many places in the literature. There are two very similar formal definitions of the

[42]More formally defined, this relation can be found in Mas-Colell et al. (1995, p. 653). Also, an account on why the expression "improve upon" should be used rather than the not uncommon "block" is provided by Shapley (1973). We will follow his advice throughout.

[43]The contract curve is the locus of all Pareto-efficient allocations, for a given initial endowment, within the so-called Edgeworth–Bowley box.

core. And even though only one is commonly used in the literature, we state both, for the sake of completeness. The first, the *inequality core*, is the one generally referred to as the core. It can be found for example in Peleg and Sudhölter (2003, p. 25) or Forgó et al. (1999, p. 225):

Definition 2.45 (Inequality Core). The core of a coalitional game (N,v), denoted $C(N,v)$, is given by all allocations satisfying the following:

$$C(N,v) = \{\mathbf{x} \mid \mathbf{x} \in \mathbb{R}^n,\ x(N) = v(N),\ x(S) \geq v(S) \ \text{ for all } \ S \subseteq N\}. \qquad (2.67)$$

The name arises because $C(N,v)$ is the solution to a number of inequalities. Equation (2.67) is the least redundant way of defining the inequalities core. Because $x(S) \geq v(S)$ holds for singletons and implies $x(N) = v(N)$, we could drop the latter equation and replace $\mathbb{R}^n$ with the set of imputations $I(N,v)$ in (2.67). It is often found in the literature, but we refrain from it. The above definition coincides with Definition 2.44.

The second notion is the *domination core*, defined via the relation of "domination through":

Definition 2.46 (Domination Core). The core of a coalitional game (N,v), denoted $C(N,v)$, is given by all allocations satisfying the following:

$$C(N,v) = \left\{\mathbf{x} \in I(N,v) \mid \nexists\, \mathbf{y} \in I(N,v) \ \text{ and } \ \nexists\, S \subseteq N \ \text{ such that } \ \mathbf{y} \succ^S \mathbf{x}\right\}. \qquad (2.68)$$

The less formal counterpart of Definition 2.46 is Definition 2.43. Note that the domination and the inequalities core need not coincide, but whenever the latter is nonempty, they do. This is due to the use of strict rather than weak inequalities which allows for more allocations with the core property in the latter case. So by inequalities, a game could have a nonempty core, while the same game has an empty core in terms of domination.[44] In our subsequent treatment, we relate exclusively to the inequality concept from (2.67).

In essence, three conditions are imposed on allocations $\mathbf{x} \in \mathbb{R}^n$ in order to exhibit the core property: The first and second, derived from $x(N) = v(N)$, state that an allocation in the core must be *feasible* and *efficient* at the same time, which amounts to *Pareto efficiency*. The third condition, $x(S) \geq v(S)$ for all $S \subseteq N$, is often called *individual* and *coalitional* rationality, depending on the size of S. It ensures for all coalitions, from singletons with $|S| = 1$ to the grand coalition $S = N$, that none of them is allocated less than it could achieve on its own. As already noted above, the set of core-allocation candidates can be narrowed down to the set of imputations, $I(v)$.

In the light of Definition 2.45, the set of allocations qualifying as elements of the core are the solution to a system of linear inequalities: For each of the $2^n - 2$

---

[44] For detailed conditions on this coincidence, see Chang (2000, p. 458) or Rafels and Tijs (1997, p. 492).

nonempty and proper subsets $S \subsetneq N$, the restriction $x(S) \geq v(S)$ represents a closed half-space in $\mathbb{R}^n$, denoted $H_+(\mathsf{X}_S, v(S))$. The characteristic vector X_S, see Definition 2.40, determines the slope (or direction) of the hyperplane bounding the half-space. The worth of coalition S, $v(S)$, equals the distance of this hyperplane from the origin.

Next, the efficiency restriction $x(N) = v(N)$ coincides with the hyperplane $H(\mathsf{X}_N, v(N))$. Together with the n half-spaces $H_+(\mathsf{X}_{\{i\}}, v(\{i\}))$ corresponding to $x(\{i\}) \geq v(\{i\})$ for all $i \in N$, this defines the $(n-1)$-dimensional simplex on which we operate. Because in finite-dimensional real vector spaces, all hyperplanes and half-spaces are convex sets, and because the intersection of arbitrarily many convex subsets of such a vector space is still convex, the core of a game (N, v) is a compact and convex polyhedron of dimension less than or equal to $n-1$.[45] These half-spaces and hyperplanes always have the same slopes, regardless of the game, because their respective gradients vectors, the characteristic vectors, are identical across games. It is only their individual distance to the origin of $\mathbb{R}^n$ which changes, as given by the values of the function v. According to Shapley and Shubik (1973, pp. 50 f.), the visualization via half-spaces and hyperplanes lead to the baptism of this concept as "core": Each half-space cuts from the simplex of imputations the part which is dominated through its respective coalition. Whatever (possibly) remains on the inside after all coalitions have been considered, is then called the "core" of the simplex.

From the above the following characterization can be made:

Proposition 2.6. *The core of a coalitional game, $C(N, v)$, is subset of the set of imputations $I(N, v)$: $C(N, v) \subseteq I(N, v)$. Furthermore, the core is compact (i.e. closed and bounded) and convex.*

Proof. For an empty core there is nothing to prove. For a nonempty core, each of its elements is an efficient allocation. Rearranging the efficiency condition to $x_j = v(N) - \sum_{i \neq j} x_i$ shows that in a given game (N, v) a vector of dimension $n-1$ suffices to define an efficient allocation to n players.

We also know core allocations to be individually rational, which invites a representation of all such allocations on a simplex of dimension $n-1$:

$$\Delta^{n-1} = \left\{ \mathbf{x} \in \mathbb{R}^n | \sum_{i \in N} x_i = v(N) \text{ and } x_i \geq v(\{i\}) \text{ for all } i \in N \right\}. \tag{2.69}$$

This simplex, as it happens, coincides with the set of imputations $I(N, v)$. With all the boundaries of the simplex determined by the n weak inequalities $x_i \geq v(\{i\})$ and real-valued $v(\{i\})$, it is closed and bounded, hence compact. The core, obliged to fulfill more than just those n requirements, is therefore a subset of this simplex and also of $I(N, v)$. Consequently, the core is bounded as well. The core is closed, because all inequalities of the form $x(S) \geq v(S)$ are weak and so the half-spaces

[45] If the core consists only of the empty set, these conditions are trivially fulfilled.

they describe are closed, a property that is preserved over arbitrary intersections of closed sets. So the core is also compact.

The core is convex, because it is the intersection of $2^{|N|} - 2$ convex half-spaces and a convex hyperplane, where the property of convexity again is preserved over arbitrary intersections. ■

We have now established the most current definitions of the core and characterized the set of resulting allocations. The latter exhibits the properties of boundedness and convexity, which will prove to be very convenient. Next, we turn to the question of existence, i.e. whether a nonempty core can be the result of every coalitional game, or whether certain requirements must be met.

2.4.3.3 The Existence of the Core

This section is concerned with the existence of the core in coalitional games. By definition, the core always "exists", as it may as well just be the unspecific empty set: $C(N,v) = \emptyset$. So whenever we speak of existence, we refer to a nonempty core, i.e. the existence of allocations that satisfy (2.67). We start this section with an example for a game with an empty core and then introduce different requirements for existence. For illustrative purposes and to develop some intuition, we begin with 3-player games and work our way via a necessary condition for general games to the *Bondareva–Shapley Theorem*, which is the reference criterion for games to have a nonempty core.

Example 2.6 illustrates the possibility of coalitional games (N,v) with empty cores:

Example 2.6. Given is a coalitional game (N,v) with player set $N = \{1,2,3\}$ and a characteristic function v. The latter assumes the values $v(\emptyset) = 0$ and $v(\{i\}) = 0$ for all $i \in N$, $v(\{1,2\}) = 3\frac{1}{2}$, $v(\{1,3\}) = 4$, $v(\{2,3\}) = 3$, and $v(N) = 5$. It is readily checked that v satisfies superadditivity and is essential. Also, as it turns out, we have an empty core, $C(N,v) = \emptyset$. Because the value assigned to the grand coalition N is equal to the sum of any efficient allocation, it does not suffice to satisfy the aggregate interests of the 2-player coalitions S. If each such coalition would be assigned at least as much as they can achieve on their own, feasibility is violated. This is depicted in Fig. 2.8, where the three half-spaces representing the core conditions of the 2-player coalitions have empty intersection.

To remain in the framework of this example, we continue with a set of conditions for superadditive 3-player games that ensure the nonemptiness of the core.

2.4.3.4 Core Existence in 3-player Games

Let us start directly with a sufficient condition for a nonempty core in 3-player games (N,v) that are superadditive and essential.

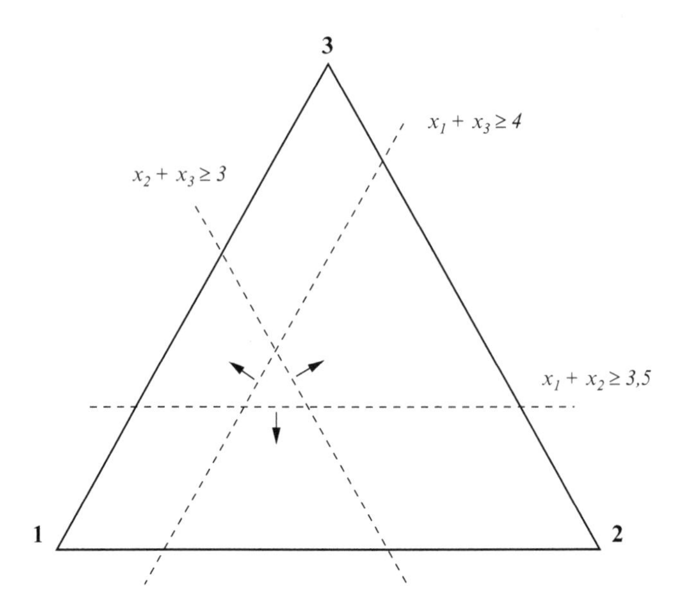

Fig. 2.8 A 3-player game with empty core

Proposition 2.7. *If a coalitional game (N, v) is superadditive and satisfies*

$$2 \cdot v(N) \geq \sum_{S \subseteq N : |S|=2} v(S), \tag{2.70}$$

then the core is nonempty, $C(N, v) \neq \emptyset$.

Proof. For reference, we spell out the core conditions for a game (N, v) with $N = \{1, 2, 3\}$. For simplicity, the stand-alone values of all players are assumed to be symmetric. This can be done without loss of generality, as long as the other conditions imposed by Proposition 2.7 are not violated.

$$x_i \geq v(\{i\}) = z \quad \text{for all } i \in N, \tag{2.71}$$

$$x_1 + x_2 \geq v(\{1,2\}) = a, \tag{2.72}$$

$$x_1 + x_3 \geq v(\{1,3\}) = b, \tag{2.73}$$

$$x_2 + x_3 \geq v(\{2,3\}) = c, \tag{2.74}$$

$$x_1 + x_2 + x_3 \geq v(N) = d. \tag{2.75}$$

Because the game is superadditive, we know that $2z$ is less than or equal to all a, b, and c, i.e. the sum of two such singleton allocations is no larger than the allocation of the respective 2-player coalition. Likewise for combinations of a singleton and a 2-player coalition, we know that d is larger than or equal to $z + a$,

$z+b$, and $z+c$. Consequently, we also have $d \geq 3z$. As we know core allocations to be efficient and so equal to d, we can conclude that there exist allocations that satisfy the individual rationality constraints (2.71) for all $i \in N$. From superadditivity it also follows that d is large enough to satisfy, one by one, all three cases of payoffs to a 2-player coalition and the remaining singleton. But does this hold for all three possibilities simultaneously in a given allocation? This question is answered by summing over (2.72)–(2.74):

$$2 \cdot (x_1 + x_2 + x_3) \geq a + b + c. \tag{2.76}$$

Because core allocations are efficient, the weak inequality in (2.75) is binding and we can replace the parenthesis on the left hand side of (2.76) with d:

$$2 \cdot d \geq a + b + c. \tag{2.77}$$

This yields exactly expression (2.70) when substituting the value function for a, b, c, and d:

$$2 \cdot v(N) \geq v(\{1,2\}) + v(\{1,3\}) + v(\{2,3\}). \tag{2.78}$$

Therefore, a superadditive 3-player game satisfying (2.70) has a nonempty core. ■

However, superadditivity is crucial for the previous result. This we reveal in the following:

Proposition 2.8. *No essential constant-sum game with three players can satisfy expression (2.70).*

Proof. In a constant-sum game, $v(N \setminus \{i\}) + v(\{i\}) = v(N)$ for all $i \in N$, and by assumption $v(\{i\}) = z$ for all $i \in N$. Substituting this into (2.70), or rather (2.78), leads to

$$2v(N) \geq v(N) - z + v(N) - z + v(N) - z = 3v(N) - 3z, \tag{2.79}$$

which rearranged yields

$$3z \geq v(N), \tag{2.80}$$

a contradiction to essentiality. ■

This result extends readily to n-player games.

2.4.3.5 Core Existence in n-player Games

Moving from 3-player games to the more general case with n players, we begin by extending Proposition 2.8 to obtain a sufficient condition for an empty core.

Proposition 2.9. *If a coalitional game (N,v) is essential and constant-sum, its core is empty: $C(N,v) = \emptyset$.*

Proof. (Forgó et al. (1999, p. 226)) Proceeding by contradiction, we suppose that the core is not empty and that allocation $\mathbf{x} \in \mathbb{R}^n$ is an element of the core: $\mathbf{x} \in C(N,v)$. Core membership of $\mathbf{x}$ requires group rationality for all coalitions $S \subseteq N$, see (2.67). Then, in particular, the following must hold for all players $i \in N$:

$$x(N \setminus \{i\}) = \sum_{j \in N \setminus \{i\}} x_j \geq v(N \setminus \{i\}). \tag{2.81}$$

Because (N,v) is constant-sum by assumption, we can rewrite the second part of (2.81) by substituting $v(N) = v(N \setminus \{i\}) + v(\{i\})$:

$$\sum_{j \in N \setminus \{i\}} x_j + v(\{i\}) \geq v(N) \quad \text{for all } i \in N. \tag{2.82}$$

The efficiency requirement of core membership for an allocation $\mathbf{x}$ allows us to replace the right hand side of (2.82) by $x(N) = \sum_{j \neq i} x_j + x_i$:

$$\sum_{j \in N \setminus \{i\}} x_j + v(\{i\}) \geq \sum_{j \in N \setminus \{i\}} x_j + x_i \quad \text{for all } i \in N. \tag{2.83}$$

The summations on both sides cancel out, regardless of i, and so summing (2.83) over all $i \in N$, we get

$$\sum_{i \in N} v(\{i\}) \geq \sum_{i \in N} x_i = x(N). \tag{2.84}$$

Because the game (N,v) is assumed to be essential, we can replace the weak inequality with a strict one, obtaining

$$v(N) > x(N), \tag{2.85}$$

which is a contradiction to efficiency of $\mathbf{x}$ and its assumed core membership. ■

But as we are rather interested in games with nonempty cores, such negative results have, other than instructive purposes, little appeal to us. More interesting are conditions that need to be met by coalitional games to yield nonemptiness of the core. Before we come to the standard reference for core existence, we look at a very general necessary condition for n-player games.

2.4.3.6 A Necessary Condition for General Games

Another necessary condition for core existence, this time for n-player coalitional games (N,v), can be deduced directly from the core conditions (2.67). Kannai (1992, p. 359) mention this as an extended form of essentiality, yet not quite superadditivity. We summarize in the following proposition:

Proposition 2.10. *If and the core of a coalitional game* (N,v) *is nonempty,* $C(N,v) \neq \emptyset$, *then*

$$\sum_{j=1}^{m} v(P_j) \leq v(N) \tag{2.86}$$

holds for all partitions P, *where* $P_j \cap P_k = \emptyset$ *for all* $P_j, P_k \in \mathsf{P}$, $j \neq k$, *and* $m = |\mathsf{P}|$.

Proof. Core membership requires efficiency, see (2.67), and so any allocation $\mathbf{x} \in C(N,v)$ satisfies $x(N) = v(N)$. With $x(\cdot)$ being an additive function, we can decompose $x(N)$ and rearrange its n summands into arbitrary partitions P. The efficiency condition can thus be rewritten to

$$x(P_1) + x(P_2) + \cdots + x(P_m) = v(N), \tag{2.87}$$

for any partition P. According to rationality $x(S) \geq v(S)$ for all coalitions $S \subseteq N$, which is also true for the elements of any given partition P. Now, replacing the elements on the left hand side of (2.87) with the (no larger) corresponding values of v leads to a weak inequality:

$$v(P_1) + v(P_2) + \cdots + v(P_m) \leq v(N). \tag{2.88}$$

This is identical to (2.86) and as such fulfilled by any allocation in the core. ■

Unfortunately, this condition is far from sufficient, as a second look at Example 2.6 reveals. Even though (2.86) is fulfilled for all five partitions of $N = \{1,2,3\}$, the core is empty. Also, it is readily verified that condition (2.86) admits several special cases, where the core is also empty: For any two element partition $\mathsf{P} = \{S, N \setminus S\}$, $S \subseteq N$ we obtain a constant-sum game if equality holds in (2.86); for the trivial partition, where $\mathsf{P} = \{\{1\},\{2\},\ldots,\{n\}\}$, an equality would signal an inessential game. Both cases have empty cores. This was shown in Proposition 2.9 for the first case, given (N,v) is also essential; the second case deserves no separate treatment, because it fulfills the requirements of a constant-sum game, given nonnegative marginal values of the players. Any possible gains arising from proper coalitions would contradict the core's feasibility constraint in (2.67) when allocated.

Having built up some intuition on what makes a coalitional game (N,v) exhibit a nonempty core, or what prevents it from having one, we now investigate a condition which is both necessary and sufficient for the nonemptiness of the core.

2.4.3.7 The Bondareva–Shapley Theorem

This section's contents lead up to a result that was established independently by Bondareva (1963) and Shapley (1967). They both state the essentially same condition for the nonemptiness of the core of a coalitional game, based on the concept of balancedness. It is therefore commonly referred to as the *Bondareva–Shapley Theorem* in the literature and has resurfaced in varying nuances. In its original form, it was the first necessary and sufficient condition for nonemptiness of the core applicable to fairly general coalitional games (N, v). In this respect, it is the standard reference in cooperative game theory, not only because of its originality, but also because its assumptions on the game are considered to be reasonable, at least compared to other results of nonemptiness.

Before we get to the theorem itself, we formally introduce the concept of balanced sets and games, and provide some intuition for both. In addition to the original sources mentioned above, treatments of this theorem can be found in the literature in varying degrees of detail. With harmonized notation, our approach is based on Weber (1994, pp. 1298 f.), Peleg and Sudhölter (2003, p. 36 ff.), Shapley and Shubik (1969, pp. 13 ff.), Owen (1995, pp. 224 ff.), Forgó et al. (1999, pp. 226 ff.), and Kannai (1992, pp. 359 ff.).

Definition 2.47. A collection $\mathscr{C} = \{S_1, \ldots, S_m\}$ of distinct, nonempty subsets of the player set N is *N-balanced*, or simply *balanced*, if there exist m nonnegative real numbers $\gamma_1, \gamma_2, \ldots, \gamma_m \in \mathbb{R}_+$ satisfying

$$\sum_{j=1}^{m} \gamma_j \, w_{S_j}(i) = \sum_{j|i \in S_j} \gamma_j = 1 \quad \text{for all } i \in N. \tag{2.89}$$

We refer to $\gamma_1, \gamma_2, \ldots, \gamma_m$ as the *weights* or *balancing coefficients* which balance collection $\mathscr{C}$, and $w_{S_j}(i)$ is the indicator function assuming values of 1 or 0, depending on i's membership to S_j.[46]

If the elements of $\mathscr{C}$ are not only distinct, but also pairwise disjoint, and exhaust N, the collection $\mathscr{C}$ is also a partition of N. Then, the balancing coefficients must be $\gamma_j = 1$ for all $j = 1, \ldots, m$ in order to satisfy (2.89), because each player $i \in N$ is contained in only one element of the partition, i.e. in one set of the collection $\mathscr{C}$. Based on this connection, balanced collections are often considered a more general form of partitions. To our knowledge, there are no intuitive procedures allowing to detect a balanced collection of sets. A geometric interpretation of balancedness is given by Hildenbrand and Kirman (1988, p. 129), which we extend in the following.

Even though the characteristic vector's entries X_S^i assumes only values of 0 and 1 for all coalitions $S \subseteq N$ and players $i \in N$, it is possible to apply it in the n-dimensional Euclidean space. Let us interpret all n characteristic vectors of singleton coalitions, $\mathsf{X}_{\{i\}}$, as the spanning unit vectors in $\mathbb{R}^n$. Define for each coalition $S \subseteq N$ the convex hull of its members' unit vectors:

$$co^S := co\left\{\mathsf{X}_{\{i\}} | i \in S\right\}. \tag{2.90}$$

Note that co^N corresponds to the $(n-1)$-dimensional unit-simplex in $\mathbb{R}^n$.

---

[46] See Definition 2.41, p. 52.

Let b_S denote the barycenter of the convex hull as defined in (2.90) for a coalition $S \subseteq N$. We will refer to it as barycenter of S. It is calculated according to

$$b_S = \frac{\sum_{i \in S} m_i \mathsf{X}_{\{i\}}}{\sum_{i \in S} m_i} = \frac{1}{|S|}\mathsf{X}_S, \tag{2.91}$$

where the second equality follows from two facts: For one, the weights[47] m_i assigned to each $i \in S$ are equal and cancel out. Also, the characteristic vectors are additive (only) across disjoint coalitions:

$$\mathsf{X}_S + \mathsf{X}_T = \mathsf{X}_{S \cup T} \iff S \cap T = \emptyset \text{ for all } S, T \subseteq N. \tag{2.92}$$

Now, a collection of sets $\mathscr{C}$ is balanced if and only if the barycenter of the grand coalition, b_N, lies in the convex hull of the barycenters of all elements of $\mathscr{C}$:

$$\mathscr{C} \text{ is balanced} \iff b_N \in co\{b_S | S \in \mathscr{C}\}. \tag{2.93}$$

We can think of the $(n-1)$-dimensional unit simplex in $\mathbb{R}^n$ as the space of imputations in an $(0,1)$-normalized n-player game. Its barycenter b_N is the allocation where each player would receive the payoff $\frac{1}{n}$. The barycenters b_S corresponding to the other sets $S \in \mathscr{C}$ are also feasible payoffs as by definition $b(S) = 1$ for all $S \subseteq N$.[48] So a balanced collection $\mathscr{C}$ can contain only coalitions with barycenters that constitute a convex hull which includes b_N. This is illustrated in Fig. 2.9.

As before, the barycenters are denoted by b_S for all $S \subseteq N = \{1,2,3\}$. For each singleton coalition $S = \{i\}$, the corresponding simplex or face is the point $\mathsf{X}_{\{i\}}$, as is its convex hull and also the latter's barycenter: $\mathsf{X}_{\{i\}} = co^{\{i\}} = b_{\{i\}}$ for all $i \in N$. For the 2-player coalitions, the convex hull of the corresponding unit vectors forms a line segment between the two members. Its barycenter is the midpoint of the line, as indicated. The barycenter of the grand coalition's convex hull (the shaded triangle), b_N, lies on the intersection of the three lines from the 2-player barycenters to the opposite vertex.[49] It can be verified graphically now that for example all possible partitions of the player set N are a balanced collection Also, e.g. $\mathscr{C} = \{\{1,2\},\{1,3\},\{2,3\}\}$ is a balanced collection. These examples are all the *minimally* balanced coalitions possible on N: No set $S \in \mathscr{C}$ can be removed if one wants to uphold the balancedness of $\mathscr{C}$. Nevertheless, we can add certain sets $S \subseteq N$ to such minimally balanced coalitions without destroying balancedness (Of course, the coefficients γ need to be changed accordingly). Arguing geometrically, this is because either the additional barycenter already lies within the convex hull of the

---

[47] The weights m_i of extreme points of the convex hull co^S are not to be confused with the weights γ_j assigned to the sets of a balanced collection.

[48] The sets co^S are $|S|-1$-dimensional unit-simplices themselves. They are called the *faces* of the simplex Δ^N. By definition, the faces are all subsets of the $(n-1)$-dimensional unit-simplex and so their barycenters b_S must also be contained in Δ^N. Also, $b(S) = \frac{1}{|S|}\sum_{i \in S} \mathsf{X}_S^i$ for all $S \subseteq N$.

[49] For reasons of clarity, we omit these lines and display the barycenter b_N, where their intersection would be.

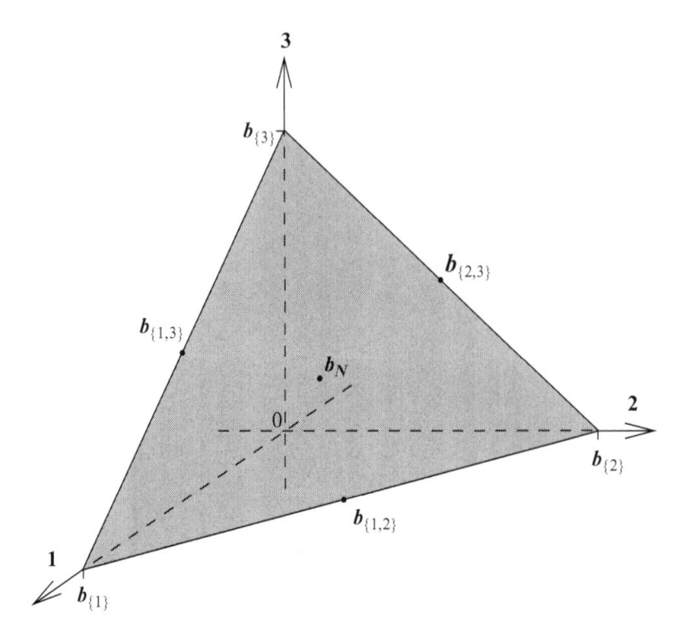

Fig. 2.9 Barycenters b_S of all faces of the unit-simplex of a 3-player game

previous ones. Or if not, by definition of convexity and convex hull, the new convex hull will be a superset of the previous one, and hence b_N will remain an element of it. Here, the graphical representation has its limits, as the balancing coefficients are not emphasized.[50]

For larger dimensions, this geometrical intuition breaks down, but some remarks can be made nevertheless: The collection of all 2-element subsets of a general set N is balanced by the weights $\gamma = \frac{1}{n-1}$, because every element $i \in N$ belongs to $n-1$ subsets of N. More general, the collection of all s-element subsets, for any $s \leq n$, is balanced. The coefficients are again symmetric for all $j = 1, \ldots, m$ and given by

$$\gamma = \frac{1}{\binom{n-1}{s-1}}, \tag{2.94}$$

because all elements $i \in N$ belong to the same number of coalitions with cardinality s in collection $\mathscr{C}$.

The reader might have noticed that so far the characteristic function has not been invoked at all, as we were concerned with the balancing of arbitrary sets, not necessarily coalitions of players. This will change now, as we carry the concept of balancedness to *balanced games*:

---

[50]Take for example the minimally balanced collection $\mathscr{C} = \{\{1,2\},\{1,3\},\{2,3\}\}$ from above. Balancing weights would be $\gamma_j = \gamma = \frac{1}{2}$ for all $j = 1,2,3$. If we add the sets $\{1\}$,$\{2\}$, and $\{3\}$ to collection $\mathscr{C}$, it is still balanced with coefficients $\gamma_j = \gamma = \frac{1}{3}$ for all $j = 1,\ldots,6$. Minimally balanced coalitions corresponding to larger player sets N can also be extended in the manner above, but also in a less symmetric way.

Definition 2.48. A coalitional game (N,v) is balanced, if for each balanced collection $\mathscr{C} = \{S_1,\dots,S_m\}$ and the corresponding set of coefficients $\gamma_1,\dots,\gamma_m$, the following is true:

$$\sum_{j=1}^{m} \gamma_j\, v(S_j) \leq v(N). \tag{2.95}$$

Relating to our previous geometrical interpretation, the weights γ_j scale the size of the convex hull of the characteristic vectors $\mathrm{X}_{\{i\}}$ of all $i \in S_j$. Because the weight equally affects all members of S_j, the barycenter b_{S_j} is moved along a line through the origin, depending on γ_j. Hence, the shape of the convex hull $co\{b_S | S \in \mathscr{C}\}$ and whether it contains b_N depends on the weights $\gamma_1,\dots,\gamma_m$, given a collection $\mathscr{C}$ that can fulfill (2.89).

We can see how expression (2.95) coincides with (2.86), whenever $\mathscr{C}$ is a partition of N and $\gamma_j = 1$ for all $j = 1,\dots,m$. If $\mathscr{C}$ is not a partition and there are possible overlaps, its elements S_j are weighted with possibly distinct γ_j to ensure (2.95) is not violated.[51] With all preliminaries in place, the main theorem can be stated as follows:

Theorem 2.1 (Bondareva–Shapley Theorem). *A coalitional game (N,v) is balanced if and only if $C(N,v) \neq \emptyset$.*

Proof. Sufficiency: Suppose $\mathbf{x} \in C(N,v)$ is a core allocation. In order for a collection of sets $\mathscr{C} = \{S_1,\dots,S_m\}$ to be balanced, corresponding nonnegative weights $\gamma_1,\dots,\gamma_m$, or $\gamma \in \mathbb{R}^m_+$, must exist, which fulfill the condition for balancedness, (2.89): $\sum_j \gamma_j w_{S_j}(i) = 1$ for all $i \in N$.

For any core allocation $\mathbf{x} \in C(N,v)$ we have $x(S) \geq v(S)$ for all $S \subseteq N$. Consequently, the latter inequalities also apply to the $j = 1,\dots,m$ sets $S_j \in \mathscr{C}$ of any balanced collection, and also their (equally) weighted sums:

$$\sum_{j=1}^{m} \gamma_j\, v(S_j) \leq \sum_{j=1}^{m} \gamma_j\, x(S_j). \tag{2.96}$$

Rearranging the right-hand side of (2.96) leads to the following chain of equalities,

$$\begin{aligned} \sum_{j=1}^{m} \gamma_j x(S_j) &= \sum_{j=1}^{m} \gamma_j \sum_{i \in S_j} x_i = \sum_{j=1}^{m} \gamma_j \sum_{i \in N} x_i\, w_{S_j}(i) \\ &= \sum_{i \in N} x_i \underbrace{\sum_{j=1}^{m} \gamma_j\, w_{S_j}(i)}_{=1 \text{ for all } i \in N} = \sum_{i \in N} x_i \\ &= x(N), \end{aligned} \tag{2.97}$$

[51] Notably, expression (2.86) was deduced from the core conditions!

which holds for all balanced collections $\mathscr{C}$ and corresponding coefficients γ. Because for an element in the core, $\mathbf{x} \in C(N,v)$, efficiency applies, $x(N) = v(N)$, we can combine (2.96) and (2.97) to yield

$$\sum_{j=1}^{m} \gamma_j \, v(S_j) \le v(N), \tag{2.98}$$

the definition of a balanced game given by (2.95). Therefore, a nonempty core, $C(N,v) \neq \emptyset$, is a sufficient condition for a game to be balanced.

Necessity : Take any balanced game (N,v) and index the elements of the power set of N, possibly with increasing cardinality[52]: $2^N = \{S_1,\ldots,S_m\}$, where $m = 2^n$ and, e.g., $S_1 = \emptyset$ and $S_m = N$. Define a matrix $\mathbf{A} \in \mathbb{R}^{mn}$ with $a_{ji} = w_{S_j}(i)$ for all coalitions $j = 1,\ldots,m$ and players $i = 1,\ldots,n$:

$$\mathbf{A} := \begin{pmatrix} a_{11} & a_{12} & a_{13} & \ldots & a_{1n} \\ a_{21} & a_{22} & a_{23} & \ldots & a_{2n} \\ a_{31} & a_{32} & a_{33} & \ldots & a_{3n} \\ a_{41} & a_{42} & a_{43} & \ldots & a_{4n} \\ \vdots & \vdots & \vdots & \ddots & \vdots \\ a_{m1} & a_{m2} & a_{m3} & \ldots & a_{mn} \end{pmatrix} = \begin{pmatrix} 0 & 0 & 0 & \ldots & 0 \\ 1 & 0 & 0 & \ldots & 0 \\ 0 & 1 & 0 & \ldots & 0 \\ 0 & 0 & 1 & \ldots & 0 \\ \vdots & \vdots & \vdots & \ddots & \vdots \\ 1 & 1 & 1 & 1 & 1 \end{pmatrix}.$$

The elements of $\mathbf{A}$ are composed entirely of 1s and 0s, with no row or column exactly alike. As each row represents some distinct coalition $S \subseteq N$, the value 1 is set for all members of this coalition and 0 for all others. Likewise, the columns represent the players $i \in N$, where we find an entry of 1 whenever the player assigned to the column is a member of the respective coalition, and an entry of 0 if he is not. Consequently, the first row (corresponding to the empty set) is composed entirely of 0s, and the last row (grand coalition) consists only of 1s. Each row is identical to the transpose of the characteristic vector X'_S of its assigned coalition S.

Next, define a vector $\mathbf{c} \in \mathbb{R}^m$, whose elements are the values of the characteristic function according to the index introduced above: $c_j = v(S_j)$ for all $j = 1,\ldots,m$. Also, let $\mathbf{1} \in \mathbb{R}^n$ be a (column) vector of dimension $n \times 1$ whose entries are all equal to 1. We will introduce now two polyhedrons, the first of which,

$$G_1(\mathbf{A},\mathbf{1}) = \{\gamma \in \mathbb{R}^m \mid \gamma'\mathbf{A} = \mathbf{1}', \gamma \geq \mathbf{0}\}, \tag{2.99}$$

[52]Because set inclusion is only a partial order 2^N cannot be ordered in a unique way according to the cardinality of its elements. But one could at least index elements with the same cardinality consecutively, starting with the singleton coalitions and finishing with N. A more systematic approach, given players were indexed themselves, would be the following: $\{\emptyset,\{1\},\{2\},\{1,2\},\{3\},\{1,3\},\{2,3\},\{1,2,3\},\ldots,N\}$. Whenever a new player, say i, is considered, he will be added as singleton first and thereafter in union with all coalitions that already exist. This method allows the collection of sets to be extended without destroying its logic. For increasing N, new elements are simply added at the end.

contains all vectors γ that would qualify as balancing coefficients according to (2.89). Each of the n entries of $\gamma'\mathbf{A} \in \mathbb{R}^n$ can alternatively be expressed as

$$\sum_{j=1}^{m} \gamma_j \, w_{S_j}(i), \tag{2.100}$$

which is equal to one for all $i \in N$. The second polyhedron is given by

$$G_2(\mathbf{A}, \mathbf{c}) = \{\mathbf{x} \in \mathbb{R}^n \mid \mathbf{A}\mathbf{x} \geq \mathbf{c}\}, \tag{2.101}$$

Each of the m entries of the weak inequality of (2.101) can be rewritten to

$$x(S_j) \geq v(S_j). \tag{2.102}$$

The left hand side of (2.102) stands for each of the m entries of $\mathbf{A}\mathbf{x} \in \mathbb{R}^m$ representing the $2^n = m$ subsets S_j of N. Accordingly, the right-hand side are the elements of $\mathbf{c}$.

For $S_j = N$, equation (2.102) reveals that the sum of the entries of each element $\mathbf{x} \in G_2(\mathbf{A}, \mathbf{c})$, given by $x(N)$, is at least as large as the gains of the grand coalition, $v(N)$, i.e. the sum of the entries of $\mathbf{c}$, given by $c(N)$. With this in mind, consider the following equation:

$$\arg \min_{\mathbf{x} \in G_2(\mathbf{A}, \mathbf{c})} \mathbf{1}' \cdot \mathbf{x} = \arg \min_{\mathbf{x} \in G_2(\mathbf{A}, \mathbf{c})} x(N) \geq v(N). \tag{2.103}$$

The scalar $\mathbf{1}' \cdot \mathbf{x}$ on the left hand side is just another way of expressing the sum of all individual payoffs as given by the additive function $x(N)$. Because the minimizing argument $\mathbf{x}$ is picked from the same set, $G_2(\mathbf{A}, \mathbf{c})$, on both sides, they must be equal.

From results on duality in linear programming,[53] we can establish

$$\arg \min_{\mathbf{x} \in G_2(\mathbf{A}, \mathbf{c})} \mathbf{1}' \cdot \mathbf{x} = \arg \max_{\gamma \in G_1(\mathbf{A}, \mathbf{1})} \gamma' \cdot \mathbf{c}. \tag{2.104}$$

By assumption, the underlying game (N, v) is balanced, and so

$$\arg \max_{\gamma \in G_1(\mathbf{A}, \mathbf{1})} \gamma' \cdot \mathbf{c} = \arg \max_{\gamma \in G_1(\mathbf{A}, \mathbf{1})} \sum_{j=1}^{m} \gamma_j \, v(S_j) \leq v(N). \tag{2.105}$$

[53] A very concise introduction to linear programming, including the fundamental results and their proofs, can be found in Owen (1995, pp. 35 ff.). Here, we make use of the theorem given ibid. pp. 40 ff. Also, Owen (1999, pp. 77 ff.) focuses on the dual relation of linear programs and shows how one can be derived from the other. At the respective extreme point of the dual programs, i.e. minimum or maximum, the weak inequality turns into an equality, which allows for expression (2.104).

By definition of the polyhedron $G_2(\mathbf{A},\mathbf{c})$, the inequality $\mathbf{Ax} \geq \mathbf{c}$ is fulfilled for all $\mathbf{x} \in G_2(\mathbf{A},\mathbf{c})$, also for the smallest argument from (2.103).

Combining (2.103) and (2.105) via (2.104) leads to the following conclusion:

$$\arg \min_{\mathbf{x} \in G_2(\mathbf{A},\mathbf{c})} x(N) = v(N). \tag{2.106}$$

If the game (N,v) is balanced, a core allocation $\mathbf{x} \in C(N,v)$ must therefore exist and $C(N,v) \neq \emptyset$. ■

As Theorem 2.1 exhibits some redundancy concerning the not necessarily minimally balanced collections, it can be stated in a more precise form:

Theorem 2.2 (Bondareva–Shapley Theorem, strong). *A coalitional game (N,v) has a nonempty core if and only if for every minimally balanced collection $\mathscr{C} = \{S_1,S_2,\ldots,S_m\}$ and corresponding weights $\gamma_1,\gamma_2,\ldots,\gamma_m$, the following is true:*

$$\sum_{j=1}^{m} \gamma_j \, v(S_j) \leq v(N). \tag{2.107}$$

We omit the lengthy proof in favor of a short intuitional sketch and refer the reader to either Owen (1995, pp. 226 ff.) or Peleg and Sudhölter (2003, pp. 40 ff.). The main benefit when switching to minimally balanced collections is the sharp decrease in number: Minimally balanced sets are limited to at most n elements and are the only kind of balanced collection to exhibit unique balancing coefficients. Because any balanced collection is the union of minimally balanced collections, the vectors of balancing weights for the latter coincide with the extreme points of the polyhedron $G_1(\mathbf{A},\mathbf{1})$ corresponding to the maximization problem from expression (2.105).[54] All other solutions to (2.105), if they exist, can be obtained as convex combinations of these extreme points.

The Bondareva–Shapley theorem provides us with a condition, balancedness, which imposed on a coalitional game (N,v), ensures the existence of a nonempty core. As a matter of fact, the opposite is true too, even though by itself, balancedness has very little intuitive value. Another condition with positive effects on the nonemptiness of the core is convexity of the underlying coalitional game. We will see in the next section that this property is sufficient and will endow the resulting core with a number of convenient features.

2.4.3.8 The Core in Convex Games

This section is concerned with the core of convex games. We first establish that convexity of a coalitional game is a sufficient condition for the core to be

[54] Each γ that solves the underlying polyhedron $G_1(\mathbf{A},\mathbf{1})$.

nonempty. We then characterize the core of a convex game via marginal vectors and decomposability of the underlying value function.

We have previously introduced convexity (Definition 2.25) as a property of a game, according to which the marginal value of every arbitrary player $i \in N$ is nondecreasing in the size of a coalition. Formally, this was expressed through

$$v(S \cup \{i\}) - v(S) \leq v(T \cup \{i\}) - v(T) \quad \text{for all } S \subseteq T \subsetneq N,\ i \notin T. \tag{2.108}$$

Under the notion of supermodularity (see Definition 2.24) this same property is given with the marginal value of coalitions instead of a player under consideration:

$$v(S \cap T) + v(S \cup T) \geq v(S) + v(T) \quad \text{for all } S, T \subseteq N. \tag{2.109}$$

Here, the same idea is expressed in terms of sets instead of individual players. The sum of the values of any two coalitions is no higher than the value of their union and of their intersection.

The requirement of convexity is not uncommon in the context of coalitional games. Regardless, it has to be handled with care. In addition to the nonnegativity of marginal values, as already given through superadditivity, convexity also states globally nondecreasing returns to scale. For all players and all coalitions, their marginal value cannot go down if the coalition they join is increasing in size. What we get in return is a nonempty core to begin with. This, among other important findings in relation with convex games, was shown by Shapley (1967). Subsequently, convex games have received much attention in the literature. Mostly seen as byproduct in applications, where the property was invoked for the ease of finding a solution, there are also some direct theoretical contributions, the most useful of which are: Ichiishi (1981), who shows among other things the converse of Shapley (1967) and Ichiishi (1983, pp. 120 ff.), providing a more general treatment on convex games. Before we proceed to trace a proof of the result in Shapley (1967), we introduce some notation.

A *permutation* is a function which maps its domain onto itself. In our case, a permutation of the set of players, denoted $\pi : N \to N$, is a reordering, where the value $\pi(i) \in N = \{1, 2, \ldots, n\}$ is the new number assigned to player i under the permutation. For a set N, there exist $n!$ possible orderings of the elements, hence there also are $n!$ permutations on N. Let the set $S_{\pi,j}$ contain all players that do not appear after j in the ordering given by π.

$$S_{\pi,j} = \{i \in N \mid \pi(i) \leq j\} \tag{2.110}$$

In other words, $S_{\pi,j}$ contains the first j players according to π.

From these, create a system of linear equations $x(S_{\pi,j}) = v(S_{\pi,j})$, or hyperplanes $H(\mathsf{X}_{S_{\pi,j}}, v(S_{\pi,j}))$, for all $j \in N$[55]:

$$\begin{aligned}
x_1 &= v(\{1\}) \\
x_1 + x_2 &= v(\{\{1\},\{2\}\}) \\
x_1 + x_2 + x_3 &= v(\{\{1\},\{2\},\{3\}\}) \\
\vdots \quad & \quad \vdots \quad \quad \vdots \\
x_1 + x_2 + \ldots + x_n &= v(\{\{1\},\{2\},\ldots,\{n\}\}) = v(N)
\end{aligned}$$

For notational simplicity and especially to uphold a somewhat traceable ordering, we have chosen the permutation where $\pi(i) = i$ for all $i \in N$ in the above system of equations. Instead of $x_{\pi(i)}$ on the left-hand side and $\{\pi(i)\}$ on the right-hand side we simply used x_i and $\{i\}$, respectively. If we solve the above system (in case of arbitrary π) for x_i^{π}, we obtain for all $i \in N$

$$x_i^{\pi} = v(S_{\pi,\pi(i)}) - v(S_{\pi,\pi(i)} \setminus \{i\}). \tag{2.111}$$

A solution is guaranteed, as the equations are linearly independent. The vector $\mathbf{x}^{\pi} \in \mathbb{R}^n$ is exactly the intersection of all $j = 1,\ldots,n$ hyperspaces $H(\mathsf{X}_{S_{\pi,j}}, v(S_{\pi,j}))$. As there are $n!$ permutations π, there are as many possible solutions to x_i^{π} in (2.111), where $\mathbf{x}^{\pi}$ is the marginal worth vector for the permutation π (see Definition 2.39). What we want to show next is that if the core inequalities $x(S) \geq v(S)$ hold with equality for all permutations π and their respective sets $S_{\pi,j}$ for all $j \in N$, they are also fulfilled for all other $S \subseteq N$, and so $\mathbf{x}^{\pi}$ is a core allocation. The following theorem sums this up:

Theorem 2.3. *The core $C(N,v)$ of a convex game (N,v) is nonempty.*

Proof. Pick any proper subset $T \subsetneq N$, for which the player j under permutation π is the first player just not to be included in T. In other words, j is the first player in set $N \setminus T$ according to π. Therefore,

$$T \cup S_{\pi,\pi(j)} = T \cup \{j\} \quad \left(= S_{\pi,\pi(j)}\right), \tag{2.112}$$

and

$$T \cap S_{\pi,\pi(j)} = S_{\pi,\pi(j)} \setminus \{j\} \quad (= T), \tag{2.113}$$

[55] This system is a proper – and binding – subset of the core inequalities $x(S) \geq v(S)$ which hold for all $S \subseteq N$ (see (2.67)).

which is more accessibly illustrated in the following way:

$$N = \left\{ \overbrace{\underbrace{1,2,3,\ldots,j-1}_{},j}^{S_{\pi,\pi(j)}\setminus\{j\}=T},j+1,\ldots,n \right\}. \tag{2.114}$$

with $S_{\pi,\pi(j)}=T\cup\{j\}$ under $1,2,3,\ldots,j-1,j$.

By assumption of (N,v) being convex, we substitute (2.112) and (2.113) accordingly into expression (2.109), i.e.

$$v(S_{\pi,\pi(j)}\setminus\{j\})+v(T\cup\{j\}) \geq v(S_{\pi,\pi(j)})+v(T). \tag{2.115}$$

Noting from (2.111) that $x_j^\pi = v(S_{\pi,\pi(j)}) - v(S_{\pi,\pi(j)}\setminus\{j\})$ yields

$$x_j^\pi \leq v(T\cup\{j\}) - v(T). \tag{2.116}$$

Adding the marginal values x_i^π for all players $i \in T$ to both sides and rearranging, we obtain

$$x^\pi(T\cup\{j\}) - v(T\cup\{j\}) \leq x^\pi(T) - v(T), \tag{2.117}$$

where $x^\pi(T)$ is the known additive function (see (2.53)). If we repeat this argument with stepwise bigger sets that include each time the "next" player according to permutation π, we can establish a chain of weak inequalities with $n-|T|+1$ elements, whose first and last are associated in the following way:

$$x^\pi(N) - v(N) \leq x^\pi(T) - v(T). \tag{2.118}$$

Because the left hand side of (2.118) must be equal to zero for any core allocation and because T was picked arbitrarily, the allocation $\mathbf{x}^\pi$ is indeed in the core: $\mathbf{x}^\pi \in C(N,v)$. This shows that for convex games the core is nonempty. ∎

Having shown that an arbitrary marginal worth vector $\mathbf{x}^\pi$ is an element of the core in convex games, the following is a logical consequence:

Corollary 2.3. *The core $C(N,v)$ of a convex game contains all marginal worth vectors $\mathbf{x}^\pi$:*

$$\mathbf{x}^\pi \in C(N,v) \text{ for all } \pi \in \Pi_N.$$

Proof. As the nonemptiness of the core was shown in the proof of Theorem 2.3 with an arbitrary marginal worth vector $\mathbf{x}^\pi$, it must hold true for all $n!$ permutations π on the player set N. ∎

We now know that the core of a convex game contains all marginal worth vectors. Another interesting result is presented in Shapley (1971, p. 22). It characterizes the core, if the underlying convex game is decomposable. In order to illustrate this, we take a slight detour for two general remarks on convex, decomposable games.

Proposition 2.11. *A coalitional game (N,v) which is decomposable on a partition P, is also convex, if and only if every component is convex, or each restriction of v on $P \in \mathsf{P}$, denoted $v|_P$, is convex.*

Proof. Suppose every component of the game (N,v) is convex, then a expression (2.109) holds:

$$v(S_j \cap T_j) + v(S_j \cup T_j) \geq v(S_j) + v(T_j) \text{ for all } S_j, T_j \subseteq P_j,\ j = 1, \ldots, m. \quad (2.119)$$

Every pair of sets $S, T \subseteq N$ whose elements originate in more than one element of partition P, can be broken down into the respective subsets $S_1, \ldots, S_m$ and $T_1, \ldots, T_m$, some of which might be empty. Denote $S_j := S \cap P_j$ and $T_j := T \cap P_j$ for all $j = 1, \ldots, m$ with $m = |\mathsf{P}|$. Because (N,v) is decomposable[56] with respect to P, we calculate that

$$v(S) = \sum_{j=1}^{m} v(S_j) \text{ and}$$

$$v(T) = \sum_{j=1}^{m} v(T_j), \text{ as well as}$$

$$v(S \cap T) = \sum_{j=1}^{m} v(S_j \cap T_j) \text{ and}$$

$$v(S \cup T) = \sum_{j=1}^{m} v(S_j \cup T_j).$$

Hence, adding (2.119) over all components preserves the inequality and thus the convexity of the game (N,v) itself. This proof works in both directions. ■

Proposition 2.12. *A convex game (N,v) is decomposable, if and only if*

$$v(N) = v(P_1) + v(P_2) + \cdots + v(P_m) \quad (2.120)$$

is fulfilled for some partition $\mathsf{P} = \{P_1, P_2, \ldots, P_m\}$ with $m \geq 2$ nonempty elements.

Proof. Assuming (2.120) holds for P, from the convexity of v we can establish for some arbitrary $S \subseteq N$ the system of m inequalities depicted below. The inequalities are based on supermodularity as given in (2.109), where the first expression in each line is simplified, because as elements of a partition, all P_j's are pairwise disjoint:

$$\left(S \cup P_1 \cup \cdots \cup P_{j-1}\right) \cap P_j = S \cap P_j \text{ for all } j = 1, \ldots, m. \quad (2.121)$$

[56] Decomposable games were introduced in Definition 2.31, p. 43.

The system then takes the following form:

$$\begin{aligned}
v(S \cap P_1) + v(S \cup P_1) &\geq v(S) + v(P_1) \\
v(S \cap P_2) + v(S \cup P_1 \cup P_2) &\geq v(S \cup P_1) + v(P_2) \\
v(S \cap P_3) + v(S \cup P_1 \cup P_2 \cup P_3) &\geq v(S \cup P_1 \cup P_2) + v(P_3) \\
\vdots \qquad\qquad & \quad \vdots \qquad\qquad \vdots \\
v(S \cap P_m) + v(S \cup P_1 \cup \cdots \cup P_m) &\geq v(S \cup P_1 \cup \cdots \cup P_{m-1}) + v(P_m).
\end{aligned}$$

A summation over the m above inequalities yields

$$\sum_{j=1}^{m} v(S \cap P_j) + v(N) \geq v(S) + \sum_{j=1}^{m} v(P_j), \tag{2.122}$$

where we can replace $v(N)$ according to (2.120) and simplify to

$$\sum_{j=1}^{m} v(S \cap P_j) \geq v(S). \tag{2.123}$$

Because convexity implies superadditivity, (2.123) must hold with equality, like the one given in Definition 2.31. The other direction is much easier: Set $S = N$ in (2.47) to obtain the result immediately. ■

The previous two propositions highlight the two sides of one coin: Proposition 2.11 states that a decomposable game is convex if and only if all its components are convex. The other, Proposition 2.12, assumes a convex game and links its decomposability to a necessary and sufficient condition regarding additivity of the component-restricted value function. The next result, the characterizetion of the core through decomposability of the underlying value function, can now be brought forward.

Theorem 2.4. *The core of a decomposable convex game is the Cartesian product of the cores of the component games:*

$$C(N, v) = C(P_1, v|_{P_1}) \times C(P_2, v|_{P_2}) \times \cdots \times C(P_m, v|_{P_m}). \tag{2.124}$$

This holds for all partitions P, *for which the game* (N, v) *is decomposable.*

Proof. We proceed in two parts: First, we show that any allocation that is in the Cartesian product of the component cores is also in the core. Then we illustrate the converse, according to which every element that is not in the "Cartesian core" can also not be a member of the core.

Suppose the characteristic function v is additive on the partition $\mathsf{P} = \{P_1, P_2, \ldots, P_m\}$ and suppose $\mathbf{x} \in \times_{j=1}^{m} C(P_j, v|_{P_j})$, i.e. the allocation $\mathbf{x}$ is an element of the Cartesian product of component cores. Consequently, for all $j = 1,\ldots,m$ components and all coalitions $S \subseteq N$, we know that

$$x(S \cap P_j) \geq v(S \cap P_j), \tag{2.125}$$

as well as

$$x(P_j) = v(P_j). \tag{2.126}$$

Because the game (N,v) is decomposable over P and hence $v(S) = v(S \cap P_1) + \ldots + v(S \cap P_m)$ for all $S \subseteq N$, expressions (2.125) and (2.126) each sum to

$$x(S) \geq v(S) \tag{2.127}$$

and

$$x(N) = v(N), \tag{2.128}$$

and therefore $\mathbf{x} \in C(N,v)$.

Now, we assume the allocation $\mathbf{x}$ not to be an element of the Cartesian product of component cores: $\mathbf{x} \notin \times_{j=1}^{m} C(P_j, v|_{P_j})$. In this case, at least one component j must exist for which one or both types of core conditions are violated:

$$x(P) < v(P) \;\text{ for some }\; P \subseteq P_j \quad \text{or} \quad x(P_j) > v(P_j), \tag{2.129}$$

The first case in (2.129) violates the general core condition for coalition P, and so $\mathbf{x} \notin C(N,v)$. If the second case in (2.129) applies for some component j, then either

$$x(N) > v(N) \quad \text{or} \quad x(N \setminus P_j) < v(N \setminus P_j), \tag{2.130}$$

violating either feasibility or a rationality condition for $N \setminus P_j$. The latter inequality follows from the additivity of $x(\cdot)$, where $x(P_j) = x(N) - x(N \setminus P_j)$ and from (2.120), according to which $v(P_j) = v(N) - v(N \setminus P_j)$. Again, $\mathbf{x} \notin C(N,v)$. ∎

This important result shows more than how the core of a decomposable game can itself be split up according to the components of the decomposition. The reader might have noticed that convexity was not required in the previous theorem at all. This is due to the fact that we deliberately omitted the second part as stated by Shapley (1971): There, he specifies the dimension of the core of a convex and decomposable game to be $n - m$, where $m = |\mathsf{P}|$ is the number of elements of the finest decomposition possible for v. If v is not decomposable at all, then the core of a convex game has the full dimensionality $n - 1$. The reasons for omitting a detailed treatment of the above are the technical requirements which have no apparent connection to our subsequent work and would have forced us to reproduce large sections of Shapley (1971).

Another interesting feature of the core-decomposability should be noted though. If a coreless game can be decomposed into components, one of which is convex, then we know to be able to find a nonempty core for this component. However, this component core is overridden in the undecomposed game. This might allow for some interesting insights and will be of use later on. After highlighting these important features of convex games, we will now juxtapose the von Neumann Morgenstern solution, or stable set, and the core to highlight their relationship.

2.4.3.9 Stable Sets and the Core

In this section, we show how the core and stable sets are related to one another. We bring together the definitions of both, where we divert from our previous convention and rely on the domination core from Definition 2.46. This might seem odd, but has two apparent reasons: For one, it is much more instructive to relate the concepts when both are based on the same binary relation, as opposed to the core being inequality-based. Secondly, both core definitions coincide, whenever the domination core is nonempty. In case it is empty, there is not much to relate to anyway.

We start out by introducing some additional notation based on the binary relation of domination through $\succ^S$ as introduced in Definition 2.37, p. 51. It will be extended to entire sets of imputations. For some imputation $\mathbf{x} \in I(N,v)$ and coalition $S \subseteq N$, define

$$\mathrm{dom}_S\,\mathbf{x} = \{\mathbf{y} \in I(N,v) | \, \mathbf{x} \succ_S \mathbf{y}\} \tag{2.131}$$

as the *dominion of* $\mathbf{x}$ *through* S: It is the set of all imputations $\mathbf{y}$ which are dominated by imputation $\mathbf{x}$ through the effective set S.

In reference to the general relation of domination, $\succ$, (see Definition 2.38, p. 51) we define the *dominion of* imputation $\mathbf{x}$ as

$$\mathrm{dom}\,\mathbf{x} = \bigcup_{S \subseteq N} \mathrm{dom}_S\,\mathbf{x}. \tag{2.132}$$

It is the set of imputations that is dominated by $\mathbf{x}$ through any effective set. Proceeding to sets of imputations $D \subseteq I(N,v)$ is accomplished analogously:

$$\mathrm{dom}_S\,D = \bigcup_{\mathbf{x} \in D} \mathrm{dom}_S\,\mathbf{x}. \tag{2.133}$$

is the *dominion of* the set of imputations D *through* coalition S. It is given by all imputations which are dominated by any imputation $\mathbf{x} \in D$ with respect to coalition S. Finally, this extends to

$$\mathrm{dom}\,D = \bigcup_{S \subseteq N} \bigcup_{\mathbf{x} \in D} \mathrm{dom}_S\,\mathbf{x} = \bigcup_{S \subseteq N} \mathrm{dom}_S\,D, \tag{2.134}$$

simply called the *dominion of* D when we drop the restriction to a single effective set S. The previous notions were introduced to the literature by Gillies (1959, pp. 49 ff.) but can also be found elsewhere, for example Lucas (1971, p. 500) or Lucas (1992, p. 551).

It is important to note that by the strict inequality that defines $\succ^S$, it is a transitive and asymmetric relation, which is not necessarily true for $\succ$. From this it follows that the dominion of a set of imputations, $D \subseteq I(N,v)$, can include elements from D, and hence $\text{dom}D \cap D \neq \emptyset$.

The requirements (2.65) and (2.66) for the stable set can be rewritten in terms of the dominion defined above:

$$\mathfrak{S} \cap \text{dom}\mathfrak{S} = \emptyset. \tag{2.135}$$

This is equivalent to internal stability, where no two elements in $\mathfrak{S}$ can dominate one another, because the intersection of the stable set with any imputation that might be dominated by one of its members is empty. The second condition is expressed by

$$\mathfrak{S} \cup \text{dom}\mathfrak{S} = I(N,v), \tag{2.136}$$

where the connection to (2.66) is less obvious: As $\text{dom}\mathfrak{S}$ is setwise dominated by $\mathfrak{S}$, for each element $\mathbf{z} \in \text{dom}\mathfrak{S}$ there must exist some imputation $\mathbf{x} \in \mathfrak{S}$ for which $\mathbf{x} \succ \mathbf{z}$ holds. This amounts to external stability. The sets $\text{dom}\mathfrak{S}$ and $\mathfrak{S}$ form a partition of $I(N,v)$, because any element not in $\text{dom}\mathfrak{S}$ must by (2.66) be in $\mathfrak{S}$. To highlight the partitioning character of $\mathfrak{S}$ and $\text{dom}\mathfrak{S}$, (2.135) and (2.136) can be combined:

$$\mathfrak{S} = I(N,v) \setminus \text{dom}\mathfrak{S}. \tag{2.137}$$

This expression emphasizes the "circularity" of the definition of stable sets, as $\text{dom}\mathfrak{S}$ is dependent on $\mathfrak{S}$.

The core, being the set of all undominated imputations according to Definition 2.43, can be rewritten as follows in terms of dominion:

$$C(N,v) = I(N,v) \setminus \text{dom}\, I(N,v). \tag{2.138}$$

We show this by contradiction: Suppose there exists some imputation $\mathbf{z} \in I(N,v)$ and some core allocation $\mathbf{x} \in C(N,v)$ for which $\mathbf{z} \succ \mathbf{x}$ holds. By definition of domination this implies both $z_i > x_i$ for all $i \in S$ and $z(S) \leq v(S)$ for some effective set $S \subseteq N$. But since $\mathbf{x}$ is in the core and satisfies $x(T) \geq v(T)$ for all $T \subseteq N$, the consequence would be $z(S) > x(S)$, leading to $v(S) \geq z(S) > x(S) \geq v(S)$, which cannot be true. This holds for all such $\mathbf{z} \in I(N,v)$, also for those in the core.

Hence, the core is the set of imputations that are undominated, both by themselves and by any imputation outside. Here we have the difference to the concept of stable sets, which allows elements $\mathbf{x} \in \mathfrak{S}$ to be dominated by outside-imputations $\mathbf{z} \notin \mathfrak{S}$. This hints to the possibility that the core is a subset of the von Neumann Morgenstern solution. To substantiate this claim, we take another look at the notion of dominion as defined in (2.131)–(2.134).

When we interpret dominion as a function from the space of imputations to its power set, given by

$$\text{dom}_S : I \to 2^I, \tag{2.139}$$

the value of the function is the (possibly empty) set $\{\mathbf{y} \in I(N,v) | \mathbf{x} \succ_S \mathbf{y}\}$ for an argument $\mathbf{x} \in I(N,v)$. By merely changing its domain, we can extend this notion to

$$\text{dom}_S : 2^I \to 2^I. \tag{2.140}$$

Now the arguments are no singleton imputations but whole sets. The function's value is simply the union of the values of (2.139) over all imputations contained in the new argument. This function is therefore nondecreasing in $D \subseteq I(N,v)$ in terms of set inclusion, as the value for each imputation is the empty set at worst. Taking this another step, we look at a function

$$\text{dom} : 2^I \to 2^I, \tag{2.141}$$

which again is the union of (2.140) over all coalitions $S \subseteq N$ and therefore nondecreasing in terms of set inclusion. This allows for the following proposition:

Proposition 2.13. *The core $C(N,v)$ of a coalitional game (N,v) is contained in any stable set $\mathfrak{S}(N,v)$:*

$$C(N,v) \subseteq \mathfrak{S}(N,v). \tag{2.142}$$

Proof. This follows directly from (2.137), (2.138), (2.68) and the fact that dom: $2^I \to 2^I$ is a nondecreasing function in $D \subseteq I(N,v)$: Because $\mathfrak{S}(N,v) \subseteq I(N,v)$, we have dom $\mathfrak{S}(N,v) \subseteq$ dom $I(N,v)$ and consequently $C(N,v) \subseteq \mathfrak{S}(N,v)$. ■

To conclude this section, we will show that in convex games the core coincides with the von Neumann Morgenstern solution and hence that the latter is unique. This was first shown by Shapley (1971, pp. 24 ff.).

Theorem 2.5. *The core of a convex game is the (von Neumann Morgenstern) stable set: $C(N,v) \subseteq \mathfrak{S}(N,v)$ and $\mathfrak{S}(N,v) \subseteq C(N,v)$.*

Proof. In order to show that the core of a convex game (N,v) is a stable set, we will establish that for any imputation $\mathbf{z} \notin C(N,v)$ there exists an imputation $\mathbf{x} \in C(N,v)$, for which $\mathbf{x} \succ \mathbf{z}$ holds. This follows from the definition of stable sets, especially (2.65), which states no two elements of the solution can dominate one another, and from the fact that the core is always contained in a stable set, as given in Proposition 2.13. If $\mathbf{x} \succ \mathbf{z}$ is the case, then also $\mathbf{x} \in \mathfrak{S}$ and hence $C \subsetneq \mathfrak{S}$ is not possible. Let us define a function $g : 2^N \to \mathbb{R}$ with $g(\emptyset) = 0$ and

$$g(S) = \frac{v(S) - z(S)}{|S|} \quad \text{for all } S \subseteq N,\ S \neq \emptyset. \tag{2.143}$$

Suppose that for some coalition S^*, the (well-defined) function g attains its maximum and $g(S^*) \geq g(S)$ for all $S \subseteq N$. As $\mathbf{z}$ is not in the core by assumption, there must be some $S \subseteq N$ for which $z(S) \geq v(S)$ is violated. This allows to conclude

$g(S^*) > 0$ and hance excludes the empty set from being arg max for g, i.e. $\emptyset \neq S^*$. Now let us construct a desired core allocation $\mathbf{x}$. Each of its n components is given by

$$x_i = \begin{cases} z_i + g(S^*) & \text{if } i \in S^* \\ c_i & \text{otherwise} \end{cases}, \quad \text{for all } i \in N. \tag{2.144}$$

Here, c_i denotes the ith component of a core allocation $\mathbf{c}$, more specific $\mathbf{c} \in C_{S^*} = C(N,v) \cap H(\chi_{S^*}, v(S^*))$, from the intersection of the core with the hyperplane $H(\chi_{S^*}, v(S^*))$.[57] C_{S^*} is the part of the core, where $x(S^*) = v(S^*)$, i.e. coalition S^* is allocated exactly its own value. Summing (2.144) over the payoffs $i \in S^*$, we obtain:

$$x(S^*) = \sum_{i \in S^*} \Big(z_i + g(S^*)\Big) = z(S^*) + |S^*| \cdot \frac{v(S^*) - z(S^*)}{|S^*|} = v(S^*). \tag{2.145}$$

Equations (2.144) and (2.145) show that $\mathbf{x} \succ \mathbf{z}$, as $x_i > z_i$ for all i from the effective set S^*, and $x(S^*) \leq v(S^*)$. As a whole, the allocation $\mathbf{x}$ is also feasible as we shall immediately show:

$$x(N) = x(S^*) + x(N \setminus S^*) = v(S^*) + c(N \setminus S^*) = c(N) = v(N). \tag{2.146}$$

The last equality follows from the fact that $\mathbf{c} \in C_{S^*}$ is a core allocation. We replaced $v(S^*)$ with $c(S^*)$, because $|S^*| \cdot g(S^*)$ is the difference between $z(S^*)$ and what S^* could obtain on its own by $v(S^*)$. As any core allocation on the hyperplane $H(\chi_{S^*}, v(S^*))$, for example $\mathbf{c} \in C_{S^*} = C(N,v) \cap H(\chi_{S^*}, v(S^*))$, yields exactly the amount $v(S^*)$ to S^*, it must hold that $v(S^*) = c(S^*)$.

The last step is to ensure that $\mathbf{x}$ also satisfies the core inequalities $x(S) \geq v(S)$ for all $S \subseteq N$. For this, we take some $T \subseteq N$ which we partition into sets Q and R, where $Q = T \cap S^*$ and $R = T \setminus S^*$. Then, $x(T)$ can be taken apart and reformulated as follows:

$$x(T) = x(Q) + x(R) = x(T \cap S^*) + x(T \setminus S^*). \tag{2.147}$$

The $x(T \cap S^*)$ we rearrange according to (2.144), and because $T \setminus S^* \subseteq N \setminus S^*$, we know $x(R) = c(R)$, yielding first part of (2.148).

$$\begin{aligned} x(T \cap S^*) &= z(Q) + |Q| \cdot g(S^*) + c(R) \\ &\geq z(Q) + |Q| \cdot g(Q) + c(T \cup S^*) - c(S^*) \\ &\geq v(Q) + v(T \cup S^*) - v(S^*) \\ &\geq v(T). \end{aligned} \tag{2.148}$$

[57] In a convex game, there exists no $S \subseteq N$ for which $C_S = \emptyset$, hence also $C_{S^*} \neq \emptyset$. This is shown in Shapley (1971, p. 18 f., p. 22).

The weak inequality in the second line follows from the maximum-property of S^* where $g(S^*) \geq g(S)$ for all $S \subseteq N$. Then we replace the first two terms on the right-hand by $v(Q)$, analogous to (2.145). Also, because $\mathbf{c} \in C_{S^*}$ is a core allocation, the core inequalities must hold. This yields the weak inequality in the third line. The last inequality follows from convexity (see (2.109)) of the game (N,v) substituting $T \cap S^*$ for Q. This shows that $x(T) \geq v(T)$ for arbitrary coalitions $T \subseteq N$.

Therefore, for each imputation $\mathbf{z} \notin C(N,v)$, we can find some $\mathbf{x} \in C(N,v)$ such that $\mathbf{x} \succ \mathbf{z}$. This shows that the core and the von Neumann Morgenstern solution coincide whenever the game (N,v) is convex. ■

Corollary 2.4. *The core of a convex game (N,v) is the unique stable set.*

Proof. This follows directly from Theorem 2.5 in conjunction with Corollary 2.2, by which no stable set can be a proper subset of another stable set. ■

Having related core and stable sets, our main treatment of the core and set-valued solution concepts is almost concluded. The last concept we consider is an attempt to tweak the core conditions in order to ensure the nonemptiness of the core. It is known as the least-core.

2.4.3.10 The Least-Core

Concluding the chapter on set-valued solution concepts, and as a possible bridge to the point-valued counterparts, we briefly want to touch a totally different approach to make sure the solution, or rather the core, is nonempty: Instead of imposing restrictive assumptions on the game (N,v) itself, we can simply modify or loosen the membership requirements of the solution concept. One example is the ε-core, where the core inequalities are relaxed in order to yield a nonempty core.[58] Different notions of ε-cores were introduced by Shapley and Shubik (1966) in the context of monetary economies and then generalized in Shapley and Shubik (1973, pp. 50 ff.). We elaborate only on the most common notion.

In reference to the definition of the core (see (2.67), p. 59), the *strong ε-core*, or simply ε-core is given by

$$C_\varepsilon(N,v) := \{\mathbf{x} | \mathbf{x} \in \mathbb{R}^n,\ x(N) = v(N),\ x(S) \geq v(S) - \varepsilon \ \text{ for all } \ S \subsetneq N\}. \quad (2.149)$$

In this approach, the core conditions are loosened by increasing a nonnegative ε. Multiple interpretations for the ε-core exist: For one, ε could be seen as the fixed cost which arises from forming or organizing a coalition S. Alternatively, it could be considered the threshold to act. Allocations $x(S)$ where deviations within an ε around (or better: below) $v(S)$ occur, are not worthwhile improving upon, even though the coalition S could do better by itself. Sometimes ε is also referred to as a

[58] This of course only makes sense for games where otherwise $C(N,v) = \emptyset$.

measure of how distant from a nonempty core a given game is and where this core would be located.[59]

The actual least-core is defined as follows:

Definition 2.49. The *least-core* of a coreless game, $C_{\varepsilon^*}(N,v)$, is the ε-core at the minimal value of ε, denoted ε^*, such that $C_\varepsilon(N,v) \neq \emptyset$.

Such an ε^* is guaranteed to exist, because the underlying core concept is defined in terms of intersections of closed half-spaces with the simplex of imputations, itself a compact and convex set. Increasing ε moves the half-spaces towards the boundaries of the (original) simplex of imputations. The latter do not expand with ε, as we kept $x(N) = v(N)$ in (2.149). For an illustration, see the following example:

Example 2.7. Continuing from Example 2.6, where an empty core resulted, we now calculate ε^* for the least-core and illustrate some ε-cores for $\varepsilon > \varepsilon^*$. Recall the characteristic function v with values $v(\emptyset) = 0$ and $v(\{i\}) = 0$ for all $i \in N$, $v(\{1,2\}) = 3\frac{1}{2}$, $v(\{1,3\}) = 4$, $v(\{2,3\}) = 3$, $v(N) = 5$. As it stands, $C(N,v) = \emptyset$, which was illustrated in Fig. 2.8. In order to find ε^*, we have to solve the following system of equations:[60]

$$\begin{aligned} x_1 + x_2 &= 3\frac{1}{2} - \varepsilon \\ x_1 + x_3 &= 4 - \varepsilon \\ x_2 + x_3 &= 3 - \varepsilon \\ x_1 + x_2 + x_3 &= 5. \end{aligned}$$

This yields $\varepsilon^* = \frac{1}{6}$, as well as the locus of intersection of the three lines where the core constraints are binding: $\mathbf{x}_{\varepsilon^*} = (x_1, x_2, x_3) = (\frac{13}{6}, \frac{7}{6}, \frac{10}{6})$. Here, the least-core is a singleton, $C_{\varepsilon^*}(N,v) = \{\mathbf{x}_{\varepsilon^*}\}$. This is illustrated in Fig. 2.10, along with the ε-cores for different values of ε, given by $\varepsilon_4 = 1$, $\varepsilon_3 = \frac{3}{4}$, $\varepsilon_2 = \frac{1}{2}$, and $\varepsilon_1 = \frac{1}{4}$. Their respective ε-cores are represented by the intersection of the (upside-down) equilateral triangles and the simplex.

Another closely-related notion is the so-called *weak ε-core*. In addition to the strong ε-core, it takes into account the size of the coalition. The modification of the core inequalities is then achieved by multiplying ε with the cardinality of S:

$$C_\varepsilon(N,v) := \{\mathbf{x} | \mathbf{x} \in \mathbb{R}^n,\ x(N) = v(N),\ x(S) \geq v(S) - \varepsilon \cdot |S| \text{ for all } S \subsetneq N\}. \tag{2.150}$$

[59] Note that by definition no sign of ε is fixed. Naturally, for games with empty core, one would assume it to be positive, but it can also be used with a negative value to strengthen the core inequalities for games where $C(N,v) \neq \emptyset$. In this regard one could think of the ε-core as an instrument to narrow down the multiple choices of allocations a core might provide.

[60] This is actually a convenient shortcut, as none of the adapted core-inequalities $x(\{i\}) \geq v(\{i\}) - \varepsilon$ are binding in this example. In general, the solution is a linear program including all $2^n - 1$ core constraints.

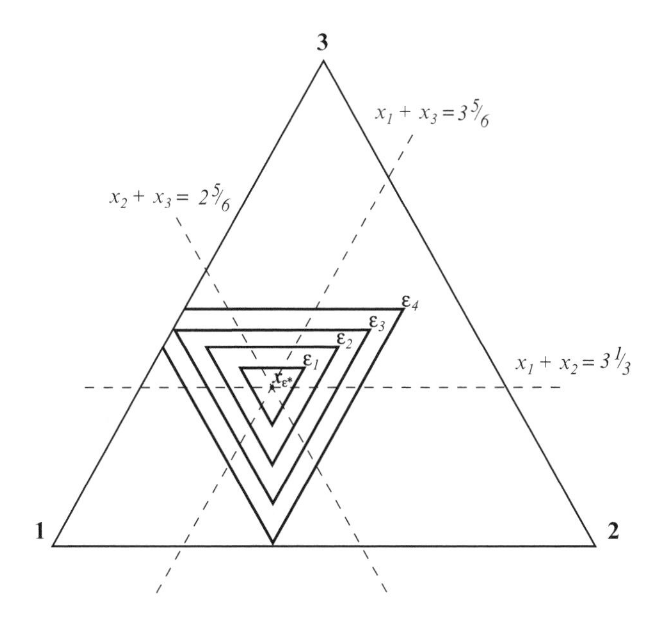

Fig. 2.10 The least-core and ε-cores for different values of ε

However, the least-core is not necessarily a singleton as it was in Example 2.7. And even though the ε-core expands symmetrically around the corresponding least-core, it might be misleading to interpret the allocation(s) in the least-core as equitable, because its location within the set of imputations need not center around the barycenter of $I(N, v)$.

So regardless of the assumptions imposed on the game or of alterations to the solution concept to ensure nonemptiness, the core-style solutions can generally not be expected to result in singleton outputs. If we really insist on single allocation solutions, we must consider different solutions concepts, which are more punctual by nature. These point-valued concepts are widely known as allocation rules and will be subject of the subsequent sections.

2.4.4 Allocation Rules as Point-valued Solutions

From the previous sections we have learned that the core and related solution concepts do not necessarily provide singleton results. As a matter of fact, they usually do not. In some contexts it maybe enough to justify some outcome by showing that it lies within what is proposed by the core. In other situations though, one might prefer a solution concept that proposes a unique allocation, with no margins for interpretation at all.[61] In the latter case it comes natural to look at

[61] Lucas (1971, pp. 519 f.) classifies solution concepts according to their coverage.

so-called *allocation rules*, which assign to each coalitional game one and only one allocation as the game's solution. Such allocation rules are usually based on certain axioms which suggest properties that the rule has to satisfy. We will list some of those below. Our main focus lies on the most prominent of these allocation rules, the *Shapley value*, introduced in Shapley (1953b), as well as one of its derivatives, the *Weighted Shapley value*.[62] Both concepts assign the average of the marginal values of the players upon joining a coalition as resulting allocation, the latter under consideration of a weights system on the players. Before we go into detail, we define the concept of an allocation rule in its most general form and then list some desirable or common properties of these types of solution concepts.

In its most general form, an allocation rule, or *operator*, or *value* is a mapping that assigns to each coalitional game (N, v) a unique vector of payoffs corresponding in dimension to the cardinality of the player set. In conjunction with allocation rules, we usually denote games by their characteristic functions v or the corresponding $(2^n - 1)$-dimensional vectors, $\mathbf{v} \in \mathbb{R}^{2^n - 1}$.

Definition 2.50. An *allocation rule* is a function from the space of coalitional games with player set N to the Euclidean space $\mathbb{R}^n$, assigning each player $i \in N$ a unique payoff:

$$\Phi : \mathbf{V}_N \to \mathbb{R}^n. \tag{2.151}$$

Next, we compile a number of possible properties that allocation rules may exhibit. Again, we are by no means exhaustive but aim to deliver the most basic assumptions, especially with respect to what is to follow. This should also help to develop intuition. We have chosen no particular order but roughly aggregate the properties in two groups: First, those that only affect allocations to the players within a given game, and secondly, those that consider the change in payoffs when the underlying game is altered. Properties of allocation rules appear in the literature in various places, we draw mainly from Monderer and Samet (2002), as well as Young (1985) and Kalai and Samet (1985), especially for monotonicity-related issues.

Definition 2.51. An allocation rule $\Phi : \mathbf{V}_N \to \mathbb{R}^n$ is *efficient*, if for all games $\mathbf{v} \in \mathbf{V}_N$ the resulting allocation exhausts the value assigned under v to the grand coalition N:

$$\sum_{i \in N} \phi_i(\mathbf{v}) = v(N). \tag{2.152}$$

Under this property, the allocation rule Φ distributes the whole amount $v(N)$ to the players $i \in N$. Technically, using the aggregate value as orientation relates to a feasible allocation only, without any further restrictions on the individual payoffs.[63] The next axiom slightly narrows down the possibility of allocating any gains from cooperation.

[62] Especially the Weighted Shapley value will reveal a close relation to the core, even though their theoretical foundations differ completely.

[63] This axiom must not be confused with Axiom 2 in Shapley (1953b), to which the author misleadingly refers as "efficiency". He actually applies the "carrier" property, which we treat next.

Definition 2.52. An allocation rule $\Phi : \mathbf{V}_N \to \mathbb{R}^n$ satisfies the *carrier property*, if for all games $\mathbf{v} \in \mathbf{V}_N$ and all carriers T of $\mathbf{v}$, the value created by this carrier is also allocated entirely to its players $i \in T$:

$$\sum_{i \in T} \phi_i(\mathbf{v}) = v(T). \tag{2.153}$$

The value assigned to any carrier of a game must be exhausted under Φ among the players of this carrier only. Because any superset of a carrier is again a carrier, this property implies efficiency. It is also the dual to the following property:

Definition 2.53. An allocation rule $\Phi : \mathbf{V}_N \to \mathbb{R}^n$ satisfies the *null-player property*, if for all games $\mathbf{v} \in \mathbf{V}_N$ and all null-players $i \in N$ the following holds:

$$\phi_i(\mathbf{v}) = 0. \tag{2.154}$$

All players that do not belong to the minimal carrier of game $\mathbf{v}$ are null-players and are consequently allocated nothing under either of the previous two properties, carrier and null-player. Also, the null-player property can be considered a special case of the dummy player property with constant marginal values of zero:

Definition 2.54. An allocation rule $\Phi : \mathbf{V}_N \to \mathbb{R}^n$ satisfies the *dummy player property*, if for all games $\mathbf{v} \in \mathbf{V}_N$ and all dummy players $i \in N$ with constant marginal values $d_i(S) = v(\{i\})$ for all $S \subseteq N \setminus \{i\}$, the individual allocations equal the stand-alone value:

$$\phi_i(\mathbf{v}) = v(\{i\}). \tag{2.155}$$

Any player whose marginal value is constant, must be assigned his stand-alone value under Φ. The consequence of this can again be summarized in a property:

Definition 2.55. An allocation rule $\Phi : \mathbf{V}_N \to \mathbb{R}^n$ is a *projection*, if for all additive games $\mathbf{v}$ all players are allocated their stand-alone values:

$$\Phi(\mathbf{v}) = (v(\{1\}), v(\{2\}), \ldots, v(\{n\}))'. \tag{2.156}$$

In an additive game, the value of any coalition $S \subseteq N$ is the sum of the stand-alone values of its members: $v(S) = \sum_{i \in S} v(\{i\})$ for all $S \subseteq N$. Consequently, as all players are dummy players, their allocation under Φ is their stand-alone value.

Definition 2.56. An allocation rule $\Phi : \mathbf{V}_N \to \mathbb{R}^n$ is *symmetric* or *anonymous*, if for all games $\mathbf{v} \in \mathbf{V}_N$ and permutations $\pi \in \Pi_N$ the following is true:

$$\phi_{\pi(i)}(\mathbf{v}_\pi) = \phi_i(\mathbf{v}), \tag{2.157}$$

where $\mathbf{v}_\pi$ denotes the vector corresponding to the game $\mathbf{v}$ with entries are adjusted according to permutation π.

Under Φ, a relabelling of players does not influence their payoff, because the allocation rule works on the basis of a player's characteristics as mirrored in $\mathbf{v}$, or rather on his marginal values, and not on his name or numbering.

This property can be expressed without permutations, too.[64] Given any two players i and j, for which the function d_i from (2.32) satisfies

$$d_i(S) = d_j(S) \text{ for all } S \subseteq N, \tag{2.158}$$

it must be true that $\phi_i(\mathbf{v}) = \phi_j(\mathbf{v})$. If the marginal values are identical for players i and j and they can be substituted without effect on the value function, they also receive an identical allocation.[65]

Definition 2.57. An allocation rule $\Phi : \mathbf{V}_N \to \mathbb{R}^n$ is *positive*, if for all monotonic games $\mathbf{v}$ all players $i \in N$ are allocated nonnegative payoffs:

$$\phi_i(\mathbf{v}) \geq 0. \tag{2.159}$$

In a monotonic game, the marginal value of no player is ever strictly negative, and given the convention $v(\emptyset) = 0$, neither should be his allocation under Φ.

We now turn to the second group of properties, where we compare allocations across games.

Definition 2.58. An allocation rule $\Phi : \mathbf{V}_N \to \mathbb{R}^n$ is *additive*, if for all games $\mathbf{u}, \mathbf{v} \in \mathbf{V}_N$ the following holds:

$$\Phi(\mathbf{u} + \mathbf{v}) = \Phi(\mathbf{u}) + \Phi(\mathbf{v}). \tag{2.160}$$

Hence, whenever distinct games are combined, the allocation in the new game is just the sum, player by player, of the allocations in the original games. This implies that such a merging of games does not open additional chances for cooperation among the players, as this could result in changes to the aggregate value function beyond the mere summation.

On first glance, one is tempted to assume that additivity can only be applied whenever the characteristic functions u and v are both defined over the same domain 2^N. This need not be the case at all, as a common domain can easily be constructed from games where the underlying set of players is not identical, possibly even disjoint: Let $N := R \cup T$, where R is the domain of v and T that of u. Define for any $S \subseteq N$ that $v(S) := v(S \cap R)$ and $u(S) := u(S \cap T)$. This amounts to having domain N for both characteristic functions, where all players $i \in N \setminus R$ are null-players in game $\mathbf{v}$ and likewise all $i \in N \setminus T$ in game $\mathbf{u}$.

---

[64] In this form, it is often referred to as *marginality*, see for example Young (1988, p. 270).

[65] Note that expression (2.158) covers all four cases of players i and j with respect to their membership in set S, incorporating both versions of d_i as given in (2.32) simultaneously, when applicable.

Hidden in this procedure is the fact that for games that are decomposable the value of Φ assigned to a given player over the whole game equals his value on just the component to which he belongs. This is content of the next property:

Definition 2.59. An allocation rule $\Phi : \mathbf{V}_N \to \mathbb{R}^n$ is *decomposable*, if for all games $\mathbf{v} \in \mathbf{V}_N$ which are decomposable with respect to some partition P the following holds:

$$\phi_i(\mathbf{v}|_{P_j}) = \phi_i(\mathbf{v}) \text{ for all } i \in P_j,\ P_j \in \mathsf{P}. \tag{2.161}$$

Definition 2.60. An allocation rule $\Phi : \mathbf{V}_N \to \mathbb{R}^n$ is *homogeneous*, if for all games $\mathbf{v} \in \mathbf{V}_N$ and all scalars $\kappa \in \mathbb{R}$ the following holds:

$$\Phi(\kappa \cdot \mathbf{v}) = \kappa \cdot \Phi(\mathbf{v}). \tag{2.162}$$

Thereby, if we scale all possible values of the characteristic function v, given by the vector $\mathbf{v}$, using a common multiplier κ, the allocation rule will reflect this operation true to the selected scale.[66]

Combining additivity and homogeneity leads to another very common property found in allocation rules:

Definition 2.61. An allocation rule $\Phi : \mathbf{V}_N \to \mathbb{R}^n$ is *linear*, if it is additive and homogeneous:

$$\Phi(\alpha \cdot \mathbf{u} + \beta \cdot \mathbf{v}) = \alpha \cdot \Phi(\mathbf{u}) + \beta \cdot \Phi(\mathbf{v}). \tag{2.163}$$

Next come various aspects regarding monotonicity of allocation rules. They relate specific changes in the underlying value functions to the payoffs under Φ.

Definition 2.62. An allocation rule $\Phi : \mathbf{V}_N \to \mathbb{R}^n$ is *monotonic in the aggregate*, if for all games $\mathbf{u}, \mathbf{v} \in \mathbf{V}_N$ that satisfy

$$v(N) \geq u(N) \text{ and } v(S) = u(S) \text{ for all } S \subsetneq N, \tag{2.164}$$

the following holds:

$$\phi_i(\mathbf{v}) \geq \phi_i(\mathbf{u}) \text{ for all } i \in N. \tag{2.165}$$

Then, if only the grand coalition's value is weakly higher in game $\mathbf{v}$ than in $\mathbf{u}$, while the values of all other coalitions remain the same, no player should be worse off under the same allocation rule Φ in game $\mathbf{v}$.

Definition 2.63. An allocation rule $\Phi : \mathbf{V}_N \to \mathbb{R}^n$ is *coalitionally monotonic*, if for all games $\mathbf{u}, \mathbf{v} \in \mathbf{V}_N$ that satisfy

$$v(T) \geq u(T) \text{ for some } T \text{ and } v(S) = u(S) \text{ for all } S \subseteq N,\ S \neq T, \tag{2.166}$$

[66] This property amounts to what is known as *homogeneity of degree one* for general functions: $f(\lambda \cdot \mathbf{x}) = \lambda^1 \cdot f(\mathbf{x})$, with $\lambda \in \mathbb{R}$.

the following holds:

$$\phi_i(\mathbf{v}) \geq \phi_i(\mathbf{u}) \text{ for all } i \in T. \tag{2.167}$$

This property is a generalization of the previous one, which it includes for $T = N$. Now, players in any given coalition $T \subseteq N$ can be allocated no less if T's value is at least as large in game $\mathbf{v}$ as in $\mathbf{u}$. The intuition remains the same as before. In consequence, this property even holds for individual players, when applied to all sets that contain some player i:

$$\text{If } v(T) \geq u(T) \text{ for all } T|i \in T \text{ and } v(S) = u(S) \text{ for all } S|i \notin S, \tag{2.168}$$

$$\text{then } \Phi(\mathbf{v}) \geq \Phi(\mathbf{u}) \text{ and } \phi_i(\mathbf{v}) \geq \phi_i(\mathbf{u}). \tag{2.169}$$

Monotonicity can also be restricted to single players directly:

Definition 2.64. An allocation rule $\Phi : \mathbf{V}_N \to \mathbb{R}^n$ is *strongly monotonic*, if for all games $\mathbf{u}, \mathbf{v} \in \mathbf{V}_N$, where some player i's marginal value is no less in $\mathbf{v}$ than in $\mathbf{u}$, no less is allocated to this player in $\mathbf{v}$:

$$\text{If } d_i^{\mathbf{v}}(S) \geq d_i^{\mathbf{u}}(S) \text{ for all } S \subseteq N, \text{ then } \phi_i(\mathbf{v}) \geq \phi_i(\mathbf{u}). \tag{2.170}$$

This property is even stronger than the consequence from the previous one. Now it is the marginal value of a specific player that has not decreased, as compared to the value of all coalitions a specific player belongs to, which could also result from other players' increases in marginal value. This leads to an allocation under Φ that is no less in game $\mathbf{v}$ than in $\mathbf{u}$.

At the end we present one last property. It relates an allocation rule to the core:

Definition 2.65. An allocation rule $\Phi : \mathbf{V}_N \to \mathbb{R}^n$ is *core consistent*, if for any game $\mathbf{v}$ with nonempty core, $C(N, v) \neq \emptyset$, the allocation rule coincides with a core allocation:

$$\Phi(\mathbf{v}) \in C(N, v). \tag{2.171}$$

With $\Phi(\mathbf{v})$ in the core, the allocation automatically satisfies all the requirements for core membership.

This concludes our list of potential properties of allocation rules. They will resurface in subsequent parts whenever allocation rules are considered. It depends on the individual rule and the properties assumed a priori. Sometimes a combination of properties even implies others, as we shall see. The first use we make of them is in the next section, concerning the axiomatic derivation of the Shapley value.

2.4.5 The Shapley Value

This section is devoted to the Shapley Value, probably the single most influential solution concept in cooperative game theory. It was the first point-valued solution receiving wide acceptance in the literature, which is also manifest in its denomination: Even though the concept was introduced quite neutrally as "*A value for n-person games*" in Shapley (1953b), it became known soon as the Shapley value in reference to its creator.[67] The literature related to the Shapley value is probably as close to "uncountable" as it gets; few other topics in economics were and are being paid so much attention in the literature. The reasons for this are manifold: On the one hand the Shapley value has been adapted to accommodate different situations or games. A prominent example for this is the *Myerson value* (see Myerson 1977) where a graph structure among the players influences cooperation. On the other hand, the Shapley value and its derivatives reoccur steadily in applications, most prominently for cost allocation purposes in accounting (see Shapley 1981 or Young 1994).

Aside from the original source (Shapley 1953b) we base our subsequent study mostly on Winter (2002), Monderer and Samet (2002), Moulin (1991, Chap. 5), as well as Chun (1989, 1991), Friedman (1991), and Owen (1995). Beyond that, a section on the Shapley value can be found in almost any economics textbook featuring the theory of cooperative games. This is the case as far back as Luce and Raiffa (1957) and as recent as Mas-Colell et al. (1995) But even books with more emphasis on others matters, see for example Ok (2007), cover the Shapley value on the grounds of its far reach within the science of economics.

What is so remarkable about the value is the fact that it is derived from only three axioms. All three are intuitively very plausible and not too restrictive as to prevent the concept from being widely accepted. In addition, the Shapley value results in a unique allocation and is guaranteed to exist for all coalitional games. Hence, it is a remedy to what we criticized in relation to the previously treated solutions concepts.

We proceed in the following order: After stating and commenting the axioms, we derive the operator through an *additive* approach, based on unanimity games. We also state the *marginal* approach, which is much more intuitive and relies less on technicalities.[68] Then, we prove the main theorem that establishes the operator from the three axioms. The section is closed with an example on core consistency (or the possible lack thereof) and considerations regarding strategic equivalence.

One striking feature of the Shapley operator is its reliance on only three axioms. We now list these axioms, relying on the nomenclature previously introduced in Sect. 2.4.4. Originally, Shapley (1953b) defined his value operator on the class

[67] Interestingly, in Footnote 35 of Harsanyi (1956, p. 157) the author claims to have discovered independently an allocation rule identical to that of Shapley (1953b), while on leave in Australia.

[68] The reader only interested in an intuitive foundation of the Shapley operator is welcome to skip the technicalities of the additive approach. These arise right after the axioms below and continue until (2.202) on p. 99.

of superadditive games, a further restriction.[69] With the necessary changes to the axioms, we relax this assumption.

Axiom 1: Symmetry As given in Definition 2.56, this axiom proposes the value to ignore the identity of the players. Their allocations under the operator is based exclusively on their marginal values, as we will see below, and a change of ordering has no real effect on the payoffs, given the characteristics of a player.

Axiom 2: Carrier *["Efficiency" in Shapley (1953b)]* This axiom was previously defined in Definition 2.52. As value created within any carrying coalition must be allocated exhaustively among its players, null-players also receive nothing under the Shapley operator.

Axiom 3: Additivity *["Law of Aggregation" in Shapley (1953b)]* This property, according to Definition 2.58, allows the merger of coalitional games under the Shapley operator, by summing the individual allocations under the operator from the original games.

The point of origin in Shapley (1953b) to derive the functional form of the operator are unanimity games. Applying the three axioms, it is extended to regular coalitional games. We trace Shapley's original course of action, but much more detailed, where he skips steps we deem crucial.

The first assertion we want to show is that for nonnegative constants $\gamma \in \mathbb{R}_+$, when applied to a unanimity game $\mathbf{u}_T$ with $T \subseteq N$, the operator yields

$$\phi_i(\gamma \cdot \mathbf{u}_T) = \begin{cases} \frac{\gamma}{|T|} & \text{if } i \in T \\ 0 & \text{otherwise} \end{cases}. \tag{2.172}$$

This is derived using the first two axioms: All members of the carrier T of the unanimity game are equally necessary and sufficient as a whole to reap the gains from $\mathbf{u}_T$. Because each one of them is equally necessary, they are interchangeable and receive the same amount (Axiom 1) under Φ. Because no player outside of carrier T has any influence on the value function, those players receive nothing (Axiom 2) under Φ. To show this more formally, we highlight both cases of (2.172):

Case 2.1. $i \in T$

Regardless of whichever permutation π on N, the players that constitute the carrying coalition T in $\mathbf{u}_T$ remain the same. Therefore, any two players $i, j \in T$ receive equal allotments under Φ, as postulated by Axiom 1:

$$\phi_i(\gamma \cdot \mathbf{u}_T) = \phi_j(\gamma \cdot \mathbf{u}_T). \tag{2.173}$$

[69] This has been dropped in the subsequent literature. Its origin was most likely due to the fact that von Neumann and Morgenstern (1947, pp. 241 ff.) define superadditivity as a fundamental property of a characteristic function v. This again stems from the fact that they treat exclusively zero-sum games.

From Axiom 2 we know that all value created by carrier T is distributed exhaustively among its members $i \in T$,

$$\sum_{i \in T} \phi_i(\gamma \cdot \mathbf{u}_T) = \gamma \cdot \underbrace{u_T(T)}_{=1} = \gamma, \tag{2.174}$$

and, because players i and j in (2.173) were picked arbitrarily from T, we can rewrite (2.174) to

$$\sum_{i \in T} \phi_i(\gamma \cdot \mathbf{u}_T) = |T| \cdot \phi_i(\gamma \cdot \mathbf{u}_T) = |T| \cdot \frac{\gamma}{|T|} = \gamma. \tag{2.175}$$

This confirms the first case of (2.172).

Case 2.2. $i \notin T$
From Axiom 2 it is quickly derived that $\phi_i(\mathbf{v}) = 0$ in any game (N, v) in which player i is not member of all carriers: If T is a carrier, so is $T \cup \{i\}$. Because Axiom 2 holds for all carriers, it must be true by definition of the latter that

$$\sum_{j \in T} \phi_j(\mathbf{v}) = v(T) = v(T \cup \{i\}) = \sum_{j \in T \cup \{i\}} \phi_j(\mathbf{v}). \tag{2.176}$$

The middle equality holds because for every coalition S such that $i \in S$, we have $v(S) = v(T \cap S)$ as well as $v(S) = v((T \cup \{i\}) \cap S)$. Consequently, $\phi_i(\mathbf{v}) = 0$, which holds not only for general games $\mathbf{v}$, but also for the unanimity game $\mathbf{u}_T$, confirming the second case of (2.172).

On the way to apply Shapley's value to general games, we want to remind the reader that any game $\mathbf{v}$ can be expressed as a linear combination of unanimity games:

$$\mathbf{v} = \sum_{\emptyset \neq T \subseteq N} \gamma_T \cdot \mathbf{u}_T. \tag{2.177}$$

Each of the $2^n - 1$ rows constituting (2.177) is given by

$$v(S) = \sum_{T \subseteq N} \gamma_T \cdot u_T(S) \quad \text{for all } \emptyset \neq S \subseteq N. \tag{2.178}$$

We now assign every unanimity game $\mathbf{u}_T$ its own coefficient $\gamma_T \in \mathbb{R}$:

$$\gamma_T = \sum_{R \subseteq T} (-1)^{|T|-|R|} \cdot v(R). \tag{2.179}$$

For a proof of (2.178), we refer to our previous treatment of unanimity games, especially to Proposition 2.1, p. 34.

Shapley (1953b) also remarks that it could be "*easily shown*" (without actually doing so) that $\gamma_T = 0$ whenever T is not contained in every carrier, in particular the minimal carrier of the game (N, v). As we believe this not quite so straightforward, and more importantly, highly instructive for what follows, we show this in a short excursion:

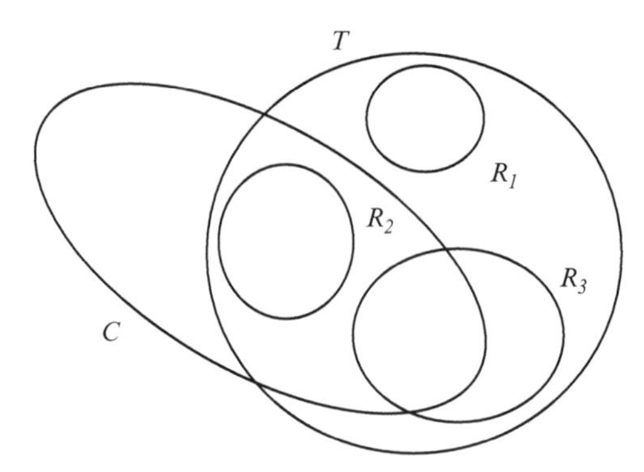

Fig. 2.11 Three possibilities for sets $R \subseteq T$

Claim. The constant $\gamma_T = 0$, if T is not a subset of the minimal carrier of a coalitional game (N, v).

Proof. For notational simplicity, we denote – for this excursion only – the smallest carrier by C. We also refrain from denoting the cardinality of sets by the respective letters in lower case, as can lead to confusion with overlapping sets.

Now, for any given set T which is not contained in the carrier, $T \not\subseteq C$, but has nonempty intersection with it, $T \cap C \neq \emptyset$, we know that $v(C \cap T) = v(T)$. With γ_T defined according to (2.179), three cases arise for the sets $R \subseteq T$: When $R \cap C = \emptyset$, obviously $v(R) = v(C \cap R) = 0$. Also, for $R \subseteq C$, we have $v(R) = v(C \cap R)$, and finally, the same holds for sets R with nonempty intersection with T inside and outside the carrier C: $R \cap (C \setminus T) \neq \emptyset$ and $R \cap (T \setminus C) \neq \emptyset$, then $v(R) = v(C \cap R)$. In Fig. 2.11 the three cases are denoted R_1, R_2, and R_3, respectively: For every set of type 2 with $R_2 \subseteq C$, there exist $\binom{|T \setminus C|}{|R_1|}$ type 1 sets with cardinality $|R_1|$ which we can use to build a set of type 3. In total, for each type 2 set R_2, there exist

$$\sum_{|R_1|=1}^{|T \setminus C|} \binom{|T \setminus C|}{|R_1|}$$

sets to augment. Note that for all such augmented sets, the value remains the same, i.e. $v(R_2 \cup R_1) = v(R_2)$ for all combinations of R_2 and R_1, because C is a carrier. In cases where γ_T is equal to zero, we need to show that

$$(-1)^{|T|-|R_2|} \cdot v(R_2) + \sum_{|R_1|=1}^{|T \setminus C|} \binom{|T \setminus C|}{|R_1|} \cdot (-1)^{|T|-|R_2|-|R_1|} \cdot v(R_2 \cup R_1) = 0 \quad (2.180)$$

holds. Precisely because $v(R_2 \cup R_1) = v(R_2)$, we can cancel out the value functions and include the first term in the summation:

$$\sum_{|R_1|=0}^{|T \setminus C|} \binom{|T \setminus C|}{|R_1|} \cdot (-1)^{|T|-|R_2|-|R_1|} = 0. \quad (2.181)$$

This, again, is a specific way of writing the binomial coefficient. Because this whole excursion is based on sets T not contained in every carrier, $|T \setminus C| > 0$ and (2.181) is true. As this procedure was based on arbitrary sets (within their assumed properties), the proof is complete. ■

We are now ready to extend the Shapley operator to general games, i.e. games which do not enjoy the property of symmetry within the carrying coalitions common to unanimity games. Axiom 3 allows us to write the following:

$$\sum_{T \subseteq N} \phi_i(\gamma_T \cdot \mathbf{u}_T) = \phi_i\left(\sum_{T \subseteq N} \gamma_T \cdot \mathbf{u}_T\right). \tag{2.182}$$

By (2.172), this is a summation whose elements are either $\gamma_T / |T|$ whenever $i \in T$, or zero, when $i \notin T$.[70] Rewriting (2.182) accordingly leads to

$$\sum_{\substack{T \subseteq N \\ i \in T}} \frac{\gamma_T}{|T|} = \phi_i\left(\sum_{T \subseteq N} \gamma_T \cdot \mathbf{u}_T\right), \tag{2.183}$$

which, after another substitution from (2.177) will result in

$$\phi_i(\mathbf{v}) = \sum_{\substack{T \subseteq N \\ i \in T}} \frac{\gamma_T}{|T|}. \tag{2.184}$$

The expression for the Shapley operator in terms of unanimity coefficients is given by (2.184). As these coefficients are not computed readily, we aim to transform this expression into something more handy. To obtain what is referred to as the formula for computing the Shapley value requires a few steps. First, we replace γ_T with (2.179):

$$\phi_i(\mathbf{v}) = \sum_{\substack{T \subseteq N \\ i \in T}} \frac{1}{|T|} \sum_{R \subseteq T} (-1)^{|T|-|R|} \cdot v(R). \tag{2.185}$$

Next, we reorder the inner summation: The first group contains all the subsets R, for which $i \in R$, the second those, where $i \notin R$:

$$\phi_i(\mathbf{v}) = \sum_{\substack{T \subseteq N \\ i \in T}} \frac{1}{|T|} \left[\sum_{\substack{R \subseteq T \\ i \in R}} (-1)^{|T|-|R|} \cdot v(R) + \sum_{\substack{R \subseteq T \\ i \notin R}} (-1)^{|T|-|R|} \cdot v(R)\right]. \tag{2.186}$$

Now let us consider, how many supersets $T \supseteq R$ exist for a given set R. We distinguish the two cases whether player i is contained in R or not.

[70] The undefined case where $|T| = 0$ is excluded.

Suppose $i \in R$: If $r = t$, this leaves but only one possible set T, one which coincides with R. For a larger set T, say $t = r+1$, we already have $n-r$ possible players to choose from and hence also have this number of possible distinct supersets T. For some general cardinality t, we have $\binom{n-r}{t-r}$ possible supersets. The upper bound is given by $t-r \leq n-r$, i.e. the number of all players not in R. The total amount of supersets $T \supseteq R$ (of any cardinality) is then calculated through

$$\sum_{t=r}^{n} \binom{n-r}{t-r}. \tag{2.187}$$

We proceed similarly with sets R such that $i \notin R$ taking into account two innate differences: First, there is no weak superset $T \supseteq R$ with equal cardinality $t = r$, because by assumption $i \in T$ but $i \notin R$. Second and consequently, for any T where $t = r+1$, we have no choice for this additional player: $T = R \cup \{i\}$. Therefore, the largest possible set $R \subseteq N$ in this scenario is such that $R = N \setminus \{i\}$. Taking this into account, the total number of supersets $T \supseteq R$ of all cardinalities is

$$\sum_{t=r+1}^{n} \binom{n-r-1}{t-r-1}. \tag{2.188}$$

Inserting (2.187) and (2.188) into (2.186), yields the following expression for the Shapley value:

$$\begin{aligned} \phi_i(\mathbf{v}) = & \sum_{\substack{R\subseteq N \\ i\in R}} \sum_{t=r}^{n} \frac{1}{t} \binom{n-r}{t-r} (-1)^{t-r} \cdot v(R) \\ & - \sum_{\substack{R\subseteq N \\ i\notin R}} \sum_{t=r+1}^{n} \frac{1}{t} \binom{n-r-1}{t-r-1} (-1)^{t-r-1} \cdot v(R). \end{aligned} \tag{2.189}$$

Note that the negative sign for the second term is compensated for by $(-1)^{t-r-1}$, where the exponent is reduced by one.[71] Comparing all possible sets $R \subseteq N$ with $i \in R$ with those $R \subseteq N$ for which $i \notin R$ (including the empty set), we notice that there are just as many subsets containing player i as there are without. Both outer summations of (2.189) have equally many summands, for each subset containing player i, there is one just like it but without i. If we pair up corresponding sets by matching an R with $i \in R$ and its counterpart $\widetilde{R}$ with $\widetilde{R} = R \setminus \{i\}$, expression (2.189) is further simplified. Because $r = \widetilde{r} + 1$, for each such pair, (2.187) and (2.188) coincide, and we obtain

$$\phi_i(\mathbf{v}) = \sum_{\substack{R\subseteq N \\ i\in R}} \sum_{t=r}^{n} \frac{1}{t} \binom{n-r}{t-r} (-1)^{t-r} \left[v(R) - v(R \setminus \{i\})\right]. \tag{2.190}$$

[71] Even though the binomial coefficient was changed to accommodate for possible combinations where $i \notin R$ but $i \in T$, this does not apply to the second exponent in (2.186).

Disregarding the difference of the value functions in square parentheses for the moment and focussing on the inner sum,

$$\sum_{t=r}^{n} \frac{(-1)^{t-r}}{t} \binom{n-r}{t-r}, \tag{2.191}$$

note that

$$\frac{1}{t} = \int_0^1 x^{t-1} dx, \tag{2.192}$$

which can be substituted into (2.191):

$$\int_0^1 \sum_{t=r}^{n} (-1)^{t-r} \binom{n-r}{t-r} x^{t-1} dx. \tag{2.193}$$

Because $x^{t-1} = x^{r+t-r-1} = x^{r-1} \cdot x^{t-r}$, we can first write

$$\int_0^1 x^{r-1} \sum_{t=r}^{n} (-1)^{t-r} \binom{n-r}{t-r} x^{t-r} dx, \tag{2.194}$$

and, followed by combining $(-1)^{t-r}$ and x^{t-r}, we obtain

$$\int_0^1 x^{r-1} \sum_{t=r}^{n} \binom{n-r}{t-r} (-x)^{t-r} dx. \tag{2.195}$$

Adapting the binomial theorem,[72] we are finally left with

$$\int_0^1 x^{r-1} (1-x)^{n-r} dx, \tag{2.196}$$

which, as we will now show, using integration by parts, turns into

$$\frac{(r-1)!(n-r)!}{n!}. \tag{2.197}$$

[72]The binomial theorem is given by

$$(x+y)^n = \sum_{k=0}^{n} \binom{n}{k} x^{n-k} y^k,$$

while our adaptions are the following: $x \to 1$, $y \to -x$, $n \to n-r$, and $k \to t-r$.

The integration by parts is done in multiple steps, $n-r$, to be exact. If we consider

$$\int_a^b u(x)v'(x)\,dx = [u(x)v(x)]_a^b - \int_a^b u'(x)v(x)\,dx$$

with (2.196) as left-hand side, we obtain

$$\int_0^1 (1-x)^{n-r}x^{r-1}\,dx = \left[(1-x)^{n-r}\frac{x^r}{r}\right]_0^1 - \int_0^1 (n-r)(1-x)^{n-r-1}\frac{x^r}{r}\,dx, \quad (2.198)$$

which again contains an integral expression on the right-hand side. Re-applying integration by parts, we get

$$\begin{aligned}\int_0^1 (n-r)(1-x)^{n-r-1}\frac{x^r}{r}\,dx &= \\ &= \left[(n-r)(1-x)^{n-r-1}\cdot\frac{x^{r+1}}{r(r+1)}\right]_0^1 \\ &\quad - \int_0^1 (n-r-1)(n-r)(1-x)^{n-r-2}\cdot\frac{x^{r+1}}{r(r+1)}\,dx, \end{aligned} \quad (2.199)$$

and after doing so again for the remaining integral,

$$\begin{aligned}&\int_0^1 (n-r-1)(n-r)(1-x)^{n-r-2}\cdot\frac{x^{r+1}}{r(r+1)}\,dx = \\ &= \left[(n-r-1)(n-r)(1-x)^{n-r-2}\cdot\frac{x^{r+2}}{r(r+1)(r+2)}\right]_0^1 \\ &\quad - \int_0^1 (n-r-2)(n-r-1)(n-r)(1-x)^{n-r-3}\cdot\frac{x^{r+2}}{r(r+1)(r+2)}\,dx. \end{aligned} \quad (2.200)$$

Following a total of $n-r$ such steps, the process comes to an end:

$$\begin{aligned}&\int_0^1 \prod_{i=0}^{n-r-1}(n-r-i)\cdot(1-x)^{n-r-(n-r)}\cdot\frac{x^{r+n-r-1}}{\prod_{i=0}^{n-r-1}(r+i)}\,dx = \\ &= \left[\prod_{i=0}^{n-r-1}(n-r-i)\cdot(1-x)^{n-r-(n-r)}\cdot\frac{x^{r+n-r}}{\prod_{i=0}^{n-r}(r+i)}\right]_0^1 \\ &\quad - \int_0^1 \prod_{i=0}^{n-r}(n-r-i)\cdot(1-x)^{n-r-(n-r)-1}\cdot\frac{x^{r+n-r}}{\prod_{i=0}^{n-r}(r+i)}\,dx. \end{aligned} \quad (2.201)$$

In this last step, the integral remaining has a zero-valued coefficient for $i=n-r$, and hence the whole expression is equal to zero. So, from each of the $n-r$ steps we are left with a summand to be evaluated at 0 and 1. Doing so at 0 reveals that each term is equal to 0, because $r > 0$ and so the fraction is always well-defined. Evaluating at 1 leaves but one summand which is not equal to zero: It is the term from the last iteration, (2.201), because $0^0 = 1$. Rearranging yields exactly expression (2.197).

Finally, replacing the inner sum of (2.190) with (2.197) results in the most commonly known formula for the computation of the Shapley value. For reasons of recognizability we also changed labelling of the sets and their cardinality from R and r to S and s:

$$\phi_i(\mathbf{v}) := \sum_{\substack{S \subseteq N \\ i \in S}} \frac{(s-1)!\,(n-s)!}{n!} \left[v(S) - v(S \setminus \{i\})\right]. \tag{2.202}$$

Because the above derivation is rather complicated and lacks intuitional support in some places, the Shapley operator is usually characterized using another approach. It is called the *marginal* characterization and can be illustrated directly by means of the common expression (2.202). Doing so is extremely intuitive and can be brought forward with only the most basic mathematical framework. If one accepts the assumption that all $n!$ possible orderings of the player set N are equally likely, the Shapley value ϕ_i can be seen as the average marginal value of player i to all coalitions $S \subseteq N \setminus \{i\}$. For some set S with $i \in S$, the expression

$$\frac{(s-1)!(n-s)!}{n!} \tag{2.203}$$

equals the probability of player i joining any coalition of cardinality s as the last player. In the numerator $(s-1)!(n-s)!$ gives the number of possible orderings of the $s-1$ and $n-s$ players belonging to $S \setminus \{i\}$ and $N \setminus S$, respectively, before and after player i who is fixed in the sth spot. In the denominator, we have the number of all possible orderings of the player set N. As the expression in (2.202) sums over all sets S with $i \in S$, it calculates the average value of the marginal values of a given player.[73] These marginal values are each weighted with the probability of player i being the last to join the coalition S, which changes with the cardinality of S. Because we see all orderings on N as equally likely, coalitions $S \subseteq N$ with equal cardinality also have equal probability of occurring. The marginal contribution of i is given by the difference in square brackets in (2.202). This interpretation is helpful in the proof of the main theorem in Shapley (1953b):[74]

[73] One could also drop the restriction $i \in S$ in (2.202), as all marginal contributions are identical to zero for coalitions not including player i.

[74] Shapley (1953b) originally defined his operator for superadditive games. To drop this restriction, changes to the axioms are necessary, which we incorporated in Theorem 2.6.

Theorem 2.6. *For a coalitional game (N, v), the unique value operator Φ satisfying the axioms of symmetry, additivity, efficiency, and dummy is given by equation (2.202).*

Proof. We will first show that Φ conforms the axioms listed and then verify its uniqueness. The axiom of symmetry is fulfilled because in calculating the value for some player i, we take into account all $n!$ possible orderings on the player set N. Therefore a relabelling or change in position of an arbitrary player does not affect the final outcome; it only changes the order in which we consider the $n!$ permutations.

The axiom of additivity is satisfied by a similar logic: For any two games, the weight attributed to a coalition S by (2.203), and therefore the weight of i's marginal value to S (not the value itself) are equal. As the Shapley operator adds all possible marginal values of i, the sum of the weighted values of this player is equal to the weighted sum of the values.

To see the validity of the axiom of efficiency, consider the following: In any given ordering of the player set N, the sum of all players' marginal values add up to $v(N)$ by definition: $\mathbf{1}' \cdot \mathbf{x}^{\pi} = v(N)$ for all $\pi \in \Pi_N$. The Shapley value for player i is the (equally weighted) average or mean of all his $n!$ possible marginal values. The sum off all players' Shapley values is then also equal to $v(N)$.[75]

Because a null-player i is defined to exhibit marginal values equal to zero for all coalitions $S \subseteq N \setminus \{i\}$, the average of all his possible $n!$ marginal values is also equal to zero. This satisfies the dummy axiom.[76] All four axioms can be deduced from the functional form of the Shapley operator.

To show the uniqueness of the Shapley value, we revert to its derivation via unanimity games: By construction, there is only one operator satisfying the axioms of symmetry, efficiency, as well as the null-player axiom for unanimity games. This is a byproduct of the two steps taken to justify (2.172).[77] Furthermore, we have shown that any coalitional game $\mathbf{v}$ is a linear combination of unanimity games (see Proposition 2.1). Because of the axiom of additivity, the Shapley value of a game $\mathbf{v}$ is equal to the sum of the Shapley values of the unanimity games that constitute $\mathbf{v}$. As the operator is unique on the latter, it must be so on $\mathbf{v}$. ■

From the additive characterization of the Shapley operator it follows readily that it is homogeneous: Scaling all coefficients γ_T with some constant is carried over

[75] Think of an $n! \times n$ matrix, where every row represents a different marginal vector. The Shapley value of player i picks exactly one entry of each row, to be summed up and averaged with $\frac{1}{n!}$. The sum of all players Shapley values is nothing else but the sum of all $n!$ marginal vectors, which each sums to $v(N)$, averaged with $\frac{1}{n!}$.

[76] Alternatively, to follow Shapley (1953b), the carrier axiom is also fulfilled by (2.202). By the definition of a carrier, no player who is not in every carrier can have an impact on the value function. His marginal values are therefore zero for all coalitions he joins and so the expression in square brackets in (2.202) is identical to zero.

[77] There, we used the carrier axiom. But in a unanimity game it is clear that $u_T(S) = u_T(N)$ holds for all sets S with $T \subseteq S \subseteq N$ and all players $i \in N \setminus T$ are null-players.

directly to the resulting coalitional game and hence to the resulting allocation. This is even easier to see from the operator's functional form (2.202). Therefore, the Shapley operator is also linear, which has some interesting implications for strategic equivalence.

Strategic equivalence for two games $\mathbf{v}$ and $\mathbf{w}$ was previously defined as

$$v(S) = a \cdot w(S) - \sum_{i \in S} c_i \quad \text{for all } S \subseteq N, \tag{2.204}$$

with $a \in \mathbb{R}_{++}$ and $c_i \in \mathbb{R}$ for all $i \in N$. Because of the operator's linearity,

$$\phi_i(\mathbf{v}) = a \cdot \phi_i(\mathbf{w}) - c_i, \tag{2.205}$$

where c_i could even be replaced with $\phi_i(\mathbf{c})$ for all $i \in N$, if we interpret the additive function $c(S) = \sum_{i \in S} c_i$ as a game itself (an inessential one, with only dummy players).

Another important consideration is to set the allocation given by the Shapley operator in relation to the core of a coalitional game. If the core is empty, then the two will trivially not coincide. But even when the core is nonempty, the Shapley vector need not be an element of the former in general games. This is shown in the following example:

Example 2.8. Take a coalitional game (N, v) with player set $N = \{1,2,3\}$. Its characteristic function is given by $v(\emptyset) = 0$ and $v(\emptyset) = v(\{i\}) = v(\{1,3\}) = 0$ for $i = 1,2,3$, as well as $v(\{1,2\}) = v(\{2,3\}) = v(N) = 1$. Then, the core collapses to a singleton, $C(N,v) = (0,1,0)$, and the Shapley value is given by $\Phi(\mathbf{v}) = (1/6, 2/3, 1/6)$, clearly two different allocations.[78]

One step towards core consistency of the Shapley operator is achieved in superadditive games, because its allocations are always and element of the set of imputations, $I(N,v)$.

Proposition 2.14. *The Shapley value of a superadditive game* $\mathbf{v}$ *is always an imputation:*

$$\Phi(\mathbf{v}) \in I(N,v). \tag{2.206}$$

Proof. By definition, the Shapley value is efficient and hence $\mathbf{1}' \cdot \Phi(\mathbf{v}) = v(N)$ for all coalitional games. Individual rationality is deduced from superadditivity. If a game is superadditive (see Definition 2.26, p. 40) the following holds also true:

$$v(S \cup \{i\}) - v(S) \geq v(\{i\}) \;\; \text{for all} \;\; S \subseteq N \setminus \{i\}, \, i \in N. \tag{2.207}$$

[78] The $(0,1)$-normalization has no influence on the outcome of either solution concept, it is merely applied for convenience. The core element $(0,1,0)$ is straightforward to explain: The coalitions $\{1,2\}$, $\{2,3\}$, and N are entitled to an allocation of 1 in the core. Single players have a stand-alone value of zero and since player 2 is the only member of all profitable coalitions, he will get the whole payoff, thereby (and only thereby) not violating any of the core constraints.

Because the combinatorial expression (2.203) is equal to 1 when summed over all sets $S \subseteq N$ with $i \in S$,

$$\phi_i(\mathbf{v}) \geq v(\{i\}) \cdot \sum_{\substack{S \subseteq N \\ i \in S}} \frac{(s-1)!\,(n-s)!}{n!} \quad \text{for all } i \in N, \tag{2.208}$$

all $i \in N$ are assigned at least their stand alone value. ∎

Certain core membership is finally attained when considering the class of convex games. Shapley (1971, pp. 23 f.) shows that in any convex game, the Shapley value is the barycenter of the vertices of the core. Since the core is convex in any game, the Shapley value must be an element of the core. We omit the lengthy and technical proof of this result, as it has very little instructive value.

In our treatment of the Shapley operator, we have focussed on two approaches of characterization: First, the additive and then the marginal approach. The former approach relies on the axioms stated in Shapley (1953b) and extends the operator from unanimity games to general games. Similar treatments can be found in almost any textbook on (cooperative) game theory, consult e.g. Owen (1995), Luce and Raiffa (1957), Mas-Colell et al. (1995), Moulin (1991), and Myerson (1991). In the non-textbook literature it is strongly represented, too, Winter (2002) provides a recent and comprehensive survey.

The second, the marginal characterization of the Shapley value is more intuitive, as it is based on the marginal values of the players. This is incorporated in the omnipresent formula (2.202), which itself has put forth not only a range of other operators, but also different methods of computing the Shapley value. As examples, we mention various probabilistic value approaches, as in Monderer and Samet (2002) or Weber (1988), or different combinatorial methods of calculating the Shapley value, see e.g. Rothblum (1988).

There also exist numerous alternative characterizations of the Shapley Value, where some or all of the axioms are replaced. Examples can be found, among others, in Chun (1989), Chun (1991), and most notably in Hart and Mas-Colell (1989).

This concludes the section on the original Shapley value. In the next section, we look at a modified version which allows us to adjust the relative importance of the players by assigning weights to them. Doing so, we incorporate differences among players which are not captured in the value function, or rather, in their marginal values.

2.4.6 The Weighted Shapley Value

The Shapley operator assigns each player a value by calculating the average of their individual marginal values. The operator processes only the information provided by the characteristic function v, more precisely the marginal values of the players. This affords opportunity for a valid point of criticism, because the players are being treated symmetrically in terms of added value (not to be confused with the property of symmetry). But many situations can be thought of, where players do exhibit payoff-relevant differences due to their size, bargaining power, or their particular contribution to making the added value possible. These are not expressed through the characteristic function.

Shapley (1981, pp. 135 f.) gives a very intuitive example in the realm of cost functions: The cost of a business trip with stops in multiple cities (which serve as the players) is allocated using weights that correspond to the relative length of the stay in each city and not only on their marginal costs. In our environment of surplus sharing one could think of a hierarchy of investors to a given undertaking, who will be allocated surplus not only in relation to their marginal values, but also according to their seniority, given by a weight system. This weight system could reflect the risk assumed by each investor, the lead time required to withdraw his money, or simply to differentiate between length of investment.

To overcome this problem, Shapley (1953a) introduced an allocation rule which became known as the *Weighted Shapley value*. As the name indicates, it assigns each player a weight in relation to which the marginal values are then allocated. The idea behind this concept can be illustrated along the lines of unanimity games, or more generally partnerships: All members of some coalition S are necessary to create value and all proper subsets $T \subsetneq S$ are powerless, i.e. they cannot create or add any value, regardless of whom they cooperate with (among themselves or in $N \setminus S$). Yet, it is possible, that within this coalition S, the contributions to cooperation of the individual members are not identical.[79] This, as mentioned above, is not "measured" by the value function. Ceteris paribus, the standard Shapley operator would allocate the gains from the partnership equally among its members. In order to take into account possible differences among the players, a weight system is applied in this derivative of the Shapley operator.[80] After the original work, Shapley (1953a), the most notable treatment of the Weighted Shapley value is probably given in form of an axiomatization by Kalai and Samet (1987). Other influential publications are Monderer and Samet (2002) and Chun (1991). Along their lines, we develop our exposition.

As before, we define the weighted operator on the basis of unanimity games $\mathbf{u}_S$, which we then extend to the general case, $\mathbf{v}$. The heart of this concept is a lexicographic weight system $\omega := (\lambda, \Sigma)$. The first element is an n-dimensional vector $\lambda \in \mathbb{R}^n_{++}$, the second an ordered partition of the set of players $\Sigma = (S_1, S_2, \ldots, S_m)$.[81] The set of all weights on n-player games is given by Ω_N, or simply Ω, when no confusion can arise.

---

[79] A word of warning is given by Owen (1968), as he points out that for some types of games (a majority game in his example), the weights can skew the allocations given by the Weighted Shapley operator into counterintuitive directions.

[80] In order to enable zero-allocations to players who are not null-players, we skip the so-called *positively* Weighted Shapley value (see also Shapley 1953b).

[81] As usual, $S_i \cap S_j = \emptyset$ for all $i \neq j \in \{1, 2, \ldots, m\}$ and $\bigcup_{i=1}^m S_i = N$. As the partition Σ is ordered, its elements appear not in a set, but as entries of an m-dimensional vector.

Denote by $\Phi^{\omega}: \mathbf{V}_N \to \mathbb{R}^n$ the weighted Shapley value, a mapping from the space of games into the n-dimensional real numbers. Then:

$$\phi_i^{\omega}(\mathbf{u}_S) = \begin{cases} \dfrac{\lambda_i}{\sum\limits_{j\in\widehat{S}} \lambda_j} & \text{if } i \in \widehat{S} \\ 0 & \text{else} \end{cases}, \tag{2.209}$$

with $\widehat{S} = S \cap S_k$, where $k = \max\{j | S_j \cap S \neq \emptyset\}$.

The above is the allocation assigned by Φ^{ω} to player i in unanimity game $\mathbf{u}_S$. Although the individual weights are all strictly positive, $\lambda_i > 0$ for all $i \in N$, the ordered partition Σ allows a grouping of players where some might be allocated nothing, even when they belong to the carrying coalition S of the underlying unanimity game $\mathbf{u}_S$. Only the players belonging to $\widehat{S}$, i.e. the carrier S **and** the "highest" element S_k of the ordered partition Σ, are allocated a strictly positive amount, while all players $i \notin \widehat{S}$ receive a payoff of zero, even as members of S. In addition, despite (2.209) being defined on the unanimity game $\mathbf{u}_S$, the players $i \in \widehat{S}$ need not be allocated identical payoffs, but rather in relation to their weights λ_i.[82]

From (2.209) we deduce that the Weighted Shapley operator is homogeneous of degree zero with respect to its weights λ, i.e.

$$\phi_i^{(k\cdot\lambda,\Sigma)}(\mathbf{u}_S) \equiv k^0 \cdot \phi_i^{(\lambda,\Sigma)}(\mathbf{u}_S) \quad \text{for all } k \in \mathbb{R}_{++}.$$

This conveniently allows us to normalize the weights such that $\mathbf{1}' \cdot \lambda = 1$. In order to obtain the weighted Shapley value for general games $\mathbf{v}$, a linear combination is applied, analogue to the procedure given in (2.182)–(2.184):

$$\phi_i^{\omega}(\mathbf{v}) = \sum_{T\subseteq N} \phi_i^{\omega}(\gamma_T \cdot \mathbf{u}_T) = \sum_{T\subseteq N} \gamma_T \cdot \phi_i^{\omega}(\mathbf{u}_T). \tag{2.210}$$

Substituting from (2.209), where nonzero values are assigned to $i \in \widehat{S}$ only, this reduces to

$$\phi_i^{\omega}(\mathbf{v}) = \sum_{\substack{T\subseteq N \\ i\in\widehat{S}}} \gamma_T \cdot \frac{\lambda_i}{\sum_{j\in\widehat{S}} \lambda_j}. \tag{2.211}$$

Unfortunately, the computation of the Weighted Shapley value for general games, $\Phi^{\omega}(\mathbf{v})$, does not lend itself to a comparably straightforward expression as it was possible for the ordinary operator $\Phi(\mathbf{v})$ through (2.202). This difficulty arises,

[82]For the trivial partition $\Sigma = (N)$, the sets S and $\widehat{S}$ always coincide, and the result is the positively Weighted Shapley value, where no player (unless a null) can end up with a zero-allocation. If, in addition, the weights are all equal (and strictly positive), i.e. $\lambda_i = \lambda_j$ for all $i, j \in N$, then the outcome is the ordinary Shapley value.

because now the value also depends on the underlying weight system ω which defines $\widehat{S}$ through Σ. Appropriate procedures for calculation are provided in the appendix of Harsanyi (1959) and Maschler (1982), for example.[83]

A readily computable expression in terms of marginal contributions can be derived via the probabilistic definition of the Weighted Shapley value. We follow the approach of Kalai and Samet (1987, pp. 208 ff.) which also includes the proof that their expression indeed corresponds to the Weighted Shapley operator. First, we extend some of the notation introduced before. We denoted a permutation on the set of players N by $\pi \in \Pi_N$. Now, for permutations on subsets $S \subseteq N$, we write $\pi_S \in \Pi_S$, and by $\pi_\Sigma \in \Pi_\Sigma$ we denote a permutation of the ordered partition Σ in such a way, that a reordering takes place only within the elements of Σ, but not across. Instead of applying a permutation on the whole set of players N, we use one for each element of the ordered partition Σ.[84] Based on the strictly positive weights $\lambda \in \mathbb{R}^n_{++}$, we can establish a probability distribution P_λ over some permutation π_S for all $S \subseteq N$. For a given permutation π_S, we write

$$p_\lambda(\pi_S) = \prod_{\pi_S(i)=1}^{s} \frac{\lambda_{\pi_S(i)}}{\sum\limits_{\pi_S(k)=1}^{\pi_S(i)} \lambda_{\pi_S(k)}}. \tag{2.212}$$

Extending this notion to probability distributions over weight systems ω, denoted P_ω, we need to calculate the probability of an ordering on the player set N within the boundaries given by $\Sigma = (S_1, S_2, \ldots, S_m)$. Because the elements of the partition Σ are by definition pairwise disjoint, the probabilities $p_\lambda(\cdot)$ of an ordering π_{S_j} are independent for each S_j. Hence,

$$p_\omega(\pi_\Sigma) = \prod_{j=1}^{m} p_\lambda(\pi_{S_j}) \tag{2.213}$$

is the probability for a permutation π_Σ on the set of players N with respect to partition Σ. Then, the Weighted Shapley value for player i is defined as

$$\phi_i^\omega(\mathbf{v}) := \sum_{\pi \in \Pi_\Sigma} p_\omega(\pi) \cdot \left[v(S_{\pi,\pi(i)}) - v(S_{\pi,\pi(i)} \setminus \{i\})\right], \tag{2.214}$$

where $S_{\pi,i} = \{j \in N \,|\, \pi(j) \leq i\}$ is the set of all players up to and including player i under the ordering π. As mentioned above, this expression is a variant of the original

[83]Roughly, the idea is the following: First the stand-alone value is allocated to all singleton coalitions. Then the sum of two stand-alone values is subtract from the value of the respective coalitions of size two. The same is repeated with whatever might be left and then taken to coalitions of size three and so on, until the grand coalition is reached.

[84]Whenever no confusion can arise, we write simply π, dropping the subscript that associates the permutation with the set over which it is defined.

Shapley equation (2.202), when $\Sigma = (N)$ and the weights $\lambda_i = \lambda > 0$ for all $i \in N$ are equal. As it stands, (2.214) consists of $\prod_{j=1}^{m} |S_j|!$ summands, as there are $|S_j|!$ possible orderings for each element of the partition Σ.

With this approach, the Weighted Shapley value can also be interpreted as an operator, where the probabilities of the possible orderings of the player set N are changed according to ω. Whereas the ordinary Shapley value treats all $n!$ permutations on N as equally likely, the weighted operator even excludes those outside the confinements of the ordered partition Σ.[85] So $p_\omega(\pi_N) = 0$ for all $\pi_N \notin \Pi_\Sigma \subseteq \Pi_N$. In contrast, nonzero probabilities $p_\omega(\pi_\Sigma)$ have the following characteristics:

For each element $S_j \in \Sigma$, the corresponding relative frequency of an ordering $p_\lambda(\pi_{S_j})$, as specified in (2.212), is calculated by multiplying the relative weights of the players $i \in S_j$ The last player under π_{S_j}, say i, has his absolute weight λ_i divided by the largest denominator, i.e. $\sum_{i \in S_j} \lambda_i$, but this does not necessary translate into the smallest relative weight. The first player's relative weight, by definition, has a value of one and is therefore the maximal relative weight. For the players in between, the relative weights, given λ, depend on the ordering π_{S_j}.

For a given S_j, the highest $p_\lambda(\pi_{S_j})$ is attained, when the ordering π_{S_j} is such that $\pi(k) < \pi(l)$ if and only if $\lambda_k \leq \lambda_l$ for all $k, l \in S_j$. The ordering π most likely under $p_\lambda(\cdot)$ is one where the value of each weight λ_i is nondecreasing in its position according to π.[86] Intuitively, the probability of an ordering is higher, the lower the weights of the players up front. Or, the other way around, an ordering with a (relatively) high probability is likely to have many players with low weights in the front positions. With respect to a given player, a high weight increases his chances of appearing late in a given ordering, or more precisely, a high weight the probability of such an ordering.

Hence, the orderings $\pi_\Sigma \in \Pi_\Sigma$ have strictly positive probabilities, and so Π_Σ is the support of the probability measure p_ω. Π_Σ is the smallest subset Z of the set of all permutations on N, $Z \subseteq \Pi_N$, for which $p_\omega(Z) = 1$. Given a weight system ω, the value of $p_\omega(\pi_\Sigma)$ reaches a maximum whenever π_Σ specifies an ordering where the value of the weights is nondecreasing within the elements of the ordered partition $\Sigma = (S_1, S_2, \ldots, S_m)$.

With a workable expression of the Weighted Shapley value in place, we again want focus on the connection between the Weighted Shapley value and the core of

---

[85] As Σ restricts the possibility of permutating the player set N, but does not allow for any additional orderings, it follows that $\Pi_\Sigma \subseteq \Pi_N$.

[86] For an explanation, reconsider expression (2.212): The numerator being identical for any ordering $\pi_S \in \Pi_S$, we have to focus on minimizing the denominator. Without loss of generality, suppose that the weights increase with index l, i.e. $\lambda_l \leq \lambda_{l+1}$, and that the ordering π_S is also such that $\pi_S(l) < \pi_S(l+1)$. Now pick some player k, who is neither first nor last under π_S, and for which $\lambda_k > \lambda_{k+1}$. (If such a player doesn't exist, then $p_\lambda(\pi_S) \equiv p_\lambda$ for all $\pi_S \in \Pi_S$ and we have nothing to show.) Now, replace π_S with π'_S, which are identical, except that player k and $k+1$ are reversed. Then, in the denominator of (2.212), the kth element of the product of sums is strictly larger under π'_S than under π_S, with all other elements equal. Therefore, $p_\lambda(\pi_S) > p_\lambda(\pi'_S)$.

a coalitional game, especially when the latter is convex. We begin with the core $C(N,v)$ of general coalitional games:

Proposition 2.15. *For every element in the core of a coalitional game, there exists a system of weights for which the Weighted Shapley value coincides with this core element:*

$$\textit{For all } \mathbf{x} \in C(N,v), \ \exists\, \omega \ \textit{such that } \Phi^{\omega}(\mathbf{v}) = \mathbf{x}. \tag{2.215}$$

We omit the proof of this result and refer the reader to Monderer et al. (1992, p. 37). In consequence, Proposition 2.15 establishes that the core is a subset of the set of all weighted Shapley values: $C(N,v) \subseteq \Phi^{\Omega}(\mathbf{v})$, where Φ^{Ω} is defined as follows:

$$\Phi^{\Omega}(\mathbf{v}) = \{\mathbf{x} \in \mathbb{R}^n \| \exists\, \omega_N \in \Omega \text{ with } \Phi^{\omega}(\mathbf{v}) = \mathbf{x}\}. \tag{2.216}$$

For a coalitional game (N,v), all core allocations can be reproduced through the Weighted Shapley value with an appropriate weight system. That the other direction in not true can be verified through Example 2.8, where we provided a coalitional game in which the original Shapley value and the core had empty intersection. Taking a weight system ω, where the weights are strictly positive and identical for all players, $\lambda_i = \lambda$ for all $i \in N$, and the ordered partition such that $\Sigma = (N)$, the weighted operator coincides with the original one. Hence, $\Phi^{\omega}(\mathbf{v}) \notin C(N,v)$ is well possible for a general coalitional game. The class of games where this can be ruled out is under examination next.

The relation between the Weighted Shapley value and core is even more precise, when we narrow in on the class of convex games. In such games, it is possible to set a well defined boundary for the set of all Weighted Shapley values on a given coalitional game. Before we go on, we want to remind of some important characteristics of the core in convex games: From Corollary 2.3, p. 75, we know that all marginal worth vectors $\mathbf{x}^{\pi}$ are elements of the core.[87] Conversely, Ichiishi (1981, p. 286) proves that if all the marginal worth vectors are elements of the core, the game is convex.[88]

We demonstrate now that in convex games the core coincides with the set Φ^{Ω}, i.e. the set of allocations the Weighted Shapley value can produce with all possible systems of weights. A more detailed proof of this, which also delivers the other direction, can be found in Monderer et al. (1992, pp. 35 ff.). We focus only on one direction, and, by relying on two facts regarding marginal vectors and the core, we can reduce the necessary technical toolkit by a considerable amount.

[87] More precisely, Shapley (1971, pp. 19 f.) demonstrates that all the marginal worth vectors are the (only!) vertices of the core of a convex game.

[88] These results are refined by Rafels and Ybern (1995) and van Velzen et al. (2002) who show that only a certain set or number of marginal worth vectors have to be identified as elements of the core for the game to be convex.

Proposition 2.16. *The core of a convex game, $C(N,v)$, and the set of all Weighted Shapley values on* $\mathbf{v}$, $\Phi^{\Omega}(\mathbf{v})$, *coincide:*

$$C(N,v) = \Phi^{\Omega}(\mathbf{v}).$$

Proof. Instead of proceeding via double-sided set-inclusion, which is common for proofs of this kind, we characterize both sets, from which it then follows that they are identical.

Let us begin with the core: From Shapley (1971) we know that the $n!$ marginal worth vectors coincide (exhaustively) with the vertices of the core $C(N,v)$. Hence, all marginal vectors are extreme points of the core, and they are the only extreme points, so, all vertices are extreme points of the core. Also, as the core is the intersection of finitely many half-spaces and a hyperplane, it cannot have extreme points other than the vertices $\mathbf{x}^{\pi}$. In this situation, the Krein–Milman theorem allows us to conclude that the convex hull of all extreme points of the core is equal to the (closed convex hull of the) core[89]:

$$co\{(x^{\pi})_{\pi\in\Pi_N}\} = co\, C(N,v) = C(N,v). \tag{2.217}$$

The second equality follows from the convexity of the core itself.

We show now that $\Phi^{\Omega}(\mathbf{v})$, the set of all Weighted Shapley values on a game $\mathbf{v}$, is also limited to the convex hull of the marginal worth vectors $\mathbf{x}^{\pi}$. Each allocation $\mathbf{x}^{\pi}$ can be generated by the Weighted Shapley operator according to the formula given in (2.214). There, the expression in square brackets is nothing else than the $\pi(i)^{\text{th}}$ entry of the marginal worth vector $\mathbf{x}^{\pi}$, as initially defined in (2.111). Now, when all probability mass is concentrated on the one permutation π specifying $\mathbf{x}^{\pi}$, we have $p_{\omega}(\pi) = 1$, and, accordingly $p_{\omega}(\pi') = 0$ for all $\pi' \neq \pi$. It turns out that in order to reduce the support of p to just the element π, the weights λ are irrelevant, as long as they are strictly positive. All that matters is the ordered partition $\Sigma = (S_1, S_2, \ldots, S_m)$: Take $m = n$, so that each element of Σ contains exactly one player. The order of players must be such that $\pi(i) = j$ for all S_j with $j = 1, \ldots, n = m$. Checking equations (2.212) and (2.213) quickly reveals that a value of zero is assigned to all orderings other than π (because they are not in the support of p), whereas $p_{\omega}(\pi) = 1$. These are, again, the extremal points in terms of allocations that can be constructed under the Weighted Shapley operator in game $\mathbf{v}$. This bound follows from (2.214) and especially the construction of (2.213).

We continue to establish all allocations between neighboring marginal vectors. These, pair by pair, contribute to the boundary of the convex hull of all $\mathbf{x}^{\pi}$. Take

[89] The interested reader will find a more formal and sophisticated treatment of the Krein–Milman theorem, along with a proof, in Royden (1988, p. 207). We use its assertion that, in finite-dimensional vector-spaces, the convex hull of all extreme points of a compact convex set K is equal to the closed convex hull of the set K itself: $co\{exK\} = co\, K$.

for example the neighboring pair, where players k and $k+1$, with $k \leq n-1$, are swapped in order.[90] The ordered partition Σ must be of the form,

$$\Sigma = (S_1, S_2, \ldots, S_k, \ldots, S_{n-1}), \tag{2.218}$$

where all elements S_j, $j \neq k$, contain only one player, but $S_k = \{k, k+1\}$. This limits the support of p_ω to two permutations, say π and π', which are identical except for traded places of players k and $k+1$ in the latter. The probabilities given by (2.213) are then

$$p_\omega(\pi) = \frac{\lambda_{k+1}}{\lambda_k + \lambda_{k+1}} \quad \text{and} \quad p_\omega(\pi') = \frac{\lambda_k}{\lambda_k + \lambda_{k+1}}, \tag{2.219}$$

which add to one and can each take values in the open unit-interval, as $\lambda \in \mathbb{R}^n_{++}$. With the appropriate choice of weights, any point in the interior of the convex hull of $\mathbf{x}^\pi$ and $\mathbf{x}^{\pi'}$ can be expressed as a convex combination of $\mathbf{x}^\pi$ and $\mathbf{x}^{\pi'}$:

$$\text{For all } \mathbf{y} \in \mathit{int}\ co\{\mathbf{x}^\pi, \mathbf{x}^{\pi'}\} \quad \exists\, \lambda_k, \lambda_{k+1} > 0 \mid p_\omega(\pi) \cdot \mathbf{x}^\pi + p_\omega(\pi') \cdot \mathbf{x}^{\pi'} = \mathbf{y}. \tag{2.220}$$

It is also possible to create the endpoints $\mathbf{x}^\pi$ and $\mathbf{x}^{\pi'}$ by convex combinations, but not with the partition Σ as specified in (2.218). This follows directly from the definition of the probabilities given in (2.219), as $p_\omega(\pi) > 0$ and $p_\omega(\pi') < 1$.

This entire procedure can be extended to include more than just two players in the element S_k and leads to the basic *Minkowski Theorem*.[91] The latter allows us to infer for any $\mathbf{x} \in co\{\mathbf{x}^\pi\}_{\pi \in \Pi_N}$ there exist $n!$ nonnegative numbers $p_\omega(\pi)$ which sum to one, and for which $\mathbf{x} = \sum_{\pi \in \Pi_N} p_\omega(\pi) \cdot \mathbf{x}^\pi$ holds. That is, every element of the closed convex hull of the marginal vectors can be represented by a convex combination of these marginal vectors.[92] Because the probabilities defined in (2.212) and (2.213) each assume strictly positive values in the unit interval and sum up to one over their support (regardless of the underlying weight system), they are equivalent to the coefficients of a convex combination of marginal worth vectors, i.e. of the extreme points, given by $\mathbf{x}^\pi$. We can, therefore, by appropriate choice of a weight system ω,

[90] In principle this procedure also works for the "neighboring" players n and 1, but we want to avoid an excursion into so-called *modulo n*, an operation that (informally speaking) closes the ordering of players to a circle with neither beginning nor ending.

[91] Minkowski's Theorem, as given in Phelps (2001, p. 1), states: "If X is a compact convex subset of a finite-dimensional vector space E, and if x is an element of X, then x is a finite convex combination of extreme points of X."

[92] Actually, in this context, not all marginal vectors are required, according to Carathéodory's Theorem: "If X is a compact convex subset of an n-dimensional space E, then each x in X is a convex combination of at most $n+1$ extreme points of X." See Phelps (2001, p. 7).

represent any allocation in the closed convex hull of all marginal vectors $\mathbf{x}^{\pi}$.[93,94] This shows that for convex games, the core $C(N,v)$ and the set of all Weighted Shapley values on $\mathbf{v}$, $\Phi^{\Omega}(\mathbf{v})$, coincide. ■

The preceding result is quite striking, especially because the two solution concepts differ so vastly, both in approach and construction, as well as in definition and computation.

For a given game, the core can by definition result in a set-valued solution, while the Weighted Shapley value only yields a singleton solution for each weight system. Nevertheless, it is always possible to find a weight system, whose corresponding result can be backed up by a core allocation (for convex games this is also the boundary). This might make the Weighted Shapley value a preferred solution concept, as (some of) its solutions can always be backed up by another concept. Even more opportune, this other concept is the core, whose elements, by definition, cannot be improved upon by any possible coalition. As such, the core elements lend themselves to support the Weighted Shapley vectors with which they coincide: Why go for another solution, if this solution already is one which puts all sets of players in a position that is at least as good as what they could do for themselves. Two points of criticism arise from this: For one, usually more than one core allocation exists, so there are many other possibilities for allocations to which the same reasoning would apply, or rather, against which this reasoning offers no remedy. Secondly, and more importantly, calibrating a Weighted Shapley value to a given allocation usually involves weight systems that lack any intuitional relation to the players' characteristics ignored by the value function.[95] But aren't those exactly the reason that are supposed to justify the Weighted Shapley value in the first place?

Nevertheless, the Weighted Shapley value is a very useful extension of its basic variant, as it allows incorporate the weights system to take into account differences among the players which are not captured by the value function of the game. In finding a weight system one has simply to be careful to be able to justify it and relate it to the players' differences that are to be accounted for.

We now turn to another class of solution concepts known as bargaining solutions. Even though they can (and will in our case) be based on cooperative games and yield

---

[93]If we do not want to leave the realm of probability spaces, we can directly stick with Choquet's Theorem: It states that any element of a compact convex set (here: the convex hull of extreme points) is the barycenter of some probability measure with support on (only but not necessary all) extreme points of the set. Details and a proof can be found in Phelps (2001, p. 14).

[94]With the choice of weight system, especially the ordered partition Σ, some caveats are given: One can include many different combinations of vertices (i.e. orderings) into the support of p_{ω}, as long as the logical structure of Σ is not violated. For any n-player game with $N=\{1,2,\ldots,n\}$, it is not possible to assign nonzero probability to orderings that swap players $\pi(i)$ and $\pi(j)$ with $\pi(i) >> \pi(j)$, without also assigning nonzero probabilities to permutations of players in between i and j. All Players from $\pi(j)$ to $\pi(i)$ have to be included in the same element of Σ. So, if i and j can be swapped under some permutation in the support, this must also be true for the players in between.

[95]This problem resurfaces in the application of the Weighted Shapley value, in Sect. 5.3.6.

unique point-valued payoff vectors, they do not rely as directly on the characteristic function in terms computation as allocation rules. We only introduce two variants, the *Nash solution* and the *Kalai–Smorodinsky solution*, along with a generalization and weighted derivatives.

2.4.7 Bargaining Solutions

For the remainder of this chapter we will be concerned with bargaining solutions, i.e. solution concepts to the so-called bargaining problem. This problem is one, where players have to agree on the distribution of a surplus. Usually referred to as *pure bargaining*, they either reach a unanimous decision, or fall back to a pre-specified disagreement point, the status quo ante. No "intermediate" solutions are considered which might put a subset of players in a better position than what is given by the disagreement point. A general introduction can be found in Mas-Colell et al. (1995, pp. 838 ff.), while Thomson (1994) provides an excellent overview.

We begin by formally specifying the bargaining problem and the bargaining solution in general, along with some basic properties the latter might satisfy. We then proceed to introduce two particular bargaining solutions. The first is the *Nash solution*, or *Nash product* (see Nash 1950), which maximizes the product of the individual payoffs. The second, the *Kalai–Smorodinsky solution* (see Kalai and Smorodinsky 1975), assigns payoffs from the relevant set of possible surplus distributions in relation to the players' individual upper limits.

2.4.7.1 The Bargaining Problem

To set the stage, we now define what is known as the bargaining problem in the literature.

Definition 2.66. A *bargaining problem* B is a pair $B := (U, \mathbf{d})$ where U is the feasible set and $\mathbf{d}$ the disagreement point for the players involved.

As the games previously treated, the bargaining problem covers a (finite) set of players N with cardinality n. The first element of the pair in Definition 2.66, U, is also often referred to as the *utility possibility set* or *prospect set*, as it contains all the allocations of payoffs that are possible. It is a subset of the n-dimensional Euclidean space, $U \subseteq \mathbb{R}^n$. The second element of the bargaining problem is the *disagreement* or *threat point*, given by $\mathbf{d} \in \mathbb{R}^n$. It represents the status quo ante, i.e. the allocation before any bargaining takes place. It is also the allocation that results when the players cannot reach an unanimous decision on how to distribute the surplus and, as such, an element of the feasible set, $\mathbf{d} \in U$. Denote by $\mathscr{B}_N$ the space of bargaining problems on the player set N.

Regarding the feasible set (or domain) U of a bargaining problem, some characteristics are assumed: For a well defined problem, U is a convex set that is furthermore closed and bounded (hence compact). We also assume the existence of some allocation $\mathbf{x} \in U$ for which $\mathbf{x} > \mathbf{d}$ holds, i.e. which proposes strictly larger payoffs for all $i \in N$ players.[96] This implies full dimensionality of U and we call such bargaining problems *non-degenerate*. Another property of the domain is its being *d-comprehensive*, or simply *comprehensive*. Accordingly, for all allocations $\mathbf{x} \in U$ it must be fulfilled that any allocation $\mathbf{y}$ with $\mathbf{x} \geq \mathbf{y} \geq \mathbf{d}$ is also an element of the domain $\mathbf{y} \in U$. That is to say, the projection of any feasible allocation in any dimension remains feasible. Denote by $\mathbf{a}(U,\mathbf{d}) \in \mathbb{R}^n$ the vector of maximal individual allocations with respect to the disagreement point, i.e. where

$$a_i(U,\mathbf{d}) = \max\left\{x_i \,\middle|\, \mathbf{x} \in U\right\} \text{ for all } i \in N. \tag{2.221}$$

A *bargaining solution* is a function f assigning to each bargaining problem $B \in \mathscr{B}_N$ an allocation $f(U,\mathbf{d})$ which lies in the feasible set: $f : \mathscr{B}_N \to U$. The properties a bargaining solution might fulfill are intuitively similar to those stated in Sect. 2.4.4 for allocation rules. Nevertheless, as the solutions are based on a different concept, technical differences arise and so we restate some of the properties, mostly without additional comment.

Definition 2.67. A bargaining solution $f : \mathscr{B}_N \to U$ is *individually rational*, if for all bargaining problems $B \in \mathscr{B}_N$ the resulting allocation assigns to each player at least his disagreement value:

$$f(U,\mathbf{d}) \geq \mathbf{d} \text{ for all } B \in \mathscr{B}_N. \tag{2.222}$$

Definition 2.68. A bargaining solution $f : \mathscr{B}_N \to U$ is *individually monotonic*, if a player i's allocation under f is no less when the feasible set increases weakly in his dimension:

$$\begin{aligned} &\text{If } U \subseteq U' \text{ and there exist players } j \in N \text{ with } a_j(U,\mathbf{d}) = a_j(U',\mathbf{d}), \\ &\text{then } f_i(U',\mathbf{d}) \geq f_i(U,\mathbf{d}) \text{ for all } i \neq j. \end{aligned} \tag{2.223}$$

This property contains more than it conveys at first sight. It seems intuitive that some player i might receive an allocation that is no less when the feasible set grows in his dimension, i.e. $a_i(U',\mathbf{d}) \geq a_i(U,\mathbf{d})$. But this property also allows for those players j whose a_j remains constant to actually be allocated less when the feasible set increases in some "other" dimension. The following property is related:

[96] This vector inequality is not to be confused with the previously defined relation of domination, even though it could be reproduced by the relation "domination through" with effective set N (see Definition 2.38, p. 51).

Definition 2.69. A bargaining solution $f : \mathscr{B}_N \to U$ is *independent of irrelevant alternatives*, if a possible enlargement of the feasible set U has no impact on the allocation under f:

$$\text{If } U \subseteq U' \text{ and } f(U',\mathbf{d}) \in U, \text{ then } f(U,\mathbf{d}) = f(U',\mathbf{d}). \tag{2.224}$$

Properties under the above name surface in many variants in economics, see Paramesh (1973), and have regularly been criticized for being too restrictive. We do not consider this criticism too appropriate, as in our case an extended feasible set always allows for solutions from within the "new" part. Then, the property is simply not applicable. What it intends to prevent is that through the extension of the feasible set, an allocation is selected as solution which was already feasible before the extension occurred.

Definition 2.70. A bargaining solution $f : \mathscr{B}_N \to U$ is *Pareto-optimal*, if it assigns to all bargaining problems $B \in \mathscr{B}_N$ an allocation $f(U,\mathbf{d})$ on which no Pareto improvement is possible within the feasible set:

$$\text{For all } B \in \mathscr{B}_N, \; \nexists\, \mathbf{x} \in U \;\Big|\; x_i \geq f_i(U,\mathbf{d}) \text{ for all } i \in N, \tag{2.225}$$

where strict equality holds for at least one $i \in N$.

Definition 2.71. A bargaining solution $f : \mathscr{B}_N \to U$ is said to be *symmetric*, if it assigns to all bargaining problems $B \in \mathscr{B}_N$ with symmetric feasible set U an allocation $f(U,\mathbf{d})$ for which $f_i(U,\mathbf{d}) = f_j(U,\mathbf{d})$ for all $i, j \in N$.

The feasible set U is *symmetric*, whenever for any given allocation $\mathbf{x} \in U$ we also have $\mathbf{x}' \in U$ for all $n!$ possible permutations on the entries of the vector $\mathbf{x}$. This implies also that $a_i(U,\mathbf{d}) = a_j(U,\mathbf{d})$ for all $i, j \in N$, i.e the maximal allocations in U are identical for all players.

Definition 2.72. A bargaining solution $f : \mathscr{B}_N \to U$ is *invariant to positively affine transformations*, if

$$f_i\left(\widetilde{U}, \beta \otimes (\mathbf{d}+\alpha)\right) = \beta_i\left(f_i(U,\mathbf{d}) + \alpha_i\right) \text{ for all } i \in N \tag{2.226}$$

holds, where $\alpha \in \mathbb{R}^n$, $\beta \in \mathbb{R}_{++}$, and $\widetilde{U} = \{(\beta_1 \cdot (x_1 + \alpha_1), \ldots, \beta_n \cdot (x_n + \alpha_n)) \,|\, \mathbf{x} \in U\}$.

Under the last property, it is possible to specify arbitrary origins through α and apply a positive scaling of the payoff-units via β. Because this is one of the standard properties assumed for bargaining solutions, the disagreement point $\mathbf{d}$ is usually normalized to $\mathbf{0} \in \mathbb{R}^n$ and then dropped from the bargaining problem, reducing the latter to its feasible set U. In a next step, we will take a closer look at two specific bargaining solutions and their weighted derivatives. We begin with the so-called *Nash solution*.

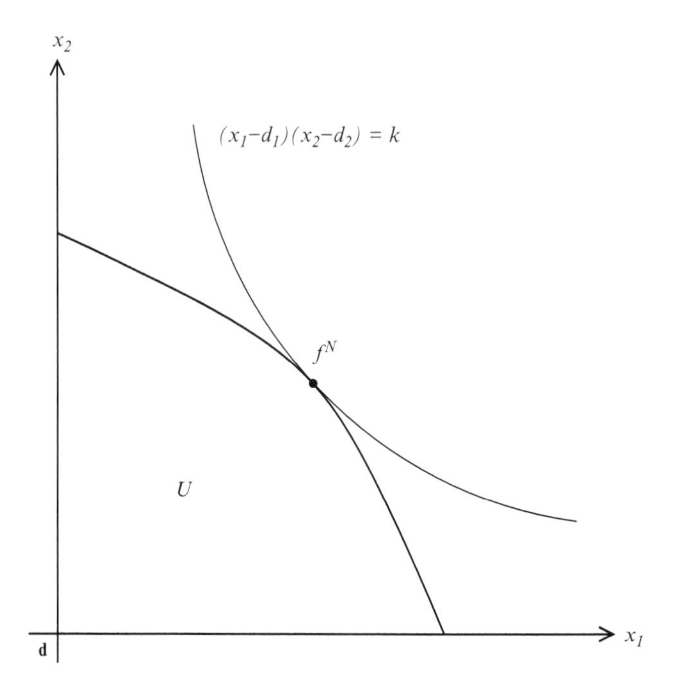

Fig. 2.12 The Nash solution f^N on the feasible set U

2.4.7.2 The Nash Solution for Bargaining Problems

Originally introduced in Nash (1950), the *Nash solution* was defined for 2-player bargaining problems. We present a generalized version for n players, which is adapted from Harsanyi and Selten (1972) and Thomson (1994, pp. 1245 ff.), but rely on a graphical illustration in 2 dimensions.

Definition 2.73. The *Nash solution* for bargaining problems, denoted f^N, is a function $f : \mathscr{B}_N \to U$ with the functional form

$$f^N(U,\mathbf{d}) = \arg\max_{\mathbf{x}\in U} \prod_{i=1}^{n} (x_i - d_i). \tag{2.227}$$

One can interpret the Nash solution as the allocation which maximizes the product of the individual allocations in case $\mathbf{d} = \mathbf{0}$, or as the product of the change in individual allocations for general disagreement points $\mathbf{d}$. As f^N is closely related to the geometric mean, it reaches its maximum for identical coefficients which exhaust the overall surplus that can be distributed.

Since the function f^N is continuous and its domain is convex and compact, a solution to (2.227) is guaranteed to exist. It is also unique, as can be deducted from its similarity to the geometric mean. We illustrate this in Fig. 2.12. Because f^N is the product of n elements, and hence commutative, it is symmetric, given that the feasible set U is so. The solution is also Pareto-optimal, because

$$\frac{\partial f^N}{\partial x_i} = \prod_{j \in N \setminus \{i\}} (x_j - d_j) > 0 \text{ for all } i \in N \tag{2.228}$$

holds with strict equality for all allocations $\mathbf{x} > \mathbf{d}$, which we assume to exist by non-degeneracy of the bargaining problem $(U, \mathbf{d})$. Hence, increasing – ceteris paribus – the allocation to a given player $i \in N$ will strictly increase the value of f^N, and so the $\arg\max f^N$, denoted by $\mathbf{x}^*$, must be on the border of U. As a consequence, the Nash solution is also independent to irrelevant alternatives: Removing elements from U which are not the $\arg\max f^N$ have no effect on the outcome. Finally, we show that f^N is also invariant to positively affine transformations:

$$\begin{aligned}\arg\max_{\mathbf{x}\in U} \prod_{i=1}^{n} \beta_i \cdot (x_i - d_i) &= \arg\max_{\mathbf{x}\in U} \prod_{i=1}^{n} \beta_i \cdot \prod_{i=1}^{n} (x_i - d_i) \\ &= \prod_{i=1}^{n} \beta_i \cdot \arg\max_{\mathbf{x}\in \widetilde{U}} \prod_{i=1}^{n} (\widetilde{x}_i - d_i),\end{aligned} \tag{2.229}$$

with $\widetilde{x}_i = x_i + \alpha_i$ for all $i \in N$ and $\widetilde{U} = \{(x_1 + \alpha_1, \ldots, x_n + \alpha_n) | x \in U\}$.

If we drop the restriction of symmetry on the solution f^N we can establish what is known as the *Weighted Nash solution*. Then, even within a symmetric feasible set U, different allocations can be realized, according to the weights assigned to the players:

Definition 2.74. The *Weighted Nash solution* for bargaining problems, denoted f^{N_γ}, is a function $f : \mathscr{B}_N \to U$ with the functional form

$$f^{N_\gamma}(U, \mathbf{d}) = \arg\max_{\mathbf{x}\in U} \prod_{i=1}^{n} (x_i - d_i)^{\gamma_i}, \tag{2.230}$$

where $\gamma \in \mathbb{R}_+^n$ are the nonnegative weights assigned to the players.

Notably, for this weighted solution, it is not necessary to introduce a weight system if one wants to assign a zero allocation to some player i.[97] Regardless of the allocation x_i, if the weight γ_i equals zero, the term $(x_i - d_i)^{\gamma_i}$ is identical to one and has therefore no influence on the value of f^{N_γ}. Another solution concept for bargaining problems, the *Kalai–Smorodinsky solution* is presented next. It is based on the maximally possible allocations within U as given by $\mathbf{a}(U, \mathbf{d})$.

---

[97] As opposed to, say the Weighted Shapley value Φ^ω, where the weight system ω contains not only individual weights but also an ordered partition of the player set.

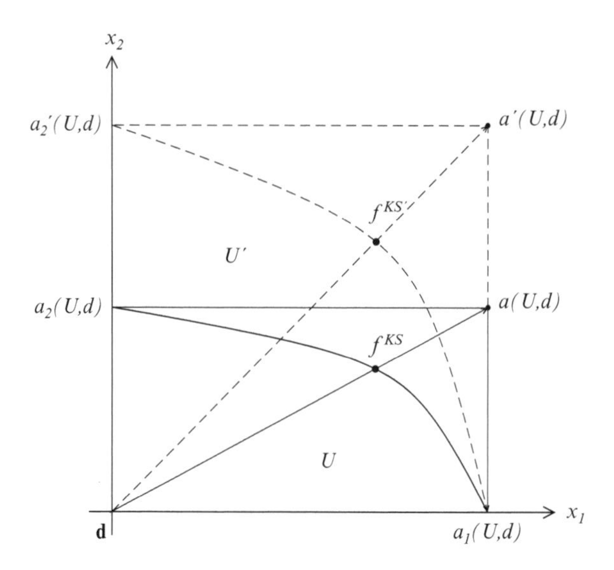

Fig. 2.13 The Kalai–Smorodinsky solution f^{KS} on the feasible set U

2.4.7.3 The Kalai–Smorodinsky Solution for Bargaining Problems

This concept was introduced to the literature in Kalai and Smorodinsky (1975), again on a 2-player basis. A version generalized to n players and a weighted variant can be found in Thomson (1994). As compared to the Nash solution, it can be decomposed into two steps: At first, the proportions in which the surplus is allocated among the players is determined according to the possible upper bound as given by $\mathbf{a}(U,\mathbf{d})$. Then, the allocation is increased in this proportion until a boundary of the feasible set is reached.

Definition 2.75. The *Kalai–Smorodinsky solution* for bargaining problems, denoted f^{KS}, is a function $f : \mathscr{B}_N \to U$ with the functional form

$$f^{KS}(U,\mathbf{d}) = \max_{k \in (0,1]} k \cdot \mathbf{a}(U,\mathbf{d}) + (1-k)\mathbf{d} \text{ subject to } f^{KS}(U,\mathbf{d}) \in U. \qquad (2.231)$$

With f^{KS} being continuous on its convex and compact domain U, an $\arg\max$ to (2.231) exists.[98] The Kalai–Smorodinsky solution is depicted in Fig. 2.13, where, for now, we ask the reader to ignore the dashed parts and all "primed" notation within. Symmetry of the Kalai–Smorodinsky solution is quickly deduced: Any relabelling of the players affects U, $\mathbf{d}$, and $\mathbf{a}(U,\mathbf{d})$ likewise, and so f^{KS} will not

[98] It should be noted here that we assume U to be comprehensive, not merely convex, but the latter property suffices for the existence of a maximum in this setting. Also, because $(U,\mathbf{d})$ is non-degenerate by assumption, k cannot be equal to 0.

change in its individual allocations. However, the prerequisite for symmetry of f^{KS}, a symmetric U is not met in Fig. 2.13. That the solution is individually monotonic follows directly from the functional form. Any increase in an $a_i(U,\mathbf{d})$ for some player $i \in N$ has a "positive" effect on the direction of the vector $\mathbf{a}(U,\mathbf{d})$, positive in the sense that the relative allocation is increased in i's favor and so his entry f_i^{KS} in the max of f^{KS}, too. This is illustrated in Fig. 2.13 through the dashed lines: If we enlarge the feasible set to U', and accordingly player 2's maximal allocation to $a_2'(U,\mathbf{d})$, he is no worse off under $f^{KS\prime}$. In our case, we deliberately changed to U', so that player 1's payoff is unaffected. But, as can be seen from the illustration, depending on the curvature of U' he could both gain and lose, even when preserving comprehensiveness.

Pareto optimality is not guaranteed for f^{KS}, but given a comprehensive U, weak Pareto optimality can be assured. We illustrate this in the following example:

Example 2.9. This example serves only to demonstrate that the Kalai–Smorodinsky solution need not result in a Pareto-efficient allocation, even if the underlying domain U is a comprehensive set. However a weakly Pareto-efficient allocation is guaranteed. Let the domain be given by the convex hull of the six vectors, $U = co\{(1,0,0),(0,1,0),(0,0,0),(1,0,1),(0,1,1),(0,0,1)\}$, and let $\mathbf{d}$ coincide with the origin, and therefore $\mathbf{a}(U,\mathbf{d}) = (1,1,1)$. Consequently, we obtain an arg max of $k = \frac{1}{2}$ where $f^{KS}(U,\mathbf{d}) = (\frac{1}{2},\frac{1}{2},\frac{1}{2})$. This allocation is an element of the border of U and as such weakly Pareto-efficient. Also an element of the border of the feasible set is the allocation $\mathbf{z} = (\frac{1}{2},\frac{1}{2},1)$, which, in addition, is (strictly) Pareto-efficient. This is illustrated in Fig. 2.14, where U is the shaded body and the vector from d to $\mathbf{a}(U,\mathbf{d})$ is partitioned in one dotted segment (within U) and one solid (outside of U). Just above the solution f^{KS} lies the allocation $\mathbf{z}$, which does not allow for any Pareto improvements.

Finally, we show by a series of equalities that also f^{KS} is invariant to positively affine transformations:

$$\begin{aligned}
&\max_{k\in(0,1]} k\cdot\mathbf{a}\left(\widetilde{U},\beta\otimes(\mathbf{d}+\alpha)\right)+(1-k)\cdot(\beta\otimes(\mathbf{d}+\alpha))\\
&\quad=\max_{k\in(0,1]}\beta\otimes(k\cdot\mathbf{a}(U,\mathbf{d})+\alpha)+\beta\otimes((1-k)(\mathbf{d}+\alpha))\\
&\quad=\beta\otimes\left(\max_{k\in(0,1]} k\cdot\mathbf{a}(U,\mathbf{d})+k\alpha+(1-k)\mathbf{d}+(1-k)\alpha\right)\\
&\quad=\beta\otimes\left(\max_{k\in(0,1]} k\cdot\mathbf{a}(U,\mathbf{d})+(1-k)\mathbf{d}+\alpha\right),
\end{aligned}\tag{2.232}$$

where $\widetilde{U}$ is specified in Definition 2.72.

Again, dropping symmetry of the bargaining solution leads to a weighted variant, denoted f^{KS_γ}. For tractability, we normalize the disagreement point so that $\mathbf{d}=\mathbf{0}$ and the argument reduces to U.

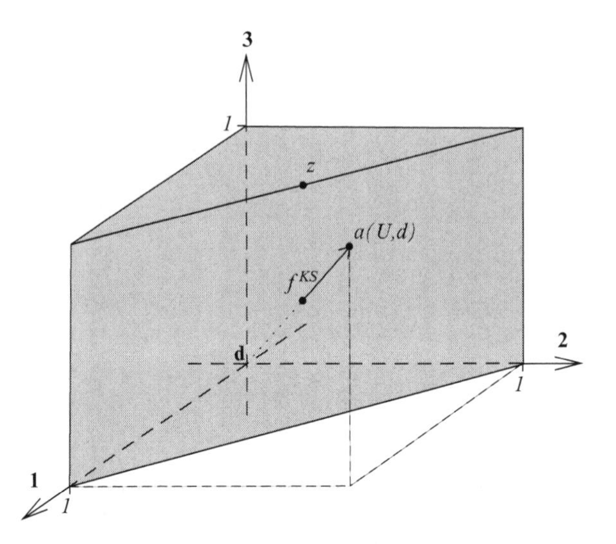

Fig. 2.14 A weakly Pareto-efficient f^{KS} on a comprehensive U

Definition 2.76. The *Weighted Kalai–Smorodinsky solution* for bargaining problems, denoted f^{KS_γ}, is a function $f : \mathscr{B}_N \to U$ with the functional form

$$f^{KS_\gamma}(U) = \max_{k \in (0,1]} k \cdot \mathbf{a}^\gamma(U) \;\text{ subject to }\; f^{KS_\gamma}(U) \in U, \tag{2.233}$$

where $\mathbf{a}^\gamma(U) = (\gamma_1 \cdot a_1(U), \ldots, \gamma_n \cdot a_n(U))$ and $\gamma \in \Delta^{n-1}$.

This weighted variant, too, allows for zero allocations via plain weights γ without introducing weight systems. In a non-degenerate bargaining problem the direction of the vector $\mathbf{a}(U)$ can be controlled entirely through the weights that scale the individual maximal allocations within U. The assumption of comprehensiveness regarding U also ensures that a projection on a certain dimension, i.e. assigning some player $i \in N$ a weight $\gamma_i = 0$, will not reduce the entries of the maximally attainable allocations a_j for the other $j \neq i$ players. In terms of weights used, the Weighted Kalai–Smorodinsky solution differs from the Weighted Nash solution. While the latter allows for general positive weights, the former is constructed to use weights picked from the unit-simplex of $n-1$ dimensions, which sum to one.

With this, we finish our consideration of solution concepts for coalitional games and even our introduction to "pure" game theory. In the next chapter, we introduce the concept of networks in economics. Based on graph theory, networks allow to visualize certain relationships between economic agents (in our case players) thus providing additional intuition in many instances. We also present different games on how such networks form, which brings us back into the realm of game theory. In this context we even consider another allocation rule, explicitly incorporating the structure of a network among the set of players.

Chapter 3
Network Theory in Economics

3.1 Overview

This chapter is intended to establish certain elements from graph theory as instruments applied within game theory. Because in game theory, the interaction between agents is modelled, it seems only natural to introduce networks that formalize the possibilities of actions taken or cooperation undertaken by and among the players. Graph theory is used to depict this network. Introducing networks in this way allows for many structural simplifications and especially an intuitive illustration of the relationships among players, once the purpose of the underlying network is specified.

The remaining chapter is structured as follows: We begin with basic notation and definitions from graph theory to be able to work with the concept of a network and certain objectees inherent to it. Next, we introduce coalitional games on such network structures, also known as *communication situations*, and elaborate on some possible properties arising from the incorporation of networks. Consequently, we also address allocation rules on such communication situations and properties of the former potentially arising specifically in this context. One such allocation rule in particular, known as the *Myerson value*, is discussed in more detail. The chapter concludes with a treatment of network formation covering formation games in strategic and extensive form, as well as stability issues of formed networks.

Our notation is mostly adapted from Goyal (2007), who provides an excellent overview on the state of network economics as a discipline within economics in general. For a more mathematical approach to graph theory, we point the reader to Wilson (1972) or Bollobás (1990).

3.2 Basic Concepts and Definitions

This section is devoted to the definition of the basic elements constituting a graph or network, as well as to some concepts used to relate to parts of the network.

P. Servatius, *Network Economics and the Allocation of Savings*, Lecture Notes in Economics and Mathematical Systems 653, DOI 10.1007/978-3-642-21096-9_3,

We begin with the graph, or network, itself. Formally, it is defined by the tuple (N,g). The first element is a (in our context) finite set of *nodes* or *vertices* of the graph, $N = \{1,2,\ldots,n\}$. The second, g, is the set of *links* or *edges* which describe the relationship between two nodes.[1] This is expressed in binary fashion, either there is a link between nodes i and j or there is not.[2] We write $g_{ij} \in \{0,1\}$ for such a potential link, and it assumes value 1 if it exists and 0 otherwise.[3] We concern ourselves only with *undirected* networks, and so $g_{ij} = g_{ji}$ for all nodes $i, j \in N$. Also, by convention we set $g_{ii} = 0$ for all $i \in N$. The set of links g can be interpreted as the collection of links existing between players in network (N,g)[4]:

$$g = \{g_{ij} | g_{ij} = 1,\ \ i,j \in N\}. \tag{3.1}$$

As the set of nodes N does generally not change in our context, we refer to a network just by the set of its links g. This notation also allows us to relate certain networks via set inclusion to one another. In network settings it is – for illustrative purposes – also common to deviate from the set-theoretic notation when removing or adding a link to a network. We use the more intuitive $g - g_{ij}$ or $g + g_{ij}$ rather than the clumsy expressions $g \setminus \{g_{ij}\}$ or $g \cup \{g_{ij}\}$. Likewise, the set of all networks on n nodes, $\mathcal{G}_N$, is simply denoted $\mathcal{G}$ where no confusion can arise.

The complete network g^c, with $|g| = \binom{n}{2} = \frac{1}{2}n(n-1)$ links, one between every pair of nodes, is defined as follows:

$$g^c := \{g_{ij} | i,j \in N\}. \tag{3.2}$$

It gives rise to the definition for the set of all networks $\mathcal{G}$ on n nodes:

$$\mathcal{G} := \{g | g \subseteq g^c\}. \tag{3.3}$$

The polar case is the so-called empty network $g^\emptyset$, for which $g^\emptyset = \{\emptyset\}$ or $|g^\emptyset| = 0$ holds. Whenever we restrict a network g to a subset of nodes $S \subseteq N$ and the corresponding links between them, we speak of a *subgraph*, or a network *restricted to S*:

$$g|_S := \{g_{ij} \in g | i,j \in S\}, \quad S \subseteq N. \tag{3.4}$$

---

[1] We do not venture in the realm of so-called *hypergraphs*, which allow more than two vertices to be connected via a single (hyper)edge.

[2] The possibility of qualitative links assuming values in between zero and one is ignored here.

[3] The notation g_{ij} can lead to ambiguities whenever the set of nodes N exceeds 11 in cardinality and its elements are indexed with natural numbers. Wherever such problems arise, we use $g_{i,j}$ to be unmistakable.

[4] Another possibility to specify (N,g) is in the form of an $n \times n$ matrix with binary entries (symmetric, in case of undirected networks) corresponding to the (non)existence of a link between nodes.

Having outlined what constitutes a network (N,g), we look next at certain relationships within a given network, and then at some typical network architectures. We begin our analysis at a single node $i \in N$. The first and most obvious question is to how many other nodes $j \in N \setminus \{i\}$ there exist links g_{ij}, for which $g_{ij} = 1$. This (possibly empty) subset of nodes is called the *neighborhood* of node i. For a specific network g, we write

$$N_i(g) := \{ j \in N | g_{ij} = 1 \} \quad \text{for all } i \in N. \tag{3.5}$$

According to this definition, node i is not a member of its own neighborhood, $i \notin N_i(g)$, and hence, if there are no links with i on one end, then $N_i(g) = \emptyset$. The elements $j \in N_i(g)$ are called the *neighbors* of i and its cardinality $\eta_i(g) := |N_i(g)|$ is the *degree* of node i in network g. Clearly, the concept of a neighbor is symmetric:

$$j \in N_i(g) \Longleftrightarrow i \in N_j(g). \tag{3.6}$$

For the corresponding set of links, $g_i(g)$, with node i at one end, we write:

$$g_i(g) = \{ g_{ij} \in g \text{ with } j \in N \setminus \{i\} \}. \tag{3.7}$$

Whenever convenient, this is reduced to g_i. The concept of a neighborhood is extended by Goyal (2007, p. 10) to the notion of a *d-neighborhood*, denoted $N_i^d(g)$, via induction:

$$N_i^d(g) = N_i^{d-1}(g) \cup \left(\bigcup_{j \in N_i^{d-1}(g)} N_j(g) \right), \quad \text{where } N_i^1(g) = N_i(g). \tag{3.8}$$

Note that $N_i^1(g) \subsetneq N_i^2(g) \subsetneq \cdots \subsetneq N_i^d(g)$ and $i \notin N_i^1(g)$ but $i \in N_i^d(g)$ for all $d > 1$. The cardinality of the difference operator,

$$\Delta N_i^d(g) := N_i^d(g) \setminus N_i^{d-1}(g), \quad d > 1, \tag{3.9}$$

provides insight on how fast a network g expands from node i on outwards. Depending on the structure and size of the network, this gives an approximation on how central the node i might be.

Along such expansions it is also interesting to consider how nodes i and j, between whom no link is present, might nevertheless be "connected": Can they be related through a series of linked nodes with i and j as endpoints? Here, we formally introduce the concept of a *path* between two nodes. Such a path exists, if there is a sequence of distinct nodes[5] $i_1, i_2, \ldots, i_t$ such that $g_{ii_1} = g_{i_1 i_2} = \ldots = g_{i_t j} = 1$.

[5]In contrast to *trails*, which require only distinct links (as opposed to distinct nodes!) and allow "crossings" and so possibly *cycles*, as well as in contrast to *walks*, which have no requirements other than the existence of links connecting i and j and allow for repetitions.

Whenever we say, i and j, both elements of some subset $S \subseteq N$, are *connected in* S, we restrict the path between them to links between nodes who are all elements of S.

There may be two or more paths between i and j in a network g, and there might also exist paths of various lengths, as counted by the number of links connecting i and j.[6] Whenever such a path exists, we say that nodes i and j are *connected.* Define as $P^g(i,j)$ the correspondence

$$P : \mathcal{G} \times N \times N \rightrightarrows 2^g. \tag{3.10}$$

It assigns to every pair of nodes all possible paths that connect them via g. By convention, every node $i \in N$ is connected to itself, and so $P^g(i,i) \neq \emptyset$ for all $i \in N$ even if $N_i(g) = \emptyset$. If P is nonempty for all pairs of distinct nodes in N, we say the network g is *connected.* If this is not the case, then the network g has more than just one *component.* The set of components, which is a partition of N is denoted N/g and given by

$$N/g := \{[i] | j \in [i] \text{ if } P^g(i,j,) \neq \emptyset\}. \tag{3.11}$$

We define components via equivalence classes based on the relation "i is connected to j" in network g. A component is therefore a maximally connected set of nodes within network g and $|N/g| = 1$ if g is connected. Even though nodes without any links also constitute a component each, by referring to "components" we usually mean the nontrivial kind with existing links among its members. An equivalent definition of the set of components is the following:

$$N/g := \{C \subseteq N | \text{for all } i \in C,\ N_i(g) \cap (N \setminus C) = \emptyset\}. \tag{3.12}$$

Note that because unlinked nodes have empty neighborhoods, the intersection of the latter with any set is the empty set again.

In this context the idea of a *minimally connected*, or simply *minimal* network arises: A network g is minimal if and only if the removal of any link $g_{ij} \in g$ will increase the number of components by one: $|N/(g - g_{ij})| = |N/g| + 1$. This notion also applies to all possible components of network g, if they can be split up into two with the removal of any link. We speak of minimal components in this case.

Trivially, this always applies if there are nodes with singleton neighborhoods, i.e. only one link. The more interesting cases though arise, when the removal of a link leads to the creation of new non-singleton components. Following Jackson and Wolinsky (1996), we will call such links *critical* links.[7]

Let us now turn to some graphical illustrations of common network structures. The two polar cases for four nodes, the complete network g^c, and the empty network,

---

[6]The shortest possible one is the *trivial* path, given by the link between i and j, if $g_{ij} = 1$.

[7]Similarly, Bilbao and Edelmann (2000) call nodes, the removal of which from the network g would lead to an increased number of non-singleton components *cutvertices*.

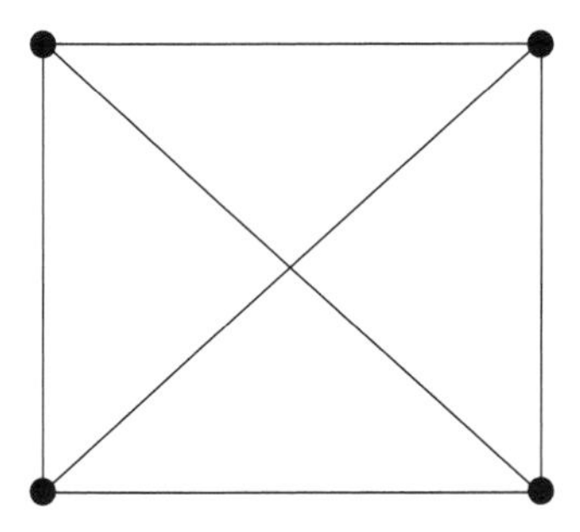

Fig. 3.1 Complete network

Fig. 3.2 Empty network

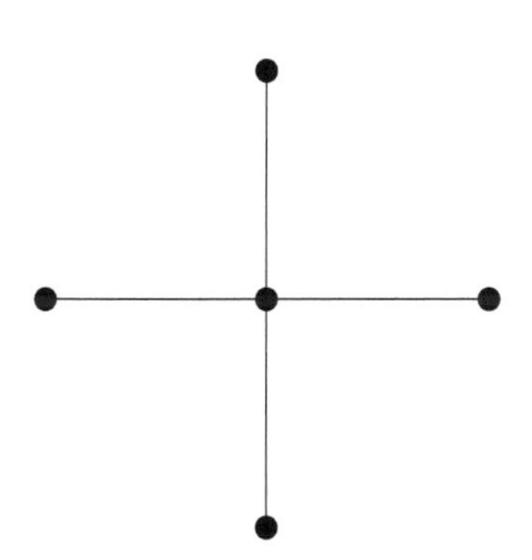

Fig. 3.3 Star network

$g^{\emptyset}$, are depicted in Figs. 3.1 and 3.2. In the former, there exist links between all four players, in the latter no links are present at all:

Figure 3.3 shows a very popular type of network, the so-called *star network*. One node is linked to all other nodes, while these are linked exclusively to the former, which then sits in a central position. This center node indeed occupies a very crucial position, as its removal would directly cause the network to be void of any links. The notion of a star network can be extended to multiple center nodes, see Fig. 3.4. It depicts an interlinked star network, where the "central" nodes on the left and the right are connected each to all four nodes in the middle.

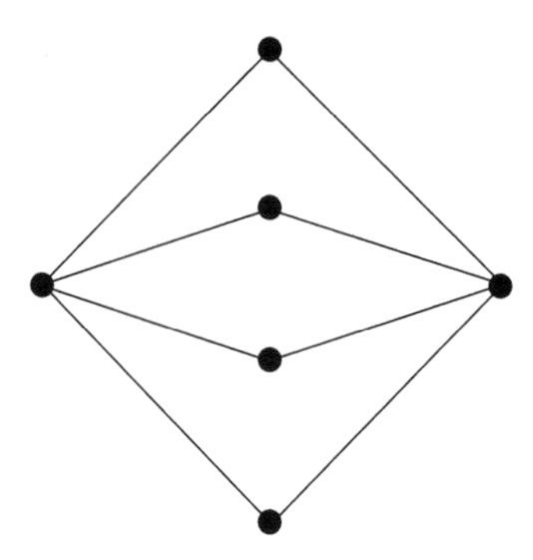

Fig. 3.4 Interlinked star with two centers

Fig. 3.5 Line network

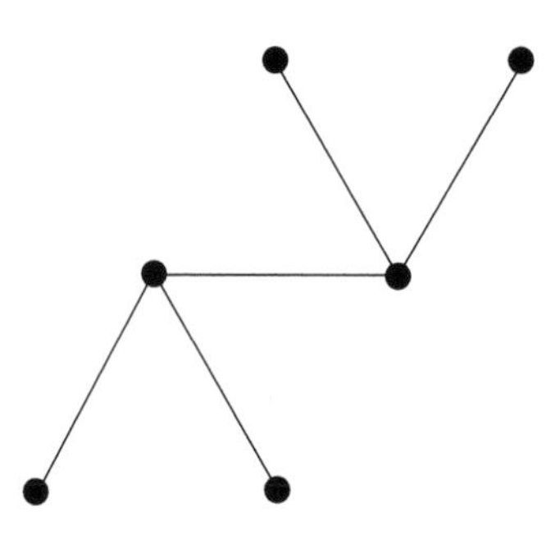

Fig. 3.6 Minimal network

Next, Fig. 3.5 contains another standard type network, the line network. And finally, in Fig. 3.6, we have an example of a minimal network, whose critical link (the horizontal one), when removed, will give rise to a network with two non-singleton components.

After this basic introduction, we now analyze how network economics can be combined with the theory of coalitional games to what is called a *communication situation*.

3.3 Communication Situations

A general coalitional game (N, v) can also be represented by a vector $\mathbf{v}$ with $2^n - 1$ entries, one corresponding to each coalition $S \subseteq N$, or rather, its value in the game. Usually, these values are calculated according to some scheme or idea how gains are generated among the cooperating players of the game and how cooperation might be restricted. In order to formalize this scheme, make it more transparent, and illustrate the effects of changes to it, a network structure (N,g) can be laid over a coalitional game (N,v). Then, the nodes of a network correspond to the respective players of the game, and the links in between, which constitute the network, illustrate to which extent cooperation is possible. If changes occur to the process of value creation through cooperation, this network structure can be altered accordingly, which makes these changes plausible and traceable, as opposed to merely changing one or more of the entries of $\mathbf{v}$.

Myerson (1977) was the first to combine a coalitional game and a network structure to illustrate how the gains from cooperation arise. Such settings have since become known as *communication situations*, and are subject of further investigation in Borm et al. (1992), Slikker and van den Nouweland (2001b), or van den Nouweland (2005), to name but a few.

Formally, a communication situation is defined by the triplet (N,v,g): A finite set of players $N = \{1,2,\dots,n\}$, a characteristic function $v : 2^N \to \mathbb{R}$, and a network g, whose set of nodes corresponds to the set N. The network g then illustrates some kind of cooperation restrictions among the players, the effects of which are measured by the characteristic function. One can combine v and g to obtain a network restricted value function $v : 2^N \times \mathscr{G} \to \mathbb{R}$, which as an argument not only includes the cooperating set of players S, but also the network structure g that might impose restrictions on cooperation. We alternate between notation $v(S,g)$ and v^g, depending on which argument we try to emphasize or hold constant. In this sense, v^g would be a network restricted value function on network g, i.e. a change in the argument $S \subseteq N$ incorporates the effects of network g. Sometimes we even refer to the network restricted value function as v, when absolutely no confusion can arise. The set of all communication situations on the player set N is denoted by $CS_N := \mathbf{V}_N \times \mathscr{G}_N$, or simply CS.

The extent to which a value function v is affected by some network g can vary. Take for example the original setting from Myerson (1977): Cooperation is fruitful only among players that are connected, i.e. among players within the same component $C \in N/g$ of network g. But within such a component C, the specific network structure $g|_C$ is irrelevant, as long as C remains a component. Consequently, the value of any coalition $S \subseteq N$ is the sum of what its members achieve in the components induced by network g on S.[8]

More restrictive settings exist as well, where value can only be created pairwise on the basis of existing links. Connectedness of two players does not have any positive influence on the value function, unless the players are also linked. Here, in addition to forming a component, the total number of links among some coalition of players S plays a crucial role in increasing gains from cooperation.

Settling in between, notions of decay are also plausible. Connectedness of players is beneficial in principle, but diminishes with increasing length of the shortest path

---

[8]This is reproduced in more detail in Sect. 3.5.

(counted in terms of links) between players. Then, even though cooperation is possible through connectedness, additional links can still have a positive impact on the gains from cooperation as measured by the network restricted value function v^g. Depending on the level of decay, we find ourselves somewhere in between the first two scenarios.

Regardless of the specification of v^g and to what extend the underlying network g affects cooperation, the two polar cases in terms of networks can be locked into position. Function v^g always coincides with the regular characteristic function v whenever $g = g^c$. With the complete network in place, no restrictions to cooperation prevail at all:

$$v^{g^c}(S) \equiv v(S) \;\text{ for all }\; S \subseteq N. \tag{3.13}$$

And, on the contrary, if the network is the empty network $g^{\emptyset}$, then no cooperation is possible at all:

$$v^{g^{\emptyset}}(S) \equiv \sum_{i \in S} v(\{i\}) \;\text{ for all }\; S \subseteq N. \tag{3.14}$$

In the context of communication situations, the marginal value of a link can be of interest, too. Formally, it is defined for all $S \subseteq N$ and all $g \in \mathcal{G}$ as

$$d_{g_{ij}}(S) = \begin{cases} v(S, g + g_{ij}) - v(S, g) & \text{if } g_{ij} \notin g \\ v(S, g) - v(S, g - g_{ij}) & \text{if } g_{ij} \in g \end{cases}. \tag{3.15}$$

It is the difference in gains a coalition S can achieve, depending on the existence of link g_{ij} in network g.

In the literature (e.g. Jackson and Watts 2002) value functions in communication situations are alternatively defined solely on the basis of links, without regards to cooperation among players. Such value functions on networks usually appear in the form $v : \mathcal{G}_N \to \mathbb{R}$. They are, after all, derived from value functions in the fashion of $v(S, g)$ but allow for more simplicity in exposition. Nevertheless, this does not suit our approach properly: By definition, it assigns a value to the whole network and not to subgraphs restricted to certain nodes $i \in S \subseteq N$ representing a cooperating coalition. We, however, want to remain in a player-based environment, where the characteristic function v^g reflects the value a coalition $S \subseteq N$ of players can realize by cooperation, given a certain network g. Here it is not the case that the network is paramount and the players are only of secondary importance. We believe this is important to maintain, in order to highlight the aspect of cooperation between players accurately.

With the communication situation established as the triplet (N, v, g), we can now introduce certain properties that arise when networks and cooperative games are combined. These naturally arise in addition to the properties already presented for general coalitional games (N, v) in Sect. 2.3.4. We begin, with no particular order in mind, with situations where the underlying network g is not changed and continue to analyze the effects of altering the network on the value function. The properties listed are complied from Jackson and Wolinsky (1996, p. 61) and Jackson (2005a, pp.132 f.), but can be found elsewhere too.

Definition 3.77. A communication situation (N,v,g) is *component additive*, if for all coalitions $S \subseteq N$ the following is true:

$$v(S) = \sum_{C \in N/g} v(S \cap C). \tag{3.16}$$

This property is closely related to the notion of decomposability for coalitional games (N,v), yet not identical. The partition required for decomposability is met by the one induced through the network g, i.e. N/g. And even though component additivity of a communication situation implies its decomposability, the converse is generally not true, because it might be possible to decompose even finer within a component $C \in N/g$. The above definition also includes the special case mentioned by Jackson and Wolinsky (1996, p. 59), where $S = N$ and component additivity is only required for the value of the grand coalition. Intuitively, component additivity poses no strict assumption. Why should coalitions be able to create a value higher than the aggregate value of its members in the respective components, when communication across components is ruled out?

Definition 3.78. A communication situation (N,v,g) is *link anonymous*, if the value created by a coalition $S \subseteq N$ under v^g depends only on the number of links within S:

$$v^g(S) = v^g(T) \text{ for all } S,T \subseteq N \text{ with } |g|_S| = |g|_T|. \tag{3.17}$$

Link anonymity imposes a certain kind of symmetry on the communication situation. But instead of symmetric players $i \in N$, the links in g (or a restriction of g on some $S \subseteq N$) are symmetric when it comes to creating value via v^g. Regardless of the size of the coalition in terms of players and linkage structure among them, only the number of existing links counts. It is similar to the previously introduced k-games (see Definition 2.23, p. 37) with $k = 2$, where only pairs of players can add value. The network g further restricts this to linked pairs of players.

Definition 3.79. A player $i \in N$ is called *superfluous* for communication situation (N,v,g), if his marginal value to any coalition $S \subseteq N$ is zero:

$$v(S,g) - v(S \setminus \{i\}, g) \equiv 0 \text{ for all } S \subseteq N,\ i \in S. \tag{3.18}$$

The superfluous player is similar to a null-player in a coalitional game. But caution is advised, because the fact that this player has no marginal value might not only rely on his particular characteristics that make him useless for any kind of cooperation. He might as well not be connected (and hence not linked) to any other player in the communication situation, but very suitable for cooperation if he was.

Definition 3.80. A communication situation (N,v,g) is *monotonic on* g, if for all $S \subseteq N$

$$v(S,g) \geq v(S,\hat{g}) \text{ whenever } \hat{g} \subseteq g. \tag{3.19}$$

Network monotonicity dictates that the marginal value of all links constituting network g is nonnegative. At worst, they have no influence on cooperation at all. Note that this applies also to links $g_{ij} \in g \setminus \hat{g}$ for which possibly one or both players i and j are not members of S. The influence of links on the value of a coalition S is not limited to those involving members of S, but to all players $i \in N$. Also consider that the statement *monotonic on* g only concerns links contained in reference network g but not those in excess of it from $g^c \setminus g$.

Definition 3.81. A communication situation (N, v, g) satisfies the property of *critical link monotonicity*, if for any critical link g_{ij} and the components C_i and C_j arising from its removal in $\widetilde{g}$, the following is true:

$$\text{If } v^g(C) \geq v^{\widetilde{g}}(C_i) + v^{\widetilde{g}}(C_j), \text{ then } \frac{v^g(C)}{|C|} \geq \max\left\{\frac{v^{\widetilde{g}}(C_i)}{|C_i|}, \frac{v^{\widetilde{g}}(C_j)}{|C_j|}\right\}, \tag{3.20}$$

where $g - g_{ij} = \widetilde{g}$, $N/g = C$, and $N/\widetilde{g} = \{C_i, C_j\}$.

Whenever we remove a critical link who contributed nonnegatively to the value of v^g, the per-capita (or average) value in either newly formed component cannot be higher than in the initial component. In other words, the removal of a critical link should not shift a big share of the value created to a new component with relatively few members. This property imposes somewhat evenly distributed marginal values on cooperation based on the links formed between players. However, a superadditive v^g will not suffice to ensure critical link monotonicity: Just suppose that from removing g_{ij} some nonempty component C_j with $v^{\widetilde{g}}(C_j) = 0$ arises. This is within the bounds of superadditivity, but the average value of C_i is strictly larger than that of $C = C_i \cup C_j$. As the definition of convexity for characteristic functions boils down to superadditivity whenever $S \cap T = \emptyset$,[9] the same example can be used to show that not even convex characteristic functions imply critical link monotonicity. This shows that the property in question is quite a strong restriction.

Definition 3.82. A link $g_{ij} \in g$ is called *superfluous* for communication situation (N, v, g), if its removal from any network $\hat{g} \subseteq g$ does not influence the value of the grand coalition:

$$v(N, \hat{g}) = v(N, \hat{g} - g_{ij}) \text{ for all } \hat{g} \subseteq g. \tag{3.21}$$

Note that by definition superfluous links exist only within the reference network g: When "disassembling" g link by link, such superfluous links could be encountered in any $\hat{g} \subseteq g$, but nothing is said about links outside of g, even when they have no influence on the value of N as well.

Definition 3.83. A link $g_{ij} \in g$ is called *strongly superfluous* for communication situation (N, v, g), if its removal does not influence the value function v^g at all:

$$v(S, g) = v(S, g - g_{ij}) \text{ for all } S \subseteq N. \tag{3.22}$$

[9] Compare with expressions (2.30) and (2.39) on pages 38 and 40, respectively.

In this notion, the strongly superfluous link has no influence under v^g on the value of any coalition in a given network g.[10] It is important to see that this property is defined for all coalitions on a fixed network g, while the previous one was in terms of the grand coalition on a range of networks $\hat{g} \subseteq g$.

Definition 3.84. A communication situation (N,v,g) is *link convex*, if the marginal contribution of an additional link g_{ij} within some component $C \in N/g$ increases with the number of links that already exist within this component. Formally, the following must be fulfilled:

$$v(C,g) - v(C,g-g_{ij}) > v(C,g-g_{lk}) - v(C,g \setminus \{g_{lk},g_{ij}\}) \tag{3.23}$$

for all networks $g \in \mathcal{G}$, components $C \in N/g$, and pairs of links $g_{lk}, g_{ij} \in g|_C$.

Because link convexity must hold for all networks g and components C induced by g on N, it also applies to any connected subset of players $S \subseteq C$, $C \in N/g$. Link convexity also implies strictly positive marginal values of links, as can be deduced from the above equation when $g_{ij} = g_{lk}$.

The last two properties refer to the network g rather than the communication situation as a whole:

Definition 3.85. The network g of a communication situation (N,v,g) is called *efficient*, if there is no other network $g' \in \mathcal{G}_N$ for which the value of the grand coalition is larger:

$$v(N,g) \geq v(N,g') \quad \text{for all} \quad g' \in \mathcal{G}_N. \tag{3.24}$$

This property is related to the superfluous link property. The repeated application of removing superfluous links from the complete network g^c will yield a minimally efficient network, but not necessarily the only efficient network. But even networks with more links than some already efficient network g can still be efficient, depending on the cost of forming links and how these are incorporated into the characteristic function. The final property includes an – as yet unspecified – allocation rule γ, one argument of which is also the network g.

Definition 3.86. The network g of a communication situation (N,v,g) is called *Pareto-efficient* with respect to allocation rule γ, if there is no other network $g' \in \mathcal{G}$ with

$$\gamma_i(N,v,g') \geq \gamma_i(N,v,g) \quad \text{for all} \quad i \in N, \quad \text{and} \quad \exists\, i \mid \gamma_i(N,v,g') > \gamma_i(N,v,g). \tag{3.25}$$

Supposing the allocation rule γ is efficient itself, the Pareto-efficient network g is also efficient.

The last property already anticipates from the next section, as it makes use of an allocation rule in a communication situation. With possible properties of (N,v,g) introduced, it now makes sense to analyze how the gains from cooperation can be distributed among the networked players.

---

[10]This of course is trivially fulfilled in any network g where $g_{ij} \notin g$.

3.4 Allocation Rules in Communication Situations

Like in coalitional games, there are allocation rules in communication situations to allocate the payoffs to the players. These, naturally, also account for the underlying networks and changes herein. We adopt the notation of Myerson (1977), who denotes network allocation rules by γ, representing a function $\gamma : CS_N \to \mathbb{R}^n$ that assigns each communication situation (N, v, g) an n-dimensional vector of payoffs, $\gamma(N, v, g) \in \mathbb{R}^n$. For a general approach to allocation rules in the context of communication situations we refer the reader to Jackson (2005a). Albeit his treatment is built on link-based value functions $v : \mathcal{G} \to \mathbb{R}$, the concepts are readily transferred into our framework, where first and foremost the players give rise to gains from cooperation, according to the underlying network structure.

Also, the procedure by which to adapt any allocation rule $\Phi : \mathbf{V}_N \to \mathbb{R}^n$ on coalitional games (N, v) to one on communication situations (N, v, g) is readily established. We only need to substitute the value function v for its network restricted counterpart v^g. In this manner, we define

$$\gamma(N, v, g) := \Phi(N, v^g) \tag{3.26}$$

as the allocation rule for an arbitrary communication situation $(N, v, g) \in CS_N$. It arises from the coalitional game (N, v) and incorporates the restrictions imposed on v through network g. For clarity, we use the letter γ to designate this special class of allocation rules. In addition to the properties allocation rules Φ can exhibit in coalitional games (N, v) (listed in Sect. 2.4.4), we now introduce a selection of properties that emerge especially in the context of communication situations (N, v, g). This list is by no means exhaustive, but rather selected to provide a useful basic overview and also to satisfy the needs of our subsequent application. The properties have been compiled from Slikker and van den Nouweland (2001a, pp. 37 ff., pp. 175 ff.), van den Nouweland (2005, pp. 69 f.), Dutta et al. (1998, p. 248), Jackson and van den Nouweland (2005, p. 424), and Myerson (1977).

Definition 3.87. An allocation rule $\gamma : CS_N \to \mathbb{R}^n$ is *component efficient* or *balanced*, if the following holds for all component additive communication situations $(N, v, g) \in CS_N$ and for all components $C \in N/g$:

$$\sum_{i \in C} \gamma_i(N, v, g) = v^g(C). \tag{3.27}$$

The property is very basic. Under γ, the value created within a component must also be allocated exhaustively among the members of the same component. This implies not only that all components cannot be allocated more than what they each create in value, but also no less, imposing some kind of (local) efficiency at the same time.

Definition 3.88. An allocation rule $\gamma : CS_N \to \mathbb{R}^n$ is *component decomposable*, if for all component additive communication situations $(N, v, g) \in CS_N$ and for all components $C \in N/g$ it is true that

$$\gamma_i(N, v, g) = \gamma_i(C, v, g|_C) \quad \text{for all} \quad i \in C. \tag{3.28}$$

According to this property, the allocation rule γ is independent of the network structure outside a component C when allocating value to the members $i \in C$ of this component. This excludes possible externalities across components N/g of a network g. Note that the properties of component decomposability and component efficiency do by themselves not imply one another!

Definition 3.89. An allocation rule $\gamma : CS_N \to \mathbb{R}^n$ is said to be *fair* or satisfying *equal bargaining power*, if the following holds for all communication situations $(N, v, g) \in CS_N$ and for all pairs of players $i, j \in N$:

$$\gamma_i(N, v, g) - \gamma_i(N, v, g - g_{ij}) = \gamma_j(N, v, g) - \gamma_j(N, v, g - g_{ij}). \tag{3.29}$$

It is important to note here that even though players i and j benefit or suffer equally from the addition or removal of link g_{ij}, this property does not imply that they split the entire marginal value of g_{ij} amongst each other![11]

Definition 3.90. An allocation rule $\gamma : CS_N \to \mathbb{R}^n$ satisfies the *balanced contributions property*, if the following holds for every $(N, v, g) \in CS_N$ and for all pairs of players $i, j \in N$:

$$\gamma_i(N, v, g) - \gamma_i(N, v, g - g_j) = \gamma_j(N, v, g) - \gamma_j(N, v, g - g_i). \tag{3.30}$$

This condition is not so easy to interpret. It seeks to confine the reciprocal effects on the allocation under γ, when a player i removes all links to his neighbors. The influence such a removal has on the allocation of some other player j, not necessarily a neighbor of i, must be symmetric under (3.30) with respect to i and j. That is, the same damage player i can do to j by removing all his links, j can do to i by proceeding likewise. Note though that this is not to be seen as a sequential process, but rather as an ex-ante "threat". The property also implies that links need the consent of both players to be established or maintained. A detailed approach to network formation follows in Sect. 3.6 below.

Definition 3.91. An allocation rule $\gamma : CS_N \to \mathbb{R}^n$ satisfies the *superfluous player property*, if for every communication situations $(N, v, g) \in CS_N$ and every superfluous player $i \in N$ the following holds:

$$\gamma(N, v, g) = \gamma(N, v, g - g_i). \tag{3.31}$$

[11] To show that Definition 3.29 holds likewise for the addition of a link g_{ij}, denote $\widetilde{g} := g - g_{ij}$ and consequently $g := \widetilde{g} + g_{ij}$ and replace accordingly in (3.29).

If a player i is superfluous for (N,v,g), we should be able to remove all his links from the network g without affecting the overall allocation γ. Note that this property alone says nothing about what a superfluous player himself is allocated.

Definition 3.92. An allocation rule $\gamma : CS_N \to \mathbb{R}^n$ satisfies the *superfluous link property*, if for all communication situations $(N,v,g) \in CS_N$ and all its superfluous links $g_{ij} \in g$, the following holds:

$$\gamma(N,v,g) = \gamma(N,v,g - g_{ij}). \tag{3.32}$$

In this situation, if a link g_{ij} has no influence on the value of the grand coalition, it should also not have any influence on the allocation of the value $v^g(N)$ through γ to the players.

Definition 3.93. An allocation rule $\gamma : CS \to \mathbb{R}^n$ satisfies the *strongly superfluous link property*, if for every communication situations $(N,v,g) \in CS$ and every strongly superfluous link $g_{ij} \in g$ the following holds:

$$\gamma(N,v,g) = \gamma(N,v,g - g_{ij}). \tag{3.33}$$

This property is based on strongly superfluous links, which have no marginal value for any $S \subseteq N$, not only for the grand coalition N, as in the previous property. Their removal should also have no influence on the outcome of the game.

Even though both previous properties only state the removal of an individual link g_{ij}, we can apply the propositions recursively, until we have stripped the initial network g of all such links.

Definition 3.94. An allocation rule $\gamma : CS_N \to \mathbb{R}^n$ is *link anonymous*, if for all link anonymous communication situations $(N,v,g) \in CS_N$ a real-valued constant $\kappa \in \mathbb{R}$ exists, such that for all players $i \in N$ the following holds:

$$\gamma_i(N,v,g) = \kappa \cdot |N_i(g)| = \kappa \cdot \eta_i(g). \tag{3.34}$$

A direct extension of link anonymity in a communication situation, this property states that the allocation to each player depends solely on his degree in g, i.e. the cardinality of his neighborhood, but neither on his position within network g nor the network's overall structure.

Definition 3.95. An allocation rule $\gamma : CS_N \to \mathbb{R}^n$ is *weakly link symmetric*, if the following holds for all communication situations $(N,v,g) \in CS_N$ and for all links $g_{ij} \in g$:

$$\gamma_i(N,v,g+g_{ij}) > \gamma_i(N,v,g) \implies \gamma_j(N,v,g+g_{ij}) > \gamma_j(N,v,g+g_{ij}). \tag{3.35}$$

A notion of minimal fairness is assigned by this property to the allocation rule γ. Whenever a link g_{ij} is added to a network, both players i,j (or neither) must be strictly better off, but not only one can profit from the link. This rules out links from which one player receives nothing while the other benefits strictly. As links from which neither side benefits are allowed under weak link symmetry, this property could be interpreted as some kind of no-envy condition.

Definition 3.96. An allocation rule $\gamma : CS_N \to \mathbb{R}^n$ satisfies the *improvement property* for all communication situations $(N, v, g) \in CS_N$ and for all $g_{ij} \in g$, if whenever there exists some player $k \in N \setminus \{i, j\}$ with $\gamma_k(N, v, g + g_{ij}) > \gamma_k(N, v, g)$, then $\gamma_i(N, v, g + g_{ij}) > \gamma_i(N, v, g)$ or $\gamma_j(N, v, g + g_{ij}) > \gamma_j(N, v, g)$, or both, holds.

The improvement property limits the possible externalities of adding a link to g, or rather, it mandates some internal profit in case of an externality. Whenever some player who is not part of a link strictly benefits from its formation, at least one of the newly-linked players must, too.

Definition 3.97. An allocation rule $\gamma : CS_N \to \mathbb{R}^n$ is *link monotonic* if for all communication situations $(N, v, g) \in CS_N$ and for all links g_{ij} the following holds:

$$\gamma_i(N, v, g + g_{ij}) \geq \gamma_i(N, v, g). \tag{3.36}$$

This property is an alteration of weak link symmetry: There, under γ, the addition of a link g_{ij} to g can never make neither player i nor j worse off. Here, though, it is possible for one player to gain strictly, while the other receives nothing. This also shows that link monotonicity does not imply weak link symmetry.[12]

The next two properties highlight both sides of the same situation: First the allocation rule, given a certain communication situation, and second, the network, given a value function and allocation rule.

Definition 3.98. An allocation rule $\gamma : CS_N \to \mathbb{R}^n$ is *network flexible*, if

$$\gamma(N, v, g) = \gamma(N, v, g^c) \tag{3.37}$$

for all communication situations (N, v, g) where the network g is efficient with respect to v^g, such that $v(N, g) = v(N, g^c)$.

Whenever a network $g \subseteq g^c$ gives rise to the same value (for the grand coalition) as the complete network, the allocation rule will distribute this value in the same manner. It ignores links which have no marginal value. In situations where the network is one that is not efficient, the addition of links that might have zero

[12] See Lemma 1 in Dutta et al. (1998, p. 248) where it is shown that the properties of component efficiency, weak link symmetry, and the improvement property, if all satisfied by a communication situation with superadditive v^g, imply link monotonicity of γ. The proof is very short: If for some players $i, j \in N$ it holds that j is strictly worse off after forming a link with i under γ, by weak link symmetry it follows that i can be no better off through this link g_{ij}. Because v is superadditive and γ component efficient, there must exist some player $k \in N \setminus \{i, j\}$ who absorbs the loss experienced by j (and possibly i) from forming g_{ij}. But this is a direct violation of the improvement property. Note that link monotonicity by itself does not imply any of the three properties that together are a sufficient condition for it.

marginal value in an efficient network, can change the allocation, even if γ is network flexible. The next property is more related to a network structure g, but it is specific to the allocation rule γ, and so we feel, it should be listed here.

Definition 3.99. A network g is called *essentially complete*, if for a given $\gamma : CS_N \to \mathbb{R}^n$ the following is true:

$$\gamma(N, v, g) = \gamma(N, v, g^c). \tag{3.38}$$

The payoffs to all $i \in N$ players must be the same under γ, whether the communication situation is based on g or the complete network g^c. We also say that for γ the networks g and g^c are *payoff equivalent*.

With this we conclude our list of possible properties for allocation rules in communication situations. In the next section, we introduce the most widely used such allocation rule, known as the *Myerson value*.

3.5 The Myerson Value

We now turn to a specific allocation rule for communication situations, known as the *Myerson value*, after its creator Roger B. Myerson who introduced the rule in Myerson (1977). In its original form, it is an extension of the Shapley value to games played on networks. Whereas the Shapley value only assigns his average marginal gains to each player (see Shapley 1953b), the Myerson value also takes into account whether and how players are connected to each other in the underlying network. This can be achieved by merely adapting a general value function to the network, before applying the Shapley operator, as we will see below. The Myerson value is by far the best known allocation rule on network structures, even before the author had been awarded what is commonly referred to as the Nobel Price in Economics[13] in 2007 for his work on mechanism design. The elegance of the Myerson operator stems from its simplicity, especially the characterization based on only two properties, component efficiency and fairness of the allocation rule, as given in Definitions 3.87 and 3.89.

Before stating his main theorem, we introduce the functional form of the network restricted value function v^g used in Myerson (1977). From it can be derived directly how the network influences the possibilities of cooperation among the players.

Define

$$v^g(S) := \sum_{C \in S/g} v(C) \quad \text{for all} \quad S \subseteq N, \tag{3.39}$$

[13]Technically, it is the *The Sveriges Riksbank Prize in Economic Sciences in Memory of Alfred Nobel*, as the Bank of Sweden instituted it in 1969 and also funded it ever since.

where S/g is the partition induced on a coalition S by network g. Then, the value of a coalition S (possibly comprising unconnected components of a network g) is calculated by adding up the values of each subcoalition of S that is maximal with respect to the components induced by g on S. So whenever a coalition $S \subseteq N$ is not connected via g, it can only achieve a value equal to the sum of the values of its components $C \in S/g$. The benefits from cooperation are then strictly limited by the network.

We can now state and subsequently prove the following theorem, both taken from Myerson (1977):

Theorem 3.7. *Given a communication situation $(N, v, g) \in CS_N$, there is a unique allocation rule $\gamma : CS_N \to \mathbb{R}^n$ satisfying the properties of component efficiency and fairness. It is defined as*

$$\gamma_i(N, v, g) := \phi_i(N, v^g) \text{ for all } i \in N, \tag{3.40}$$

where $\phi_i(N, v^g)$ is the Shapley operator payoff for player i based on the network restricted value function v^g.

Proof. We begin by showing that the fair and component efficient allocation rule γ is unique, before turning to the properties themselves.

Suppose not, and there exist two such allocation rules γ^1 and γ^2, distinct from one another. For a given communication situation (N, v, g), assume network g to be minimally connected, so that within each component $C \in N/g$ there exists a unique path in between each pair of players $i, j \in C$ with $i \neq j$. The removal of any link would turn one component into two new ones.[14] Because g is minimally connected, the removal of any link g_{ij} will lead to $\gamma^1(g - g_{ij}) = \gamma^2(g - g_{ij})$.[15] Under fairness, for both allocation rules it must hold that

$$\gamma_i(g) - \gamma_i(g - g_{ij}) = \gamma_j(g) - \gamma_j(g - g_{ij}). \tag{3.41}$$

Rearranging the above term to

$$\gamma_i(g) - \gamma_j(g) = \gamma_i(g - g_{ij}) - \gamma_j(g - g_{ij}), \tag{3.42}$$

considering $\gamma^1(g - g_{ij}) = \gamma^2(g - g_{ij})$, the right hand side of (3.42) is identical for both, γ^1 and γ^2. We can therefore set the left hand sides equal, to obtain

$$\gamma_i^1(g) - \gamma_i^2(g) = \gamma_j^1(g) - \gamma_j^2(g). \tag{3.43}$$

[14] Note that within all components $C \in S/g$ for all $S \subseteq N$ the paths in between every pair of distinct players must be unique! Also, players connected in S/g will be so in N/g.

[15] This follows from induction: Removing link after link in a minimally connected network will necessarily lead to all players being allocated their stand-alone values. Adding these links again, according to fairness, marginal values can only be allocated in a unique manner at each step.

Interestingly, (3.43) holds also for unlinked pairs of players i and j as long as they belong to the same component $C \in N/g$. This follows from the fact that (3.43) can be established for any pair of linked players and since it refers to the minimally connected network g with all links intact (and not to network $g - g_{ij}$), all pairwise comparisons are equal and the differences must be the same for all players in the same component.[16] Define $\Delta_C(g) := \gamma_i^1 - \gamma_i^2$ for all $i \in C$, where $|C|\Delta_C(g)$ is the aggregate difference of the allocation rules in component C. By component efficiency, $\sum_{i \in C} \gamma_i^1 = v^g(C) = \sum_{i \in C} \gamma_i^2$ must be fulfilled and hence,

$$0 = \sum_{i \in C} (\gamma_i^1 - \gamma_i^2) = |C| \cdot (\gamma_i^1 - \gamma_i^2) = |C| \cdot \Delta_C(g). \tag{3.44}$$

With $|C| > 0$ we conclude that $\Delta_C(g) = 0$ and no two distinct allocation rules can exist that satisfy the properties of fairness and component efficiency.

It suffices to show this on a minimally connected network, since according to fairness, adding links to any (and hence also to a) minimally connected network must benefit the newly-linked players equally. Players not directly involved in this new link can only be affected if they are in the same component C (before or after). They all benefit from the new link through v^g equally, because the values of $v^g(S)$ for all $S \subseteq C$ are at least as high as before the link was formed.

Next, we show that component efficiency is fulfilled by γ. Define for a given network g, for each component $C \in N/g$ the following *sub*game[17]:

$$u^C(T) = \sum_{R \in (T \cap C)/g} v(R) \quad \text{for all} \quad T \subseteq N. \tag{3.45}$$

Expression (3.45) is a characteristic function whose domain is the entire player set, but in terms of value only connected subsets of the component C are considered. Note that because N/g is the unique partition into components induced by network g, all players in any subset of such a component are again connected to each other. So,

$$(S \cap C)/g = S \cap C \quad \text{for all} \quad C \in N/g, \; S \subseteq C. \tag{3.46}$$

Conversely, all players connected in T are also connected in N, with

$$T/g = \left(\bigcup_{C \in N/g} (T \cap C) \right) \Big/ g, \tag{3.47}$$

[16] This is akin to saying that when adding a link to a network, all pairs of players (and hence all players) must feel the allocative difference induced by the distinct allocation rules in the same manner. Given component efficiency, this difference can only be 0 and hence the distinct allocation rules are identical after all!

[17] The expression subgame here is not to be understood in the game theoretic way as defined previously.

because $\bigcup_{C \in N/g}(T \cap C) = T$. Adding over all components, we can define the characteristic function v^g on g as follows:

$$v^g(T) = \sum_{C \in N/g} u^C(T) = \sum_{C \in N/g} \sum_{R \in (T \cap C)/g} v(R) \text{ for all } T \subseteq N. \tag{3.48}$$

Since each subgame function u^C is limited to its component C and no players $i \notin C$ are assigned any value under it, the component C is a *carrier* for u^C. This follows from the construction of u^C according to which $u^C(T \cap C) = u^C(T)$ for all $T \subseteq N$. The efficiency of the Shapley operator admits the following expression,

$$\sum_{i \in C} \phi_i \left(u^{\hat{C}}\right) = \begin{cases} u^{\hat{C}}(N) & \text{if } C = \hat{C} \\ 0 & \text{if } C \cap \hat{C} = \emptyset \end{cases}, \tag{3.49}$$

where both C and $\hat{C}$ are elements of N/g and so disjoint sets of players.

The linearity of the Shapley value, deduced from additivity in Shapley (1953b), shows for all $T \subseteq N$ and $C, \hat{C} \in N/g$ that the following applies, as we explain right below:

$$\begin{aligned} \sum_{i \in \hat{C}} \phi_i(v^g) &= \sum_{i \in \hat{C}} \phi_i \left(\sum_{C \in N/g} u^C \right) = \sum_{C \in N/g} \sum_{i \in \hat{C}} \phi_i \left(u^C(T)\right) \\ &= u^{\hat{C}}(N) = \sum_{R \in \hat{C}/g} v(R) = v(\hat{C}). \end{aligned} \tag{3.50}$$

In the first line v^g is replaced according to (3.48) and the summation rearranged. The first term in the second line follows from the efficiency of the Shapley value and the fact that in the summation over all components $C \in N/g$, the only nonzero element is the component $C = \hat{C}$. Also note that $N \cap \hat{C} = \hat{C}$ for the last summation. This establishes component efficiency of the allocation rule $\gamma(N, v, g)$.

The third part of the proof addresses the fairness property of the allocation rule. For a given network $g \subseteq g^c$, select any link $g_{ij} \in g$. Define the value function w as follows:

$$w(T) = v^g(T) - v^{g - g_{ij}}(T) \text{ for all } T \subseteq N. \tag{3.51}$$

This leaves us with two cases as to which the partition T/g will be refined after removing g_{ij} from g. For $\{i, j\} \nsubseteq T$, when either one or both players are not members of T, the partition is unaltered, $T/g = T/(g - g_{ij})$, and so

$$w(T) = \sum_{R \in T/g} v(R) - \sum_{R \in T/(g - g_{ij})} v(R) = 0. \tag{3.52}$$

Only when $\{i, j\} \subseteq T$ is $w(T) \geq 0$, since the partition of T can possibly be refined such that $|T/(g - g_{ij})| \geq |T/g|$. This last inequality is strict, whenever the removal of the link g_{ij} splits up an element of the partition of T, e.g. $R \in T/g$, refining it into two elements $R_i, R_j \in T/(g - g_{ij})$.

Additivity of the Shapley operator allows us to add the values of two independent games, here v^g and $v^{g-g_{ij}}$, player by player, when combined:

$$\phi_i(w) = \phi_i(v^g) - \phi_i(v^{g-g_{ij}}) \text{ for all } i \in N. \tag{3.53}$$

When connecting the disjoint sets R_i and R_j with the link g_{ij}, both players i and j are equally crucial to additional gains that could be realized. Because w is defined to capture only these gains and due to the symmetry of the Shapley value, we have

$$\phi_i(w) = \phi_j(w). \tag{3.54}$$

Therefore, the Myerson value is also fair, and the proof of Theorem 3.7 is complete. ■

This theorem is quite striking: The Myerson operator is characterized through only two axioms, component efficiency and fairness, which are by no means unreasonable or restrictive, and lead to a unique allocation rule. Furthermore, if we assume unconfined cooperation with $g = g^c$, the Myerson operator coincides with the Shapley operator. It is therefore also possible to characterize the latter allocation rule with only two axioms, by introducing – even if redundantly, given no restrictions when $g = g^c$ – the concept of networks and a communication situation.

In settings were connectedness suffices to reap all gains from cooperation within a coalition, the Myerson operator is certainly an interesting candidate, not least because it also satisfies the desirable properties of symmetry and carrier through its inherent relation to the Shapley value.

Despite all this, some points of criticism have been raised concerning the Myerson value, of which we consider two: The first is directly related to the value, the second rather to its property of fairness. We will illustrate them in the two subsequent examples, which are adapted from Jackson (2005b, pp. 46 ff.). There, they appear in the context of value functions on networks and not in terms of network restricted characteristic functions for coalitional games. The appropriate adjustments for them to fit seamlessly into our framework have been made here.

Example 3.10. On the player set $N = \{1,2,3\}$ as domain, consider the characteristic functions u^g and v^g. For both functions, their values for the empty set and singleton coalitions are zero, regardless of the underlying network structure. For proper coalitions, value can only be created on components, where, $u^g(S) = 1$ for all S with $|S| > 1$. Function v^g has more restrictions concerning the network. Because the original example (see Jackson 2005b, pp. 46 ff.) uses a link-based characteristic function, both a functional form and a minute description in words are more tedious to present than necessary. In detail, we have reproduced all 64 combinations for each value function in the Tables 3.1 and 3.2 below, and refer the reader to the Table 3.2 specifically, for the values of v^g.

Calculating the Myerson value for both characteristic functions u^g and v^g, given a network structure $g = \{g_{12}, g_{23}\}$, will yield the identical allocations

$$\gamma(N,u,g) = \gamma(N,v,g) = \left(\frac{1}{6}, \frac{2}{3}, \frac{1}{6}\right).$$

Table 3.1 Possible values of u^g for all coalition-network combinations

$S \setminus g$	$g^\emptyset$	$\{g_{12}\}$	$\{g_{13}\}$	$\{g_{23}\}$	$\{g_{12},g_{23}\}$	$\{g_{12},g_{13}\}$	$\{g_{13},g_{23}\}$	g^c
$\emptyset$	0	0	0	0	0	0	0	0
$\{1\}$	0	0	0	0	0	0	0	0
$\{2\}$	0	0	0	0	0	0	0	0
$\{3\}$	0	0	0	0	0	0	0	0
$\{1,2\}$	0	1	0	0	1	1	0	1
$\{1,3\}$	0	0	1	0	0	1	1	1
$\{2,3\}$	0	0	0	1	1	0	1	1
$\{1,2,3\}$	0	1	1	1	1	1	1	1

Table 3.2 Possible values of v^g for all coalition-network combinations

$S \setminus g$	$g^\emptyset$	$\{g_{12}\}$	$\{g_{13}\}$	$\{g_{23}\}$	$\{g_{12},g_{23}\}$	$\{g_{12},g_{13}\}$	$\{g_{13},g_{23}\}$	g^c
$\emptyset$	0	0	0	0	0	0	0	0
$\{1\}$	0	0	0	0	0	0	0	0
$\{2\}$	0	0	0	0	0	0	0	0
$\{3\}$	0	0	0	0	0	0	0	0
$\{1,2\}$	0	1	0	0	1	1	0	1
$\{1,3\}$	0	0	0	0	0	0	0	0
$\{2,3\}$	0	0	0	1	1	0	1	1
$\{1,2,3\}$	0	0	0	0	1	0	0	1

This is surprising, because player 2 enjoys much less pivotal status under characteristic function u^g. Given the network $g = \{g_{12}, g_{23}\}$, he is the "connecting" player in both games, but only in the communication situation v^g is this also expressed in the value function. In u^g, any player could serve his purpose, when the network would be altered accordingly.

Example 3.10 provides a case where for two communication situations (N,u,g) and (N,v,g), which differ (only) in their value function, the Myerson value is identical. The surprising payoff is the one allocated to player 2: Under u^g, any player could be in the central position of the network and contribute to the value of u^g likewise. Under v^g on the contrary, it is only player 2 who can take up this central role, and the complete network does not lead to any gains. The reason for this apparent mis-allocation is that the Myerson value only accounts for the actual network, here $g = \{g_{12}, g_{23}\}$, and the marginal values of the players within this network. Possible marginal values in other network structures are not considered. We believe this makes sense, as the Myerson value takes the network as given and does not incorporate all conceivable alterations which might have an effect on the marginal values of some players, even if there is no additional overall value.

Our next point concerns the property of fairness, which the Myerson value satisfies. Definition 3.89 states that the two players i and j at the end of a link g_{ij}, which is added to or removed from g, must be affected equally through this change in terms of their payoffs under γ. Note that fairness does not imply that they split the marginal value of the link g_{ij} in its entirety. Other players might benefit from it as well, but i and j must share their part equally. The following example shows in which situations this property might seem questionable.

Example 3.11. Consider a communication situation (N,v,g), where $N = \{1,2,3\}$ and the characteristic function v^g assumes the following values:

$$v(S,g) = \begin{cases} 1 & \text{if } 2, j \in S,\ j \in \{1,3\} \text{ and } g = \{g_{2j}\} \\ 0 & otherwise. \end{cases} \quad \text{for all } S \subseteq N. \qquad (3.55)$$

We deliberately picked player 2 to be the pivotal one. Only the coalitions containing him and some other player have nonzero values, and additionally, the underlying network g must consist of one single link between player 2 and some other player $j \in S \setminus \{2\}$. All other combinations of coalitions and networks are assigned a value of zero under v^g.[18] The Myerson value will allocate equal values to players 2 and j in such a constellation:

$$\gamma_2(N,v,\{g_{2j}\}) = \gamma_j(N,v,\{g_{2j}\}).$$

In Example 3.11 we are faced with a situation where one player is necessary for any nonzero gains and any other player cooperating with him is sufficient in order to realize those gains. Nevertheless, the marginal value is identical for the pivotal player and whomever he chooses to cooperate with, and so, under the Myerson value, their allocations are identical. The fact that the marginal value by itself is not necessarily the most appropriate measure to allocate the gains from cooperation has been pointed out before, see for example Maschler (1982). It is one of the motivations to introduce weight systems on the players to account for differences not captured by their marginal values. Such considerations support the use of weighted values such as the Weighted Shapley operator.

Despite these criticisms, we deem the Myerson value a very useful allocation rule which can be applied in a wide variety of games that incorporate a network structure among the players. Both axioms, fairness and component efficiency, seem reasonable to us, even if they might lead to the occasional outlier allocation, which intuitionally seems odd, but can be reconstructed perfectly using the axioms.

This concludes our treatment of allocation rules on communication situations, more precisely, of the Myerson operator. In the next section we examine different processes of network formation, as we so far have always taken the network structures as given, or at least spent little thought on how networks might come about.

[18] Especially the restriction on the network might seem curious, but has no qualitative influence on the allocation and serves merely the purpose to make calculations less tedious.

3.6 Network Formation

3.6.1 *Overview*

In our previous considerations of networks, especially communication situations, no mention was made regarding the formation of the network structure. So far, we took networks as given, analyzing only the effects of adding or removing links on the value function and the allocation rule. Nevertheless, the process of formation itself deserves detailed examination, as its result, the network, is the basis upon which value is created and then allocated.

The networks we are concerned with emphasize the individual nodes or players, as it is up to them to form or maintain the links between them.[19] We now take up the process of network formation, which is usually modelled along the lines of noncooperative games, where the players represent the nodes and decide whether to form a link with one another or not. As such models are always very situation-specific, we try to stay as general as possible. More details specific to our context will follow in the application, in Chap. 6. The literature on network formation is ample and relatively young: Hart and Kurz (1983) introduce strategic models of coalition formation, based on reciprocal expectations. Even though they do not directly incorporate networks, this can – albeit with very crude ones – quickly be augmented, as we will see. Aumann and Myerson (1988) describe a link formation game in extensive form, where pairs of players decide according to a predetermined sequence whether they want to be linked or not. Mutuswami and Winter (2002) employ a mechanism design framework, where players, again in turn, voice their preferred network and corresponding willingness to pay, upon which a social planner picks an efficient and balanced solution. With respect to our more cooperative approach, we direct the reader to Dutta et al. (1998) or Slikker and van den Nouweland (2001a). Other sources of interest to us are the classical models by Jackson and Wolinsky (1996) and Dutta et al. (2005). For general reference, Demange and Wooders (2005), as well as Dutta and Jackson (2003), and no less Goyal (2007) provide an excellent overview.

Our main goal is to find a formation process that terminates in a network which is maximizing the aggregate surplus achievable (through cooperation), while sharing

[19]By this we want to distinguish from the *minimum spanning tree problem*, which has been discussed extensively in the operations research literature.(See for example Graham and Hell 1985.) There, a source is supposed to supply an input to a set of agents via a physical network between them. The first goal is to find a cost-minimal network, all while supplying the desired quantity of the input to each agent. Secondly, one wishes to provide some cost-allocation method from which no group of agents wants to deviate and built their own network to the source at a lower cost. In this literature the network is more of a costly good to be provided by a central instance than a collection of bilateral relationships, which can be decided upon by the respective pairs of agents themselves.

this surplus so that no group of players has an incentive to deviate from the network, because it makes them better of. In short, we look for an efficient, yet stable network as result of the formation process.

The most fundamental distinction in network formation is whether links can be formed individually or only with the consent of both players. We differentiate here between *one-sided* and *two-sided* links, or, *unilateral* and *bilateral* link formation.[20] In the former environment, links can be established by the action of only one player, regardless of whether the to-be-linked player agrees. In many economic situations this is quite unrealistic, and usually the strongest argument for its use is modelling simplicity. Nevertheless we will see in our subsequent application that some changes will make this concept plausible under certain circumstances. Mainly though, we focus on the bilateral formation of links, where the consent of both players is necessary for a link to be established between them.[21] How exactly this link formation occurs, i.e. in what order (if any) players have to act, how many chances they get to do so, how well they are informed about what the others are doing and so on, is specific to the model employed. To this end, there also exists a range of stability or equilibrium conditions, which specify when a network formation process has come upon a "stable" network in which no individual, pair, or set of players (depending on the concept) wants to deviate from the status quo.

The next important distinction in the network formation process specifies in which order the links are formed, and at the same time how much information is available to the players about the rules of the game as reflected in the current state of the network. Here we distinguish between two approaches, the extensive, and the strategic form. In the former, links are established in sequential order, in the latter simultaneously for all players. We begin with the extensive form game of network formation introduced in Aumann and Myerson (1988). This is followed by considerations on network formation in strategic form, and finally by notions of stability of networks.

3.6.2 Network Formation in Extensive Form

Analyzing the process of network formation through an extensive form game implies that there exists a certain order in which players take their turns choosing the actions to select at a given move.[22] We present the model on bilateral link formation introduced in Aumann and Myerson (1988), where players move simultaneously

[20]This shall not be confused with *directed* links, which are employed in the context of directed networks.

[21]Such bilateral formation of links implies the possibility of unilateral severance.

[22]One could also model network formation as an extensive form one-shot game, but the possible advantages from a more detailed exposition this representation allows for (as compared to games in strategic form) would be lost.

within pairs to decide on the link between them, but the pairs themselves are chosen according to some exogenously given order.

Underlying this model is a cooperative game (N, v), according to which value is created from cooperation. What players are allocated individually depends on the allocation rule and the network to be formed. Regarding the latter, Aumann and Myerson (1988) utilize an *auxiliary linking game*, described in the following: Initially, the players are facing an empty network $g^0 := g^\emptyset = \{\emptyset\}$. A *rule of order*, which is common knowledge, specifies the sequence in which pairs of players are asked whether they want to form a link between them. If all $\binom{n}{2}$ pairs have been asked in round one, some network g^1 can be established. On the basis of g^1, the updated rule of order will go over the remaining unlinked pairs and give them another chance to form a link. If no pair of players wants to form a link, the linking game terminates. Else, it continues in this fashion until in one round no pair wants to form another link. In total, the linking game can have at most $\binom{n}{2}$ rounds, when there is only one link formed per round. Note that it is not possible for pairs of players to sever links between them at any stage of the game.

Formally, we can define the rule of order as a series of composite mappings $h_t \circ r_t$ with $t = 0, 1, \ldots, m \leq \binom{n}{2}$, where the bijection r_t, the actual rule of order, is given by

$$r_t : \left\{1, 2, \ldots, \binom{n}{2} - |g^t|\right\} \to g^c \setminus g^t. \tag{3.56}$$

Its domain varies with t and comprises at most $\binom{n}{2}$ elements, corresponding to all possible links between n players in a complete network g^c.

The binary function h_t then assigns to each pair of players the value 1 or 0,

$$h_t : g^c \setminus g^t \to \{0, 1\}, \tag{3.57}$$

depending on the players' willingness to form a link.

The network g^t is the union of all links having been formed before round t, i.e. the pairs of players that need not be considered by r_t:

$$g^t := \bigcup_{l=1}^{t} \tilde{g}^l. \tag{3.58}$$

The set of additional links resulting from a given round t is denoted $\tilde{g}^{t+1}$ and defined by

$$\tilde{g}^{t+1} := \left\{g_{ij} \in g^c \setminus g^t | h_t(g_{ij}) = 1\right\}. \tag{3.59}$$

Even though technically speaking they are, we refrain from calling the links contained in, say, $\tilde{g}^{t+1}$ a "network", because by themselves they have very little meaning in this context. We refer to it rather as a set of links.

This somewhat formal construct is actually quite easy to interpret: In the initial round, where $t = 0$, the function r_0 maps from the set of all possible pairs among

n players, $N^{\binom{n}{2}} := \{1,2,\ldots,\binom{n}{2}\}$, to the set of all possible links on n players, the complete network g^c. The order in which the pairs are asked whether or not to form a link depend on the labelling within $N^{\binom{n}{2}}$. The question regarding the actual link formation is answered by the function h_0, resulting in a value of 1 if the pair wants to form the link and 0 otherwise. The resulting network is then g^1, which includes all the links established in the first round.[23]

In the next round, round 2, the domain and the codomain of r_0 are curtailed appropriately. Because pairs of players are never asked whether they want to sever an existing link based on the updated network, r_1 now only covers the remaining, unlinked, pairs of players. The same holds for h_1, which is now only defined over those links that have not already been formed in earlier rounds.

The process is repeated and terminates as soon as all possible $\binom{n}{2}$ links have been formed and both domain and codomain of r_t are empty. The result is the complete network g^c. The process also ends, if in some round t the function h_t assumes the value zero over its entire domain, i.e. no further pairs of players want to form a link. The resulting network then is given by $g^{t+1} = g^t$. As we can see, it is sufficient to specify the function r_0, because all further $r_t, t = 1,2,\ldots,m$ keep the same principal order on the pairs, merely omitting those who already formed links.

The choices made by the pairs of players in the auxiliary linking game depend on the underlying coalitional game (N,v) and the prevailing allocation rule γ. Under the assumption of perfect information the players know their position in the game tree, and hence also the complete previous history of links formed or refused, as well as all possible payoffs. Therefore a subgame perfect Nash equilibrium in pure strategies must exist.[24] Players do not only consider the immediate effect of forming a link, but also the remaining course of the game, and with it the implications on other players' decisions. All information sets in the game tree are singleton sets, therefore each strategy resulting in an equilibrium yields a unique network g. Nevertheless, a given network g might be the result of multiple equilibrium strategies, i.e. different terminal nodes in the game tree[25] might result in the same outcome for the players in terms of network and payoffs. Aumann and Myerson (1988) call the networks resulting from equilibrium strategies *natural outcomes*. They also introduce an early notion of stability, calling a network g *stable*, if the auxiliary linking game played on it will not lead to the formation of additional links in any of the resulting subgame perfect Nash equilibria. But, as their linking game lacks the possibility of severing links, this notion must not be confused with the notions of stability which we introduce later in Sect. 3.6.4. We now illustrate the example from Aumann and Myerson (1988).

Example 3.12 (Aumann and Myerson (1988)). This simple yet instructive example on sequential network formation is based on the following specifications. The

---

[23]In the first step $g^1 = \tilde{g}^1$ holds.

[24]See Definition 2.11, p. 25, and Selten (1965).

[25]Not to be confused with the nodes of the network g!

Table 3.3 Payoff **x** resulting from network g

	g	$x = (x_1, x_2, x_3)$
1	$\{\emptyset\}$	$(0,0,0)$
2	$\{g_{12}\}$	$(30,30,0)$
3	$\{g_{23}\}$	$(0,30,30)$
4	$\{g_{13}\}$	$(30,0,30)$
5	$\{g_{12}, g_{13}\}$	$(44,14,14)$
6	$\{g_{12}, g_{23}\}$	$(14,44,14)$
7	$\{g_{13}, g_{23}\}$	$(14,14,44)$
8	$\{g_{12}, g_{23}, g_{13}\}$	$(24,24,24)$

underlying coalitional game (N, v) is defined for three players $N = \{1,2,3\}$. Its characteristic function is symmetric and assumes the values $v(S) = 0$ whenever $|S| \leq 1$, $v(S) = 60$ for $|S| = 2$, and $v(N) = 72$. The allocation rule γ applied is the Myerson value (see Sect. 3.5), which can lead to eight possible allocations, depending on the network g, as given in Table 3.3.

With an initial rule of order, where

$$r_0(1) = g_{12}, \quad r_0(2) = g_{23}, \quad \text{and } r_0(3) = g_{13},$$

we can construct the corresponding game tree, illustrated in Fig. 3.7, p. 146. The circled numbers indicate the pair of players having to move at a given node. The possible actions are always **a** for agreeing to form a link and **r** for rejecting to do so. Left-leaning edges always correspond to **a**, right-leaning ones to **r**, so only the first two edges are labelled exemplarily. To keep Fig. 3.7 from being overly crowded, we list a number from 1 to 8 instead of payoff vectors at each terminal node. These numbers correspond to the payoffs in the table above.

Through the backwards induction procedure, we find three subgame perfect Nash equilibrium outcomes, corresponding to the payoff vectors 2, 3, and 4, and a network g with one link each. In Fig. 3.7 their terminal nodes are highlighted through grey squares. These outcomes can be reached in equilibrium for any initial rule of order, as the reader is encouraged to verify.

From Example 3.12 it can be deduced that the trees for such games literally explode in extensiveness with an increasing number of players, even though there are only two actions to choose from at any move. In terms of general results for the auxiliary linking game, the literature does not provide much. Aumann and Myerson (1988, pp. 181 f.) state a theorem for the special class of weighted majority games, and even there they have to invoke many restrictions. Other attempts, for example van den Nouweland (1993), focus on the class of convex games. There, the existence of subgame perfect Nash equilibria supporting the complete network (or one with equivalent payoffs) would seem intuitive. Yet, no such results are found. Also, to reduce complexity, symmetry among players is often assumed. For results on such games with up to five players, see Slikker and Norde (2000).

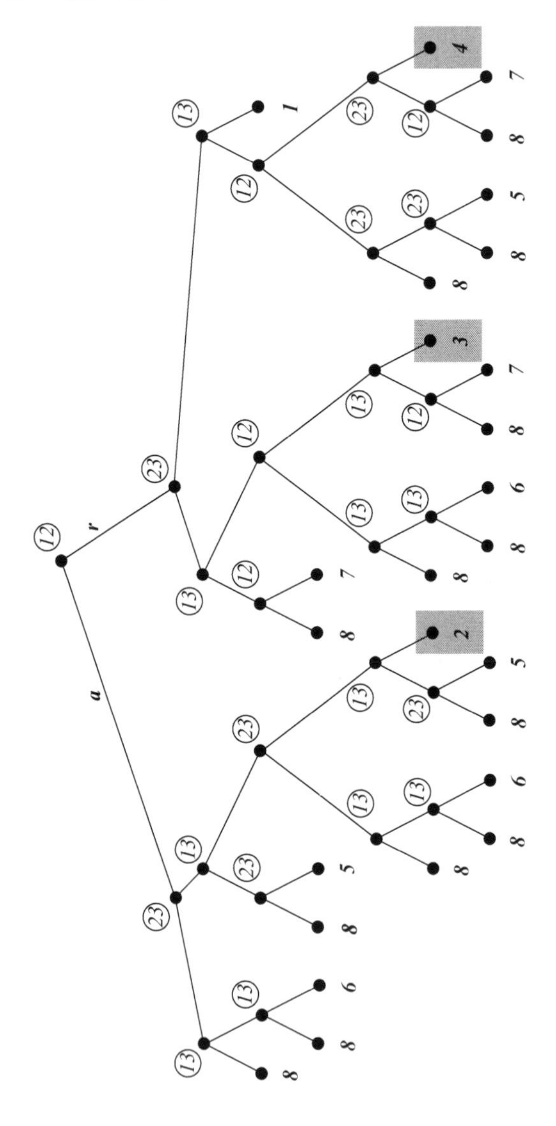

Fig. 3.7 Game tree with initial rule of order 12, 23, 13

Of course, the sequential model of network formation might not be apt for all economic situations. For one, perfect information, i.e. the publicity of link formation and rejection, might be too much to demand from an application. Also, it could be seen as questionable whether the network relations really are transitive and whether the full value of a coalition is attainable merely through its being a component. This, however, is assumed by the value function underlying the Myerson value. We will come back to sequential network formation in more detail in Chap. 6. There we address the criticisms mentioned above and make the necessary changes for our application.

3.6.3 Network Formation in Strategic Form

As opposed to the extensive form process, the strategic formation of a network is a simultaneous concept, in which all players choose their strategy at the same time, or at least have no information (when they move) about what the others might have chosen or will choose.

Based on a underlying coalitional game (N,v) and an allocation rule γ, players $i \in N$ simultaneously choose with whom they intend to form links, in order to realize gains above their stand-alone value $v(\{i\})$. Links only come into existence if two players both intend to form the link with one another, which puts us in a scenario of bilateral link formation. In the model of strategic link formation that we introduce, all players decide simultaneously with whom they would like to form a link in order to maximize their payoffs. In this step (before the allocation of payoffs according to γ) it is assumed that players cannot enter any binding agreements. Our treatment is based on Qin (1996), Dutta et al. (1998), Slikker and van den Nouweland (2001b, pp. 174 ff.), and van den Nouweland (2005), the first of which introduced the model into the literature.

Let us list the ingredients of the network formation game one by one. As mentioned above, the network formation takes place in light of gains from a coalitional game (N,v), and their allocation to the players through the allocation rule γ. Each player has exactly 2^{n-1} pure strategies $s_i \in \mathrm{S}_i = 2^{N\setminus\{i\}}$ available, where

$$s_i \subseteq N \setminus \{i\}$$

is a subset of the player set. Player i intends to form a link with some player j under strategy s_i, if and only if $j \in s_i$. Then, player j's intention regarding i are the same, if only if $i \in s_j$, upon which the link $g_{ij} = g_{ji} = 1$ is formed. If either $i \notin s_j$ or $i \notin s_i$ or both, then $g_{ij} = g_{ji} = 0$. Accordingly, a strategy combination s induces a unique network g:

$$g(s) = \left\{ g_{ij} \middle| \; i \in s_j \text{ and } j \in s_i, \text{ with } i,j \in N \right\}. \tag{3.60}$$

The network formation game in strategic form, denoted $\Gamma(N,v,\gamma)$, is given by the triplet

$$(N,\mathsf{S},f),$$

where the i-th of n components of f, $f_i : CS_N \times \mathsf{S} \to \mathbb{R}$, is a composite function based on the chosen allocation rule γ,

$$f_i(s) := \gamma_i(N,v,g(s)), \quad \text{for all } s \in \mathsf{S} \text{ and } i \in N. \tag{3.61}$$

In terms of equilibria, network formation games in strategic form can actually produce meaningful results. We begin the analysis with the standard Nash equilibrium, mainly to show why it is not sufficiently refined in our context. Then we turn to the undominated Nash equilibrium, as well as the Coalition-Proof Nash equilibrium. The approach we follow is not completely general with respect to the equilibrium network structure. Rather, it is supposed to show that complete networks or such that are essentially complete, can be supported by the equilibrium notions under consideration. This is motivated by the fact that in (superadditive) coalitional games the highest value is always assigned to the grand coalition N of all players. We make assumptions on the underlying communication situation and allocation rule, but these are rather standard and need little motivation. One that is not mentioned explicitly below is the superadditivity of the network restricted value function v^g.

Proposition 3.17. *If γ is an allocation rule satisfying component efficiency, weak link symmetry, and the improvement property, any network g can be supported in a Nash equilibrium in the game $\Gamma(N,v,\gamma)$.*

Proof. Fix, just for reference, a network g on the player set N. Assign to each player $i \in N$ the strategy

$$s_i^* = \{j \mid j \in N_i(g)\},$$

expressing i's intention to form links with all players j in his neighborhood $N_i(g)$. Strategy s^* constitutes a Nash equilibrium for the link formation game $\Gamma(N,v,\gamma)$, because no player can realize a gain by unilateral deviation. With the game being link monotonic,[26] no player $i \in N$ can gain by deviating to a strategy $s_i \subsetneq s_i^*$. Likewise, a strategy $s_i' \supsetneq s_i^*$ will not result in any additional links (and hence gains), as under s^* no strategy s_j^* for $j \notin N_i(g)$ exists, where $i \in s_j^*$. The equilibrium network resulting from s^* is equal to the reference network, $g(s^*) = g$. Individual deviations with less intended links than under s^* result in a smaller network, but an increased number of intended links of a single individual has no effects on $g(s^*)$. ■

With all networks $g \subseteq g^c$ supportable through an appropriate Nash equilibrium strategy s^*, this solution concept is not too meaningful when trying to find specific

[26]We have shown in Footnote 12, p. 133, that an allocation rule γ as specified in Proposition 3.17 is also link monotonic.

network architectures that emerge in equilibrium. We thus turn to equilibrium refinements, beginning with the undominated Nash equilibrium, as introduced in Definition 2.6, p. 17. Denote by $\overline{s}$ the strategy for which $\overline{s}_i = N \setminus \{i\}$ for all $i \in N$ and $g(\overline{s}) = g^c$ as the resulting network structure.

Proposition 3.18. *If γ is an allocation rule satisfying component efficiency, weak link symmetry, and the improvement property, the strategy $\overline{s}$ constitutes an undominated Nash equilibrium in the game $\Gamma(N, v, \gamma)$.*

Proof. Pick some player $i \in N$ and a strategy $(s_i, s_{-i}) = s \in \mathrm{S}$. Consider the networks $g := g(s_i, s_{-i})$ and $g' := g(\overline{s}_i, s_{-i})$. Because $\overline{s}_i = N \setminus \{i\}$, s_i must be a subset of it, i.e. $s_i \subseteq \overline{s}_i$. Consequently, the resulting network g cannot have more links than g' and $g \subseteq g'$. Since only i's strategy differs, all links $g_{ij} \in g' \setminus g$ are such that $j \in N_i(g')$.

Because γ is implied to be link monotonic, the addition of one or more links with i at one end will never harm i in terms of payoffs:

$$f_i(\overline{s}_i, s_{-i}) = \gamma_i(N, v, g') \geq \gamma_i(N, v, g) = f_i(s_i, s_{-i}) \quad \text{for all} \quad i \in N.$$

Hence, for no player $i \in N$ a strategy s_i exists that dominates $\overline{s}_i$, and, for $s_{-i} = \overline{s}_{-i}$ the above equation shows that $\overline{s}$ is a Nash equilibrium.[27] ■

Before we turn to the next result, we state the following lemma, which is a consequence of the fact that any allocation rule is link monotonic, if it satisfies the properties of weak link symmetry, component efficiency, and the improvement property.[28]

Lemma 3.1. *If an allocation rule $\gamma : CS_N \to \mathbb{R}^n$ satisfies component efficiency, weak link symmetry and the improvement property, then, for all networks $g \subseteq g^c$ and all pairs of players $i, j \in N$, it must be true that if there exists some player $k \in N \setminus \{i, j\}$ with $\gamma_k(N, v, g + g_{ij}) \neq \gamma_k(N, v, g)$, then $\gamma_i(N, v, g + g_{ij}) > \gamma_i(N, v, g)$ and $\gamma_j(N, v, g + g_{ij}) > \gamma_j(N, v, g)$.*

Proof. If $\gamma_k(N, v, g + g_{ij}) > \gamma_k(N, v, g)$, according to the improvement property, i or j must be better off as well, and, because of weak link symmetry of γ, they must be both.

If, on the other hand, $\gamma_k(N, v, g + g_{ij}) < \gamma_k(N, v, g)$, weak link symmetry leaves us with two cases: $\gamma_i(N, v, g + g_{ij}) > \gamma_i(N, v, g)$ and $\gamma_j(N, v, g + g_{ij}) > \gamma_j(N, v, g)$, or $\gamma_i(N, v, g + g_{ij}) \leq \gamma_i(N, v, g)$ and $\gamma_j(N, v, g + g_{ij}) \leq \gamma_j(N, v, g)$. The first gives us the desired result, so we have to show that the second leads to a contradiction; this is indeed the case, because component efficiency of γ and superadditivity of v^g would imply a player $l \in N \setminus \{i, j\}$ who would have to be better off through the formation of link g_{ij}, i.e. $\gamma_l(N, v, g + g_{ij}) > \gamma_l(N, v, g)$. But this violates the improvement property. ■

---

[27] In fact, $\overline{s}_i$ even dominates s_i.

[28] Again, see Footnote 12 on p. 133.

We can now show that all strategies that qualify for undominated Nash equilibria lead to networks that yield payoffs identical to the complete network:

Proposition 3.19. *For any strategy s leading to an undominated Nash equilibrium in the game $\Gamma(N,v,\gamma)$, the resulting network $g(s)$ is essentially complete for each allocation rule γ satisfying component efficiency, weak link symmetry and the improvement property.*

Proof. Pick an equilibrium strategy $s \neq \bar{s}$, i.e. a strategy that is distinct from the one yielding the complete network. Let us collect all players deviating from $\bar{s}$ under strategy s in the following set, which is constructed to contain the players labelled $1,\ldots,m$:

$$M := \{i \in N \mid s_i \neq \bar{s}_i\} = \{1,2,\ldots,m\}.$$

Now, we define a series of strategies $(s^0, s^1, \ldots, s^m)$ which "converges" player by player from s to $\bar{s}$, in just as many steps as there are players deviating from $\bar{s}$. Initially, we have $s^0 = s$. Now, in each of the following m steps, we replace one by one the strategies of the deviating players with those played under $\bar{s}$ and hold the strategies of all players in $N \setminus M$ (and those who reverted to $\bar{s}$ in previous steps) fixed:

$$\text{For all } k = 1,2,\ldots,m \quad s^k_k = \bar{s}_k \text{ and } s^k_j = s^{k-1}_j \text{ for } j \neq k.$$

This process leads, in the end, to $s^k = \bar{s}$, i.e. the complete network strategy. Now, for a given step k, we know that all players $j \neq k$ play the strategy they played in step $k-1$, be it s_j for the deviaters who come in line after k or $\bar{s}_j$ for the non-deviaters and those in line before k. For player k we have $s^k_k = \bar{s}_k$, or, $s^k = (\bar{s}_k, s_{-k})$.[29] Because $s_k \subsetneq \bar{s}_k$ for all $k \in M$, we can invoke the link monotonicity of γ to yield:

$$f_k(s^k) = \gamma_k(g(s^k)) \geq \gamma_k(g(s^{k-1})) = f_k(s^{k-1})$$

This equation cannot hold with strict inequality, because $s^{k-1} = (s^{k-1}_k, s^k_{-k})$, which for step $k=1$ (but player k, who depends on the indexing in M) leads to

$$f_k(\bar{s}_k, s_{-k}) > f_k(s_k, s_{-k}),$$

stating that $\bar{s}_k$ dominates s_k. This contradicts the assumption that s is an undominated Nash equilibrium. Hence for any undominated strategy s equality must prevail.

Because of weak link symmetry, it is not only player k who does not gain from additional links formed, but also the players he forms those links with. Lemma 3.1 even forbids any player not involved in the formed link to profit (or lose), if

[29] To be exact, within the strategy tuple s_{-k} the players in $\{j \in M | j > k\}$ follow s, while all others follow $\bar{s}$.

not both newly linked players are strictly better off themselves. Thus the payoffs remain unchanged for all players when stepping up from s^{k-1} to s^k and $\gamma(g(s^k)) = \gamma(g(s^{k-1}))$. The above holds for all $k = 1, 2, \ldots, m$, and so

$$\gamma(g(s)) = \gamma(g(s^0)) = \gamma(g(s^1)) = \ldots = \gamma(g(s^m)) = \gamma(g(\bar{s})).$$

This shows that the networks $g(s)$ resulting from any undominated Nash equilibrium strategy s are essentially complete, and therefore yield the same payoffs under γ as the complete network resulting from strategy $\bar{s}$. ■

Propositions 3.18 and 3.19 are also true for the concept of a Coalition-Proof Nash equilibrium. This is shown, e.g., in Dutta et al. (1998, pp. 253 f.). We omit the general consideration of Strong Nash equilibria, because their existence is not guaranteed, not even under the assumption of convexity of the underlying communication situation (N, v, g).[30] The concept will, nevertheless, be applied in Chap. 6.

Having introduced the basics for both representations of network formation games, sequential and strategic, we now turn to the related notions of network stability. Under these, networks are analyzed with respect to the incentives of players to add or remove links from the status quo.

3.6.4 On the Stability of Networks

The approach we consider now does not concentrate on the formation of networks itself. The emphasis rather lies on the "stability" of a given network, meaning whether some players or groups of players might have an incentive to deviate in a coordinated manner by adding or removing links, in order to increase their payoffs. Different notions of stability and contexts in which they are presented appear in the literature. We mainly follow Jackson and Wolinsky (1996), Jackson and van den Nouweland (2005), Goyal and Joshi (2006), and Goyal (2007).

The earliest mention of a network structure being *stable*, we find in Aumann and Myerson (1988, p. 181). But since the approach presented there does not allow for the removal of previously formed links, we need not bother going into more detail.

The most basic stability notion prevalent in the literature is that of *pairwise stability*, as introduced by Jackson and Wolinsky (1996). It is, to some extent, related to the notion of a Nash equilibrium, as it covers deviations in increments of single links.

---

[30] For this, see Example 7.4 in Slikker and van den Nouweland (2001b, p. 181).

Definition 3.100. A network g is called *pairwise stable* for a given communication situation (N, v, g) and allocation rule γ, if

1. For all $g_{ij} \in g$, we have $\gamma_i(N, v, g) \geq \gamma_i(N, v, g - g_{ij})$ and $\gamma_j(N, v, g) \geq \gamma_j(N, v, g - g_{ij})$
2. For all $g_{ij} \notin g$, if $\gamma_i(N, v, g) < \gamma_i(N, v, g + g_{ij})$, then $\gamma_j(N, v, g) > \gamma_j(N, v, g + g_{ij})$

Definition 3.100 appears quite innocuous at first. Its condition 1 states that if players i and j are both no worse off when linked in a pairwise stable network, neither has an incentive to sever this link. Condition 2 specifies that if the formation of a – presently nonexistent – link between player i and j makes the former strictly better off, it must make the latter strictly worse off. The not so obvious part is, how pairwise stability handles indifference of a player with regard to forming a link. This is hidden in a conjunction of both conditions of Definition 3.100: Following condition 1, any links from which either one or both players do neither benefit nor lose, can be part of a pairwise stable network. But can such links possibly not be part of a pairwise stable network? Let us begin with the case where one player is indifferent, while the other strictly benefits from a potential link $g_{ij} \notin g$. From the second condition, we know that if player i would be strictly better off by adding link g_{ij}, player j must be strictly worse off, if the network is supposed to be stable. So apparently, any link leading to a Pareto improvement must be part of a stable network. In this sense, the definition of pairwise stability also has limited use in network formation, as it requires the existence of certain links to be satisfied.

Unfortunately, we cannot draw any conclusions from Definition 3.100 for links g_{ij} that leave the payoffs of both players, i and j, unchanged. We summarize all this in the following proposition:

Proposition 3.20. *If a network g is pairwise stable, it includes all links g_{ij} that lead to a Pareto improvement such that*

$$\gamma_i(N, v, g) > \gamma_i(N, v, g - g_{ij}) \text{ and } \gamma_j(N, v, g) = \gamma_j(N, v, g - g_{ij}). \tag{3.62}$$

Proof. Suppose not, and there is a pairwise stable network g' and a link $g_{ij} \notin g'$ as described in (3.62). Then, player i would strictly benefit from forming g_{ij}. But from condition 2 of Definition 3.100 it follows that player j would have to be strictly worse off in order to satisfy pairwise stability, which by construction is not the case. Therefore, g' cannot be pairwise stable. ■

This notion of (pairwise) stability is further discussed in Jackson and Wolinsky (1996, pp. 66 f.), where some variants are introduced as well. Particularly interesting are side payments, allowing for pairwise stable networks, as long as players can offset for each other the potential losses and gains from linking. Basically, the Pareto-criterium is dropped and the relevant measure for stability are the aggregate gains (or losses) of the player pair forming or severing their link.

Jackson and Wolinsky (1996) also show that under reasonable assumptions no allocation rule γ exists, which, for all conceivable value functions v^g, will uphold a network g that is both pairwise stable and efficient. This is reminiscent of the communication situation from Aumann and Myerson (1988) analyzed in Example 3.12. There, the complete network and those with two links are the efficient ones. Yet, with the Myerson value as allocation rule, the network in equilibrium (which is pairwise stable) is one with a single link, and hence one that is not efficient. The theorem in Jackson and Wolinsky (1996) is the following:

Theorem 3.8. *For any player set N with $n \geq 3$, no anonymous and component efficient allocation rule γ exists, which produces for each (component additive) value function v a network g that is efficient and pairwise stable.*

Proof. Let $N = \{1,2,3\}$ and set the network restricted value function such that $v(N,\{g_{ij}\}) = 1$, $v(N,\{g_{ij},g_{jk}\}) = 1+\varepsilon$, and $v(N,g^c) = 1$.[31] Because γ is anonymous and component efficient, the allocations must be $\frac{1}{2}$ each to both linked players in a network with one link and $\frac{1}{3}$ to each player in the complete network.

The three possible efficient networks have two links each, and the player in the middle (with two neighbors), must be allocated at least as much as he is without one of the links. So, for $g = \{g_{ij},g_{jk}\}$ we have $\gamma_j(N,v,g) \geq \frac{1}{2}$. The other two players i and k are allocated each the same, because γ is anonymous, and, because it is component balanced, no more than what is left after player j has been served: $\gamma_i(N,v,g) = \gamma_k(N,v,g) \leq \frac{1}{4} + \frac{\varepsilon}{2}$.

But, as soon as $\varepsilon < \frac{1}{6}$ the allocation of both players i and k drops below $\frac{1}{3}$, which they can ensure themselves by forming a link between them. Hence, the efficient networks with two links need not be pairwise stable.[32] ■

This points to the problem that for the support pairwise stable networks, an anonymous and component efficient allocation rule can always be found, but this rule need not necessarily be efficient, too. Only if restrictions on the characteristic function or even on the allocation rule are made, can this issue be resolved. See Jackson and Wolinsky (1996, pp. 61 ff.). Also, a survey devoted to network formation with focus on stability and efficiency can be found in Jackson (2005b). The main drawback of pairwise stability is the fact that deviations from one network to another, possibly more efficient one, must proceed in increments of a single link. As soon as a simultaneous deviation by more than two players, or rather more than one link is required, pairwise stability can have its limits.

To cope with such larger changes to the network, we turn to another notion of network stability called *strong stability*, which allows for deviations with any number of links at once. The concept we present originated in Jackson and van den Nouweland (2005, p. 426). Prior to a formal definition, we need to introduce some notation:

---

[31]This implies that singleton components or coalitions cannot create any value on their own, and hence $v(N,g^{\emptyset}) = 0$, but other than that, we have not restricted the characteristic function v^g.

[32]Jackson and Wolinsky (1996) extend this proof to $n > 3$ by assigning a zero value to v^g for all networks with links to or between players other than the initial three.

Definition 3.101. A network g' is *obtainable* from network g *via deviations by* coalition $S \subseteq N$, if

1. $g_{ij} \in g'$ and $g_{ij} \notin g$ implies $i, j \in S$
2. $g_{ij} \notin g'$ and $g_{ij} \in g$ implies $\{i, j\} \cap S \neq \emptyset$

The definition embodies the basics of bilateral link formation: The addition of a link requires the consent of both players, severing a link can be realized by a single player. For a deviation brought about by coalition S, added links must be between players contained in S, while severed links require only one member of S. It is worthwhile to note that this definition implies nothing about the minimality of S, i.e. the coalition could have more members than really necessary to bring about the change from g to g'.

Definition 3.102. A network g is called *strongly stable* with respect to allocation rule γ and communication situation (N, v, g), if for any given coalition $S \subseteq N$ and all networks g' obtainable from g via S, where $\gamma_i(N, g', v) > \gamma_i(N, g, v)$ for $i \in S$, there exists a player $j \in S$ for which $\gamma_j(N, g', v) < \gamma_j(N, g, v)$. Differently, g is strongly stable, if there exists no coalition $S \subseteq N$ and some network g' obtainable by S, for which we have $\gamma_i(N, g', v) \geq \gamma_i(N, g, v)$ for all $i \in S$ with strict inequality for at least one $i \in S$.

A strongly stable network must not allow for deviations through any coalition $S \subseteq N$ to a network in which a Pareto improvement for S can be realized. The minimality issue of S arising in Definition 3.101 does not occur here, because all networks obtainable through some S must not produce this Pareto improvement, hence there are no "redundant" players in any S. Using minimal coalitions S, we can even show that strong stability implies its pairwise kin:

Corollary 3.5. *For a given allocation rule γ, any strongly stable network g is also pairwise stable.*

Proof. Take a strongly stable network g and consider a deviation to some network g', where $g' := g - g_{ij}$. This is minimally obtainable through either the singleton coalitions $\{i\}$ and $\{j\}$, or alternatively through coalition $\{i, j\}$. Now, in any deviation by i or j or both to remove g_{ij}, none can profit strictly, at least not without hurting another member of his coalition, if it exists. According to Definition 3.100, the only links that can be removed from pairwise stable networks are such that do not affect the payoffs of both players, which is covered by strong stability.

Now, consider a deviation from g to g', where $g' = g + g_{ij}$. This is minimally obtainable through $S = \{i, j\}$. Then, the definitions of strong and pairwise stability coincide directly: If, say, player i gains strictly from adding g_{ij}, player j must lose strictly. This shows that any strongly stable network is pairwise stable, too. ∎

For an excellent distinction between pairwise and strongly stable networks, Jackson and van den Nouweland (2005, pp. 430 f.) provide a very instructive example.

The stability notions on networks conclude the chapter on network theory in economics. In combination with the first chapter on game theory, we now have acquired the tools to tackle the application in the subsequent part of this work: We can analyze a communication situation, i.e. a coalitional game incorporating a network structure into the characteristic function. In this context we can allocate the gains from cooperation using various concepts, taking into account the positions of the players within this network. In a separate analysis, we can focus on how the network is actually formed, given a certain allocation rule as orientation for the players' payoffs. This follows now, in Part II, after a brief introduction to telecommunications, the area in which our application is residing.

Part II
Applications to Peering in Telecommunications

Chapter 4
Telecommunications and the Internet

4.1 Overview

This chapter provides a very basic and general introduction to telecommunications, more precisely telephony networks and the internet. It is completely nontechnical in all aspects: No formal economics appear and neither do technical specifications or standards from the world of telecommunications.

The reason for this interlude is a simple one. As the subsequent model is embedded in a telecommunications environment, the reader should have a basic idea of the functioning of telephony networks, the internet, and their relation to one another. Also, we shed some light on the sector's industry structure and its regulation. Our goal is to set the ground on which the subsequent economic analysis is based and to update those readers who have little or no knowledge of the subject.

We start with some architectural basics on telephony networks and the internet, the convergence of which leads to so-called *next generation networks*. After a historical overview of liberalization and regulation in telecommunications, we illustrate the issue of network interconnection. The chapter concludes with a treatment of voice over IP telephony and an introduction to the so-called *re-routing* problem, which is the basis for all our subsequent analysis. A very brief overview of the related literature, mostly very general, is also given.

4.2 A Brief Refresher on Telecommunications Networks

Our goal is to equip the reader with a basic understanding of the prevalent types of networks with which we are concerned: The classic telephone network, the internet, and their ongoing merger into what is known as *next generation network*, or NGN. For this, we introduce some basic vocabulary concerning telecommunications networks but stay as general as possible when it comes to technical details. We do not clutter this treatment with nomenclature on specific data transmission protocols

P. Servatius, *Network Economics and the Allocation of Savings*, Lecture Notes in Economics and Mathematical Systems 653, DOI 10.1007/978-3-642-21096-9_4,

or switching techniques, because is it unnecessary for all subsequent parts and furthermore subject to so fast a technological development and innovation that it is most likely outdated by the time this work is ready for publication. We begin by describing the components and the architecture of a classical telephone network, followed by the introduction to the internet. Then, we look at their overlap and the convergence of both types of networks, how far it has already progressed, possibly unintended, and what the future outlook is.

4.2.1 Architecture of a Classic PSTN

As stated above, we begin with the classic telephone network, also called PSTN, short for *public switched telephone network*. We neither go into technical details nor consider every possible exception that could be made regarding the network structure. Instead, we try to provide the standard image of a basic telephone network, albeit with some deviations. We rely on Hyman et al. (1987) and Linnhoff-Popien and Küpper (2002). The former treatment is more closely related to the development in the United States, but it serves our purposes well, as some of the standards and nomenclature developed there were adopted in the industry worldwide. The latter is a more general approach, which includes parts on network management and transmission protocols, both ignored here.

And even though the PSTN allows for a multitude of services[1] other than basic voice calls, we restrict our terminology as to pretend that phone calls are the only possible application. This goes without much loss of generality, because the mode of operation in such a network is identical for all possible devices connected to it.

Generally speaking, a telephone network has three components, or categories thereof:

- The terminal device: Also known as customer premises equipment or CPE, the terminal device is the telephone installed at the subscriber's location. It plays a role, both at the initiating point of a call, as well es on the end where the call terminates. Also, the terminal device acts as sender and receiver, as words spoken usually travel in both directions.
- The transmission medium: This category comprises the means used to transport the message on within the network. Usually, different sorts of cables are employed for this task, ranging (in bandwidth[2]) from copper wire over coaxial cable to optical fiber. Generally speaking, the more bandwidth a medium allows for, the more likely it is to be employed in the center of a network, where the data is concentrated.[3] Transmission media are also referred to as trunk.

---

[1]Among them, e.g., fax, data transmission via modem, and video telephony.

[2]"Bandwidth" measures the amount of data that can be transmitted in a certain interval of time.

[3]This has changed with the rise of the internet and its bandwidth-hungry applications. Nowadays coax or even optical fibre connections up to the CPE are not uncommon in some areas.

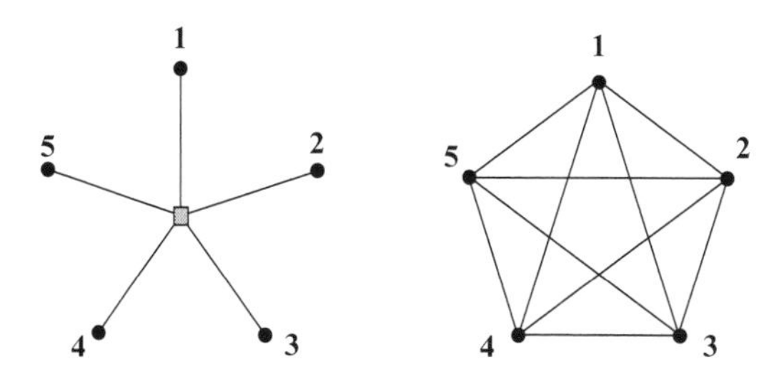

Fig. 4.1 Five subscribers connected in a star or via a fully meshed network

- The switching equipment: These are computers at specific locations within the network architecture which route calls towards their destination. Also, they possibly transform the incoming traffic into another protocol, as data will not transit through the whole network on the wire-pairs used to connect to CPE. A switch also bundles the incoming connections and sends them onto high bandwidth trunks such as optical fibre. As a switch is bi-directional, this also applies for the other direction.

The telephone network is a circuit-switched network, i.e. each call is dedicated its own circuit, from the initiator to the receiver. This can be seen as a reserved path from origin to destination for the duration of the call. When it comes to the network architecture, the setup is well arranged. As we can see from Fig. 4.1, the number of links needed differs widely, whether all subscribers are hooked up to a central switchboard (which establishes connections as needed) or whether they are connected to one another in a fully meshed network. Already in the depicted case with five subscribers, we have five connections in a switched scenario versus ten if the network is complete.[4] Aside from the higher costs it is most likely also technically impossible to connect all customers directly with one another, even on a local level. The PSTN is therefore based on a hierarchical structure composed of such star-like networks.

We show this in very abstract form in Fig. 4.2. The solid dots at the bottom of the tree are the customers, or subscribers, which are connected directly to exactly one switchboard, called their *central (end) office* or *local exchange*. It is, like all other switching facilities depicted by a grey square. The lines in between are referred to as the *subscriber loop* or *local loop*, because, typically, they happen to be a pair of copper wire.[5] The *local access transport area*, short LATA, is given by the lowest level of the network, including an end office and all its subscribers. One example in Fig. 4.2 is the grey ellipsis. Calls that originate and terminate within this confined area are handled solely by the end office represented by the shaded square. If a subscriber wants to call someone outside his LATA, which we call

[4]This difference can be computed according to $\frac{n(n-1)}{2} - n$ for $n \geq 2$ and monotonically increasing with additional subscribers.

[5]"Loop" is to be understood in the engineering context, i.e. that a pair of wires is needed in order to create an electrical circuit.

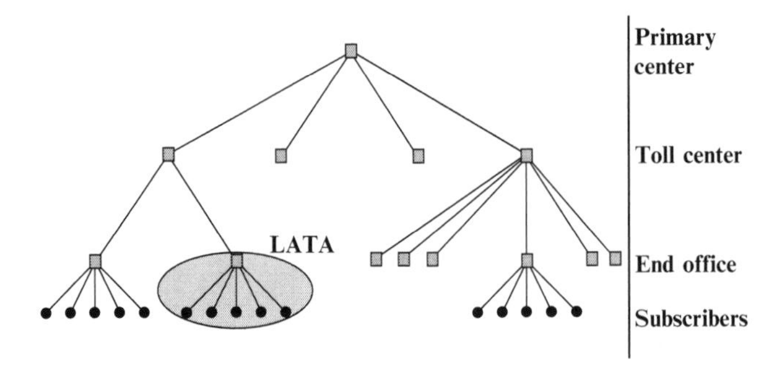

Fig. 4.2 Basic PSTN with strict hierarchical structure

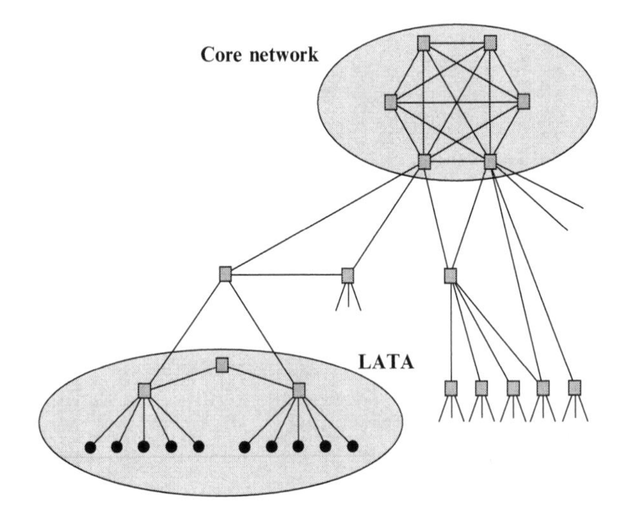

Fig. 4.3 PSTN with interhierarchical structure

simply "long distance", the end office will then route the call to a switch at a higher level. Depending on where the call destination is located within the network, more than one level of hierarchy will have to be passed. From the highest level, it is then possible to reach any subscriber in the network. How many levels of hierarchy a telephone network has depends on many factors, e.g. the geographical area covered, the number of subscribers, as well as the population density in certain areas.[6]

A less strict example in terms of network architecture is given in Fig. 4.3. Here we allow for various connections not considered in the basic example above. The first is the *tandem switch* between the end offices on the left. This increases the

[6]The classic setup in North America has five levels, or classes, of hierarchy: The lowest, class five, is the end office, followed by the toll center, a primary, and then a sectional center. At the top, we have a class one regional center. This traditionally served as rough orientation for the industry worldwide, but as the common network architecture for the United States and Canada covers a geographical area unique in size, other countries' networks have less levels. Germany's telephone network, for example, has an architecture with three such classes.

LATA to include more than just one local loop, a concept that can well be extended beyond two end offices. Such situations can be found in large cities or metropolitan areas, where one end office is not sufficient to serve all subscribers, but nevertheless, it is desirable to allow for connections among them without long distance character.

Another visible deviation from the basic framework occurs at the top. Where previously we only showed an excerpt of a network with one class three switch (primary center), we now have depicted six, who in addition are fully interconnected. Such a full-mesh configuration is often chosen for reasons of redundancy, as well as to be able to cope with high loads of traffic in all directions. This highest level part of a network is called the *backbone* or *core network*.

It is also possible to install additional trunks in lower levels of the hierarchy. If some routes show a high load of traffic, for example neighboring cities, these can be connected directly. The data is then just passed between the switches (in Fig. 4.3 those in the middle hierarchy on the left) without having to cross another level of hierarchy. Also, some switches might be connected to the core network in more than just one location, depending on traffic structure. And there also exist "short cuts", which jump levels of hierarchy if needed. We pictured the latter by the end offices on the very right, whose trunks are directly connected to the core network, possibly in addition to a "regular" connection with a class four switch.

What, in the end, a telephone network looks like in detail, is determined by many internal and external factors. Technical reasons can impose a layout, just as much as regulatory ones, when, for example, interconnection points with other network operators are mandated. But the general setup corresponds to a more or less hierarchical architecture reminding of a tree network.

4.2.2 The Internet

In this part we describe the general infrastructure of the internet as a telecommunications network. Being general, we again omit technical details regarding the means of data transmission, be it hardware or software related. Also, we do not cover possible services that can be delivered and the protocols necessary to do so. Finally, we completely ignore the internet's historical development and the organisations that contributed to it. For both, we refer the reader to Badach and Hoffmann (2007) or Pelcovits and Cerf (2003), who provide an adequate and in the former case up-to-date treatment thereof. Our focus lies instead on the infrastructure and architecture of the internet in terms of a general data-transmission (and storage) network, as compared to the PSTN.

The main difference between the PSTN and the internet is the way data is sent from one point to another. On the one hand, the PSTN is circuit switched and hence dedicates a reserved point-to-point circuit for any connection established (for its entire duration). The internet, on the other hand, is packet-switched: The stream of data is partitioned in standardized and numbered packets, each of which is labelled with the destination address. The routes these packets take in between sender and receiver are not fixed for the duration of the connection, and hence also not necessarily identical for all packets. They can be sent along the network in the most economical manner to balance the traffic load. This fundamental data transmission protocol is called the *internet protocol*, or short *IP*.[7] The internet protocol is used as basis by all other protocols regarding specific services for data-transport or applications on the internet.[8]

The unique identifiers of the "participants" of the internet, akin to phone numbers in the PSTN, are so-called *IP-addresses*. In this context, it is important to note that every device connected to the internet, not only at the end-user level, is assigned such an IP-number. This includes servers for data-storage or email transmission, computer terminals used to browse the web or to write email, as well as all the other devices that can be used or addressed via the internet. Last but not least, the *routers* are also assigned IP-numbers. These routers are located at the nodes of IP-networks, comparably to switches in the PSTN. They are the gateways between the different layers, or hierarchies, of the network.

Unfortunately though, it is impossible to describe the architecture of the internet as a hierarchical network in the way we have done before with the PSTN. This is due to the fact that what is known as the internet actually is a large number of networks on IP-basis that are interconnected. There is no single network operator for the internet, and the organisation *ICANN*,[9] which coordinates the assignment of the IP-addresses keeps no oversight with respect to network architecture. A network operator approaches ICANN to be assigned a certain range of IP-addresses for his network, but is not obliged to submit a detailed plan for what the addresses are intended. The operator also chooses freely how and where to connect his network to "the internet", i.e. to one of the networks of the internet. From an economic perspective and restricted to the US, an attempt to characterize the infrastructure of the internet in relation to geography and demography can be found in Greenstein (2005). The study reveals some similarities to what we are about to describe.

A general characterization of the internet commonly found is to subdivide it into three layers. At the top are the *backbone networks*, which provide long-distance high-bandwidth connections between selected points located in different cities or regions. They are sometimes subdivided into national and international backbones but this differentiation seems to play less and less a role, with a shift to the latter kind. One layer below come the so-called *metropolitan* or *regional networks*, which are rolled out within cities or regional areas and provide interconnection points for the networks of the lowest layer: On the one hand, they are *local area networks*, used by larger private companies, universities, public institutions, and other organisations, for which a direct connection to the metropolitan networks makes sense. On the other hand, there are *access networks*, operated by internet

---

[7]Technical details on the internet protocol, its latest standards and the services realized on its basis (so far), can be found in Badach and Hoffmann (2007).

[8]For example the *Hypertext Transfer Protocol*, commonly abbreviated to HTTP, the *File Transfer Protocol* (FTP), the *Post Office Protocol* (POP), or the *Simple Mail Transfer Protocol* (SMTP), to name but the ones most known.

[9]This acronym stands for *Internet Corporation for Assigned Names and Numbers*.

service providers, which connect private customers in their homes or smaller offices to the metropolitan or regional networks. Both local area and access networks are characterized by their connection to end-users on the individual level, who do typically not connect to any higher-layered networks directly. When it comes to the architecture, access networks are likely to bear resemblance to the LATA infrastructure of a PSTN, in the way that many subscribers are connected in a star-like fashion to some first form of concentrator. But the upper two layers follow no specific architecture and are rolled out as is best suited for their respective usage. Also, direct data-transit connections between networks of different layers are commonly found, usually to ensure sufficient bandwidth for certain services.

As the above-mentioned types of networks need not be vertically integrated and can be owned by different companies each, it is important to point out on which terms these networks connect with one another. In general, there are two kinds of interconnection agreements for IP-based networks, *transit* and *peering*: The first one is an arrangement where one network will pay the other to gain access through it to the internet. In case this paying network has other connections to the internet already, only traffic from its downstream customers[10] is being sent through the transit interconnections. These transit agreements occur within and across all layers and are not limited to networks similar in size or structure.

Under the second kind of agreement, peering, traffic is exchanged freely between the peering partners. This is usually done among (larger) networks of similar size and characteristics[11] to avoid significant imbalances. It is rarely the case that one of the peering partners has no other connection to the internet at all, or that a very large network peers with an otherwise isolated small one. The origin of the traffic, whether it comes from up- or downstream customers of a network, is irrelevant under such an agreement. Every flow of data is potentially exchanged. Peering is also often referred to as *bill & keep*.

Both kinds of agreements might be accompanied by extensive contractual frameworks specifying the conditions of the agreement with many possible contingencies. The most common are traffic-based, determining how much and what kind of traffic can be exchanged, possibly also time-dependent, and how much bandwidth must be available at least. A detailed report on such interconnection agreements can be found in Schorr et al. (2001, Chap. 7) and more specifically on traffic pricing in Gupta et al. (2005).

Based on the kind of agreement networks are interconnected with one another, another possible method of categorizing arises, the three tier system. Under this system, a network is a tier 1 network, if it can reach any IP-address without having to pay (itself) transit to any other network which might be in between. Such a network has exclusively peering relationships with other networks. A tier 2 network is one that has peering relationships, but also relies in some way on buying transit

[10]Downstream meaning traffic from a network in a lower layer according to the above definition.

[11]Where size can mean a number of things, beginning with traffic exchanged to number and type of customers or geographical area covered.

(upstream) to connect to the internet. The tier 3 network on the other hand connects to the internet exclusively on the basis of transit agreements and has no (upstream) peering agreements in place at all.

This notion somewhat coincides with the three layer approach above, as access networks or local area networks are unlikely to have upstream peering agreements. Metropolitan networks could well engage in peering with other networks of their layer, but buy transit from the top-layer backbone networks. And those in turn could only be connected to one another via peering agreements, even though transit is not uncommon between networks in countries that are separated e.g. by oceans.

Another important difference is the generality in which the internet can be used. The purpose of a PSTN is to establish connections between subscribers which sit at the very bottom of the network hierarchy. The internet in comparison has no such single purpose and offers much more flexibility. One of its main applications, the world wide web, is stored on service platforms called servers. These are typically located not on the end-user level, but in more central points of the internet, or rather upper layers, where a lot of bandwidth is available, to be able to handle as many requests at the same time as possible and to minimize the distance of traffic flow. There are also many applications, like the internet telephony substantial to this work, that effectively rely on user-to-user connections. Nevertheless user-to-services platform connections, as well as services-platform to services-platform connections, are the major part of internet traffic.

4.2.3 *Overlap and Convergence to* Next Generation Networks

As we can conclude from the preceding, there are some areas where PSTN and internet overlap. Most typically, this is the case on the local loop between the premises of the customer and the end office. Here, two access networks, one to the internet and one to the PSTN, are combined on one transmission medium using different frequencies in order not to interfere with one another. Such a setup is also referred to as an overlay network. Certainly, this already saves a tremendous amount of cost as compared to the parallel installation and operation of two entirely separate networks. Nevertheless, recent development points to so-called *next generation networks* (NGN), which rely not only on a single physical infrastructure, but also a unique and IP-based transport layer or protocol, on which all services are delivered. A more detailed account on the convergence of these services beyond telephony can be found in Dodd (1999, pp. 283 ff.). This is also covered in Picot (2006, pp. 41 ff.) who, in addition, lists some of the universal devices customers can expect in NGN.

Very generally, the International Telecommunication Union (ITU) attributes the following properties to NGN: Most importantly, the network is to exhibit package-based transfer of data, with but a single transport protocol (IP) over which a wide and unspecified range of services can be delivered. These services, via open interfaces, should be offered not only by the network operator, but also by competing service providers with unrestricted access to the physical network itself. The customers in turn should be able to choose their preferred provider for each service at will, i.e. without having to face technical restrictions. Finally, NGN should be interconnected with legacy networks like the PSTN, and all means of identifying the customers in each service (phone numbers, user names, and so on) are to be traceable to an underlying IP-address which is unique to the subscriber. The architectural structure of NGN is also much simpler, as it consists only of an access and a core network, leaving behind the multiple levels of hierarchy of a PSTN, as well as the possible unstructured combination of network types that constitute the internet.

The reasons why PSTN operators and network operators in general are upgrading to multi-purpose networks like NGN are numerous, see Sarrocco and Ypsilanti (2007, p. 15). Most pressing are certainly cost- and competition-related factors: The operation of legacy networks, and especially overlay networks is unnecessarily complex and expensive, and the ongoing decrease in revenues from voice services is closing in rapidly on the profitability of such networks. The operation of an NGN allows for the provision of multiple services on the same network architecture, with near-zero marginal costs. This opens new markets and can help against competitors, who offer similar (bundles of) services.

As far as this upgrading of networks is concerned, most network operators have already upgraded their core networks to be IP-based. This step usually goes unnoticed by the customers, because NGN-specific services can only be offered once the bottleneck, i.e. the access network, has also been upgraded.

The accelerated change of the telecommunications sector and the breadth of services available is owed in large part to the liberalization which most of the industrialized world experienced over the last two decades. It originally enabled and subsequently facilitated the shift from state-owned monopolies to the competitive landscape of today. A brief historical overview is given next.

4.3 Market Liberalization and Regulation

This section describes the change of market structure in the telecommunications sector over the past two decades. Based on Larouche (2000), Ypsilanti (2003), and Cawley (2003), we provide a quick overview of the steps taken to initiate liberalization, both by the EU and the WTO. A short note on access pricing follows thereafter.

The so-called *1987 Green Paper*, published by the European Commission,[12] was the first step towards liberalization of the telecommunications industry in the European Union. Before, each country had its own public telecommunications provider (PTO) who acted as a monopolist for infrastructure and the services

[12]Towards a Dynamic European Economy: Green Paper on the Development of the Common Market for Telecommunications Services and Equipment, COM(87)290final (30 June 1987)

hereon. These PTOs were at least partially state-owned, and sometimes even part of the national administration as an agency or ministry for telecommunications. In addition, telecommunications was largely a nationally confined business. The 1987 Green Paper changed all this: Even though it left, for the sake of network integrity, the infrastructure under monopoly, all services but public voice telephony were to be liberalized. An *Open Network Provision Framework* was demanded to standardize the relationship of the PTO and the soon-to-be competitive service providers. All firms were from now on subject to competition law, especially cross-subsidization within the PTOs was a relevant issue. This got the market moving and although general voice telephony was not yet liberalized, many companies were entering the telecommunications sector and getting ready to offer voice to the public, once it would be legal to do so. In the meantime, they would provide data or value-added services such as dial-up *Bulletin Board Services* (isolated and limited precursors to online information systems that nowadays are part of the internet) or various information or operator services accessible through the phone.[13]

The next step in European telecommunications regulation with significant impact on fixed-line telephony was taken by the European Commissions's *1992 Review*[14] which successfully recommended the full liberalization of telecommunications services (including public voice telephony) by 1998. The report also urged for a review of the policy regarding public telecommunications infrastructure by 1995. This was accomplished in the *1994 Green Paper*[15] which liberalized with immediate effect all infrastructure for the already liberalized services and by 1998 the infrastructure for public voice telephony as well, so that liberalization of services and corresponding infrastructure would go hand in hand. Following further legal action by the European Commission, the "liberalization of all telecommunications services and infrastructures was realized".[16]

Extending beyond the European Union (EU), liberalizing measures for the telecommunications sector were put in place by the World Trade Organization (WTO) as well. These were agreed upon in the post-Uruguay round negotiations

[13]As a matter of fact, the exact definition of *voice telephony* in the 1987 Green Paper already contained various loopholes through which certain voice services, e.g. calling or credit card services, as well as calls from or to mobile networks were possible. In contrast, competitive service providers could not offer services that required them to assign phone numbers to private customers.

[14]1992 Review of The situation in the Telecommunications Services Sector, SEC(92)1048final (21 October 1992)

[15]Green Paper on the Liberalisation of Telecommunications Infrastructure and Cable Television Networks of 1994: Part I – Principle and Timetable, COM(94)440final (25 October 1994), and Part II – A Common Approach to the Provision of Infrastructure for Telecommunications in the European Union, COM(94)682final (25 January 1995)

[16]This action consisted mainly of the "Consultation on the Green Paper on the Liberalization of Telecommunications Infrastructure and Cable Television Networks", COM(95)158final (3 May 1995), the "Resolution of 18 September 1995 on the Implementation of the Future Regulatory Framework for Telecommunications" [1995] OJ C 258/1, and the Directive 96/19, Full Competition Directive.

from 1994 to 1997 under the "General Agreement on Trade in Services". Simultaneous to the EU time frame, on 1 January 1998, the resulting "WTO Agreement on Basic Telecommunications Services" entered into force. With it, some 69 countries committed themselves to liberalized market access to all basic telecommunications services (including voice, data, leased lines, and so on) and infrastructure, both via own infrastructure and resale of the latter's use to market entrants. This included not only cross-border business arrangements, but also foreign direct investments.

Key elements for a fully liberalized telecommunications market are well-functioning interconnection agreements between networks. These agreements ensure the compatibility of the services offered by the network operators, which in the simplest case means that customers from one network will be able to call those of another. Especially the newly entering firms gain from such interconnection, because the number of reachable users then includes those from other providers and the incumbent, i.e. the former PTO. This in turn increases the attractiveness for actual and future subscribers. From the perspective of the incumbent, the incentive for interconnection is comparably small, as he will most likely only experience a marginal increase of subscribers from newly formed competitive providers. The incumbent is more likely to see an outflow of his subscribers to the competition.

But interconnection cannot only be mandated as such. In order to prevent the incumbent from abusing his market power by setting discriminating prices for interconnection, a thorough regulation is necessary. The fact that interconnection by itself can be a very profitable business sits at the heart of this study. Even though interconnection fees should be cost-oriented to merely compensate the network operator for terminating the call, this is often not the case and the possibility for arbitrage arises.

There are many different approaches on how incumbent networks, i.e. the ones owned by the former PTOs, should be opened to the entrants. In the light of recent technical developments and the shift to NGN, an overview is provided by Singh and Raja (2008). They stress the importance of NGN-specific regulation to assure interoperability among all such networks to facilitate the provision of services. More general, with respect to the interconnection of networks, we refer the reader to Noam (2002) who sheds light on competitive issues, citing interconnection as the key policy tool in telecommunications. Pricing for interconnection, also referred to as access pricing is the subject of Armstrong (2002). He compares different approaches of access pricing, both for horizontal as well as vertical interconnection scenarios. A comprehensive overview on regulation and interconnection is also found in Schorr et al. (2001, pp. 37 ff.), and in more economic detail in Graber (2004, pp. 241 ff.).

Our concern is mainly the interconnection arrangements between PSTN network operators, known as two-way access in the literature, and their regulation, as they apply to voice telephony in general. A comprehensive survey on such access price regulation is given in Vogelsang (2003).

Without going into any detail, we quickly mention the three most prominent approaches on how such access charges are calculated: The first to mention is *cost-based* and incorporates merely the cost of a network for terminating a call from the outside. Then, there is *Ramsey pricing*, which aims to maximize welfare, but relies on data for retail demand elasticities, which might not be available to a regulator. Thirdly, the *efficient component pricing rule* adds to the cost for termination the profits lost in retail markets due to giving access to a competitor. It also relies on elasticities of demand of the subscribers. Each has its advantages and disadvantages, depending on the necessity to correct for distortions, the availability of data and the competitive situation.

The described changes in market structure allowed competitors to offer many new services to customers, sometimes only as substitutes for already established means of communication. One of these is IP-based telephony, which we consider next.

4.4 Telephony on IP-Based Networks

In the last decade, the use of telephony based on IP-conform networks has experienced a steep growth in many different forms. Not only have many firms unified their own company-wide telephone and computer networks into one on IP-basis, covering both purposes. Also, the network operators have upgraded their core networks to IP in order to use capacity more efficiently. But most importantly for our context is the increased number of (private) subscribers, i.e. the end users, whose telephone lines are IP-based. Due to widespread availability of broadband internet access at the customers' premises, it is not uncommon, especially for new entrants into the telecommunications market, to offer telephony services that (on their end) are based purely on IP. As telecommunications firms usually do not advertise their products by the underlying technology, most such IP-subscribers are unaware that they are not connected to a legacy phone network like PSTN, since the IP-based phone networks can be a perfect substitute for the former. From now on we refer to IP-telephony by its widely used acronym *VoIP*, which is short for *Voice over IP*, or more precisely, *Voice over Internet Protocol*.

The rest of this section is organized as follows: We first give an overview of what IP telephony can refer to and then establish a working definition of VoIP for all subsequent parts. We continue to classify firms offering VoIP services according to how much of their services they provide by their own means. Finally, we introduce the *re-routing problem*, which is at the heart of the economic analysis in the following chapters.

4.4.1 IP Telephony

In its most general form, the term *IP telephony* refers to a phone call, which at some point relies on the internet protocol to convey its data. The earliest incarnations of this are long distance (or international) connections in the form of calling card services: A PSTN subscriber calls a (local) dial-up number, and in exchange for his payment information, he is connected to a number of his choice. The long distance routing of the phone call is then established on an IP-based network. Nowadays this is the case regularly as most network operators had their core networks upgraded to IP technology. In both cases, the calling and called parties need not be aware of the fact that their call is partially routed through an IP network.

Next are specific software applications for use on personal computers which allow their users to communicate with one another as if using a telephone. Such services also expanded to offering connections not only among computer users, but into the PSTN. Even the other direction is possible by assigning their customers standardized telephone numbers to be contacted through, rather than the prevalent user names, which can be chosen at will. Nevertheless, a personal computer and running the provider-specific software is still required. The most prominent example of this type of VoIP application is a software called *Skype*, which is completely internet-based.

With the advent of widely available broadband connections for end users, another possibility has arisen: Hardware-based solutions, which allow for the use of regular end user equipment (read: telephones), without a computer up and running. The devices that serve for network termination of the broadband access on the customers' premises were extended so that regular phones can be plugged in. No configuration is required and the user might not even be aware that he is not connected to the PSTN, because the way he uses the telephone is unchanged. For cost reasons, as well as for the possibility to offer multiple services at once, this approach is often chosen by new entrants on telecommunications markets.

Notably, these approaches all aim to circumvent parts of the PSTN, i.e. the network generally owned by the incumbent network operator. The last scenario also describes more or less our working definition of *VoIP*: What we are exclusively concerned with in our subsequent research are companies offering end user VoIP services. In our terminology a *VoIP firm* is any company providing to its customers a service colloquially referred to as a "basic telephone line". A standard phone number[17] is assigned to each user and no computer, proprietary software or special equipment, other than a regular telephone, is necessary for usage.[18]

A perspective on the impact of these VoIP developments on the classic telecommunications market can be found in Kelly et al. (1997, pp. 30 ff.). A recent technical overview on VoIP telephony is provided in Delley et al. (2005), especially in Ch. 12.

[17]According to the ITU-T E.164 recommendation, which contains the international public telecommunications numbering plan.

[18]By this we exclude services such as *Skype*, or *GoogleTalk*, which are merely internet-based software applications.

4.4.2 Classes of VoIP Firms

For subsequent purposes we now classify VoIP firms according to the network infrastructure they own and operate. We distinguish the following types:

- *Service Providers*: A VoIP firm without any network infrastructure at all. It relies on the customers' broadband internet access and only operates or rents capacity from a gateway server to route calls from the internet into the PSTN.
- *Core Operators*: A firm who owns a core network but relies on rented capacity in its access network.
- *Local/Regional Network Operator*: A firm operating an (access) network restricted to a certain area.
- *Network Operator*: A firm operating a full-fledged national network.

These categories are just stylized items on the possible range, and a wide variety of intermediate types can be found in most liberalized markets. Nevertheless, for our analysis, we restrict the types of VoIP firms under consideration even more, to only two: The first category contains the *service providers*. These firms rely mainly on their customers' broadband internet access. Call data is routed over the internet to a gateway location they each operate. From there, calls are handed over to the PSTN or vice versa. Because their gateway server is their only interconnection location to the PSTN, they incur costs for long distance routing of calls that terminate with customers of other firms.[19] The second includes *network operators*, i.e. companies that operate extended network structures which they either own or rent (in part) from an incumbent operator.[20] We assume the members of this group to be able to hand over calls that terminate with customers of other firms at the respective network interconnection point. In other words, their networks are extensive enough so they need not rent long distance capacity for transit from a third party and can route the call on their own infrastructure as far as technically possible. On the basis of the preceding, we can now present the re-routing problem.

4.4.3 The Re-Routing Problem

The *re-routing problem* is one endemic to VoIP telephony. It derives from insufficient interconnection on the service level of the VoIP firms. Even though their IP-based networks are physically interconnected, either directly or through the internet, the services they offer (VoIP telephony) have a range of sight limited to their own customers. We build up the re-routing problem by comparing a number of different call scenarios and the costs that accrue in each.

---

[19]For the following analysis we can also include firms in the first category that own a network, but only on a local or regional level.

[20]These are typically not firms that offer solely VoIP services, but for example cable providers or utilities that operate enhanced, IP-capable networks.

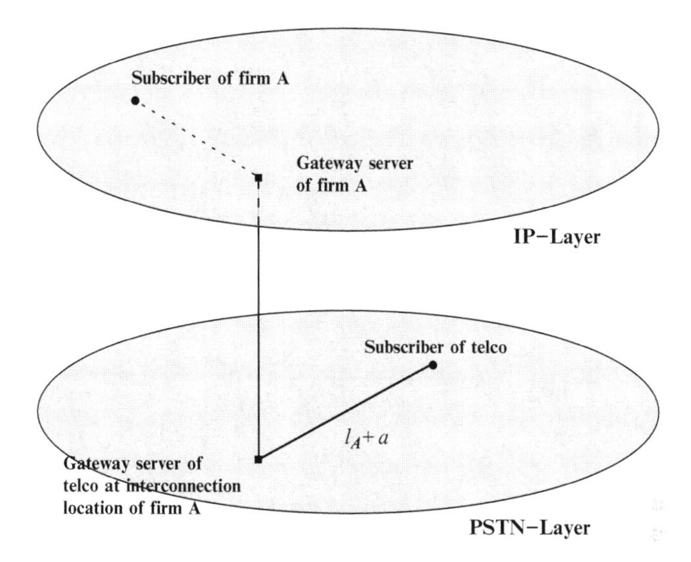

Fig. 4.4 Call routing from a VoIP to a PSTN subscriber

We begin with the simplest possibility, a call from one subscriber to another, both at the same VoIP firm. In this case, the firm is aware that the number dialled belongs to one of its own customers and hence knows the IP-address of the called party. Such a call is then easily routed between the two IP-addresses along IP-based infrastructure.

Next, take the case, where the caller is a VoIP subscriber of some firm A, and the called party is a classic PSTN user at the incumbent telecommunications firm. Firm A cannot assign the dialled number to one of its customers and treats it like any other PSTN number: It routes the call to its gateway server, where it is handed over to the PSTN. There, it is routed to its destination by the operator of the PSTN, according to the phone number. For the latter part, firm A is likely to incur long distance fees l_A, because the interconnection location housing the gateway server need not be in the same area as the called party. Also, firm A has to pay an access charge or termination fee a to compensate the PSTN operator for its costs of terminating the call on the local level.[21] This is illustrated in Fig. 4.4.

The last scenario is one where a subscriber of VoIP firm A calls a subscriber of VoIP firm B. In the absence of any cooperation among the two firms, the call is routed to the gateway server of firm A, because firm A is unaware that the called party is also a VoIP subscriber. From the phone number alone, this is not evident, and if it were, firm A would still require the corresponding IP-address to route the call accordingly. From the gateway server, the call is again routed within the PSTN by a long distance provider (at cost l_A) towards its destination, which for the PSTN-

[21] These uniform termination fees are mandated by the national regulator.

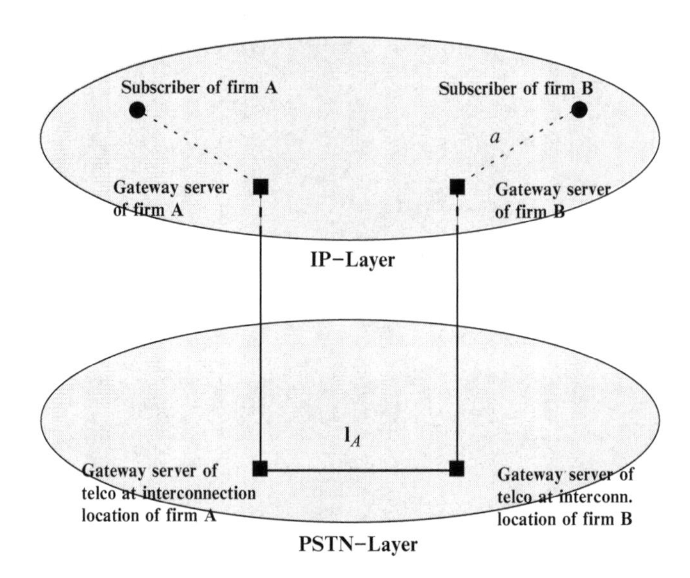

Fig. 4.5 Call routing among VoIP subscribers without peering

part coincides with the gateway server of firm B. There it is fed back onto IP-based infrastructure all the way to the called subscriber of firm B. Here, too, firm A pays the access charge a to firm B for completing the call. We depict this case in Fig. 4.5.

Now, change the scenario in the following way: Let there be a peering agreement between firms A and B. With such an agreement in place, the firms share their customers' phone numbers with one another. To each phone number is attached the respective IP-address of the subscriber, and consequently no re-routing through the PSTN is required. The call remains entirely on IP-based network infrastructure.

As depicted in Fig. 4.6, firm A incurs neither long distance transit fees l_A nor the access charges a. The first is dropped, because it is no longer necessary to route the call through the PSTN. The second falls victim to the peering agreement. This procedure implies that all VoIP firms' networks are connected to the internet (or even one another), so that a path between the subscribers is possible.

We can clearly see the existing potential for gains from peering among VoIP firms arising from operational inefficiencies: While abolishing the transfer of access charges is just a zero-sum game (informally speaking), the long distance fees that cease to apply create opportunity for some real savings for all VoIP firms involved.

The peering itself can be arranged in different ways: One is on a purely bilateral basis, where VoIP firms approach one another individually when they deem it profitable to peer. Also, central peering is possible, which allows VoIP firms to become members of a peering instance and exchange their routing tables with all other members. The peering instance could be set up by a regulating agency,

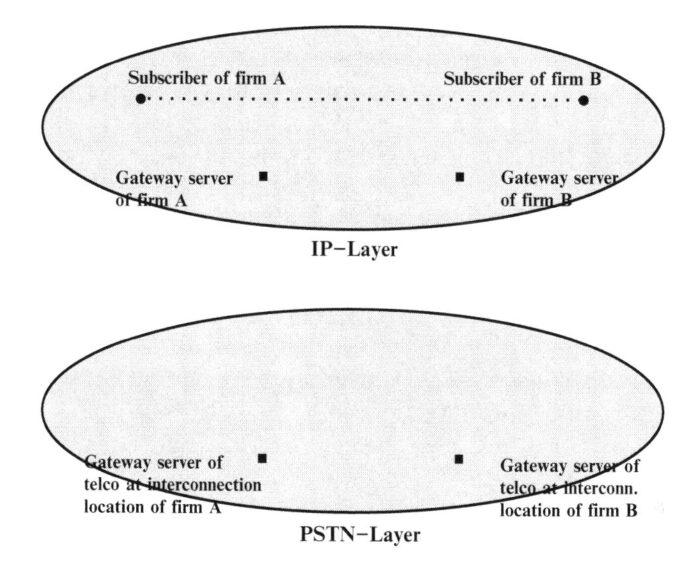

Fig. 4.6 Call routing among VoIP subscribers with peering

possibly even mandating membership, or by an organisation collectively owned by the member firms. Peering efforts are also often referred to as telephone number mapping, or short *ENUM*.[22]

It is further interesting to consider that the access fees charged for calls terminating with VoIP customers are those that apply to PSTN termination as well. The regulators that set the access fee (or a narrow range) do not yet take into account that calls to regular phone numbers can also terminate on an infrastructure other than PSTN, as there exists no regulation mandating or even concerning purely IP-based calls between VoIP firms (See Elixmann et al. 2008, p. 38 regarding the European Union). Not only does this ignore the fact that different classes of VoIP firms incur different costs for termination, but also that the underlying IP-network requires a different approach to the calculation of access prices, as its cost structure cannot be compared to that of the PSTN. Before we turn to modelling such a peering environment, we provide a brief overview of related research.

4.5 Related Research

Most of the literature on the regulation of interconnection and its pricing is revolving around the telecommunications industry, even though the basic idea applies likewise to gas, electricity, public transport, and other similar networks. Usually, there

[22]This acronym is derived from the expression ***E****.164* ***NU****mber* ***M****apping*, which is the generic term for mapping standard telephone numbers to IP-addresses.

are two standard situations considered: In the first, the network is owned by a monopolist, the formerly state-owned incumbent, and considered an essential facility, in our case to offering telephony services. Complementary to the network are the services provided hereon to its subscribers. These services can also be provided by competitors, who require so-called *one-way* access to the network in order to do so.

The second situation is one in which there exist two or more networks, i.e. the infrastructure has been replicated, albeit not on the same set of customers. Here, the competing operators seek *two-way* interconnection in order to realize network externalities, which means the ability to extend the reach of the services offered to their customers.

How we classify our research is a difficult question. Not only do we incorporate different degrees of infrastructure operated by the VoIP firms, but we also ignore the part of physical network interconnection and concentrate exclusively on the services (VoIP telephony) delivered on these networks. So, are we in a mix of both situations? Most VoIP firms operate on someone else's IP-based network to one or the other extent. Matters of traffic exchange are settled by the IP-network operators in a general, non service-specific way, via transit or peering, as we pointed out above. Now, the VoIP firms need to interconnect their services with one another, which is more the second situation, even though no relevant infrastructure has been replicated. Last of all, we do not have any essential facility problems, as the connections for calls between firms are always guaranteed for through the bypass via the PSTN.

To the best of our knowledge, no literature directly related to services interconnection exists. Consequently, we fall back on rather general sources, or such that can be adapted as to fit our purpose.

Take for example Schmidt (1999), a European Commission report on the liberalization of network industries and the role public authorities are expected to play in their deregulation. In addition, the report contains an excellent nontechnical treatment of interconnection and access charges (see pp. 132 ff.), albeit not for services but only for the networks themselves.

Similarly, Buigues and Rey (2004) provide a very broad treatment of antitrust and regulation issues in the telecommunications sector, all with a European focus. They include the economic aspects of market definition, market dominance, and network access.

Tirole (2004) argues the case why antitrust and competition policy should apply for the telecommunications sector. He also suggests that no tailoring to this industry is necessary, if a regulating agency is present.

As the internet is the network underlying all of our analysis, Pelcovits and Cerf (2003) give a helpful overview in this context with regard to regulation. They consider market entry barriers on different levels, as well as market power, and, in connection with this, switching costs. Interconnection policies are also a subject.

IP-based networks, including NGN, are considered in Alexander and Kisrawi (2005), and the regulation of communication carried out over the internet is the subject of Frieden (2003). The latter though does not treat IP-telephony at all and is more focussed on issues of users' privacy and on copyrights.

The most recent regulatory report is provided by the European Regulators Group (2009). It deals specifically with IP interconnection in NGN, including different arrangements for interconnection, as well as their possible implications on network architecture. Similar issues are covered by Marcus et al. (2008), who also highlight the coming importance of (so far mostly ignored) interconnection of services with the advent of NGN. We also refer the reader to both studies for the general motivation and benefits of regulation in the interconnection framework.

The most comprehensive and up-to-date source of regulatory issues in telecommunications, the ICT Regulation Toolkit (2008), is stored online.[23] It is aimed mainly at regulators in developing countries, intended as support for the adoption of a regulatory framework. As such, the ICT Regulation Toolkit covers all the basics and provides a sophisticated overview. The interconnection of VoIP services and arbitrage opportunities arising from it are listed, but without further analysis.

Re-routing is discussed in a setting of international calls by van den Nouweland et al. (1996) and Kelly (1996, p. 450), where it is assumed that savings can be realized by adjusting traffic flow according to peak-load traffic through different time zones.

In order to develop a feeling for the magnitude of the numbers involved, we point to OFCOM (2009), a comprehensive and up-to-date report on the Swiss telecommunications market compared with those of the EU countries. Part of the report contains pricing information on most of the elements that will appear in this study for all EU countries and Switzerland. A report similar in nature but more extensive and with focus on selected OECD countries is provided by Ypsilanti and Sarrocco (2009). All further literature that we relate to in the subsequent parts is listed where the corresponding models appear.

With this general introduction on the segment of the telecommunications industry relevant to us, we can proceed to the applications. There, we first model the situation on a national market, where interacting VoIP firms have the chance to peer with one another, possibly reducing their costs by doing so. The focus here is on these savings. In the second application, given the possible savings, we analyze under which conditions VoIP firms actually engage in peering. We consider both bilateral peering agreements as well as those realized through a central peering instance.

[23]The **I**nformation and **C**ommunication **T**echnologies Regulation Toolkit has been developed and is maintained by a cooperation of the International Telecommunication Union and the World Bank.

Chapter 5
A Model of Peering Among VoIP Firms

5.1 Overview

In this chapter we model certain aspects of the interaction among VoIP firms. To be precise, we focus on the re-routing problem as presented in Sect. 4.4. To this end, we establish a coalitional game where the VoIP firms as players can achieve gains through cooperation. These gains result from the long distance fees that are dropped under peering agreements, and we incorporate elements of network theory to model the peering relationships. Because the characteristic function takes into account the peering network, what we called a game above is actually a communication situation, see Sect. 3.3.

Subsequently, this chapter is arranged as follows: We start by describing the environment in which our model is placed, including the players, the market structure, the customers' demand, and the distribution of calls. From this, the relevant excerpts of a profit function are derived and analyzed, which in part serve as the basis for constructing the characteristic function thereafter. Finally, the game at hand is checked for desirable properties, some of which we already introduced in Sects. 2.3.3, 2.3.4, and 3.3, while some others arise only in the present context. In the next section, we apply several solution concepts and analyze the outcomes one by one. The chapter is concluded by comparing and relating these outcomes to one another.

5.2 The VoIP Peering Game

5.2.1 Basic Setup and Assumptions

The first step leading to our model is all about the environment in which interaction shall take place. We will introduce the agents in the market and how they interact with one another, which results in a certain profile of call flows, from which, finally, the profits can be derived. A few assumptions will guide us there.

P. Servatius, *Network Economics and the Allocation of Savings*, Lecture Notes in Economics and Mathematical Systems 653, DOI 10.1007/978-3-642-21096-9_5,

A point of reference for the overall setting is Laffont et al. (1998a), who model the dynamic interaction of two network operators competing in prices and access fees. Our approach is static: Not only do we consider prices fixed over our horizon, but also the market shares will not change in our model. This is because we do not focus on profit maximization through pricing and market share, but rather on the gains achieved from cooperation through peering. Particular emphasis is therefore placed on the *long distance charges*, or *transit fees*, and the *access revenue*, realized from traffic that either originates or terminates with customers of other VoIP firms.[1] Depending on wether a firm has positive or negative net access revenues, we speak either of an *access surplus* or an *access deficit*.

We analyse a national market with a finite number of n VoIP firms, represented by set the of players $N = \{1, \ldots, n\}$. These firms face a continuum of customers with mass 1.[2] Each customer's demand for calls is modelled as a function of price only, to be exact, the price p_i at which services are offered by the respective VoIP firm i. This reflects quasilinear preferences, under which the consumption of calls is neither distorted by wealth effects nor by the prices of other goods, even of close substitutes like those offered by competing VoIP firms. We do not consider this assumption overly strict in our setting, given the nature of the good in question, and its price relative to income. Also typical lock-in effects common to network industries make the switch to a competitor often a delicate and even difficult matter, or at least one that must be carefully timed. The demand function itself is real-valued,

$$q : \mathbb{R}_+ \to \mathbb{R}_{++}, \tag{5.1}$$

where $q' < 0$ exists for all $p \in \mathbb{R}_+$. We assume no finite reservation price to exist, but an upper bound to the quantity demanded, as one can only make a limited number of phone calls in a given period of time. Then, demand for calls is continuously differentiable and strictly decreasing in price p. Because the quantity demanded does not depend on the VoIP firm per se, for any pair $i \neq j$ of equal prices $p_i = p_j$, customers have identical demand for calls, whether they are subscribers of firm i or j.

The market share of each firm $i \in N$ in terms of customers is denoted α_i, with $\sum_{i \in N} \alpha_i = 1$ and $\alpha_i > 0$ for all $i \in N$. The restriction to strictly positive market shares is actually no restriction at all. As we will see below, any firm with zero market share can be dropped completely from the analysis, without any effect.

The demand a firm i faces at a given price multiplied with its market share results in the call volume, $\alpha_i q(p_i)$. In Fig. 5.1 we illustrate the important difference between market share and call volume of a VoIP firm. There are four firms, $N = \{1, 2, 3, 4\}$,

[1]Calls that either originate or terminate on the PSTN are left out of the analysis, because their share in the access revenue is unaffected by peering agreements.

[2]Without loss of generality, we assume the (Lebesgue) measure of customers subscribing to more than one VoIP firm is zero, in order to have the same mass on the customers as well as on the overall amount of subscribers to all n VoIP firms.

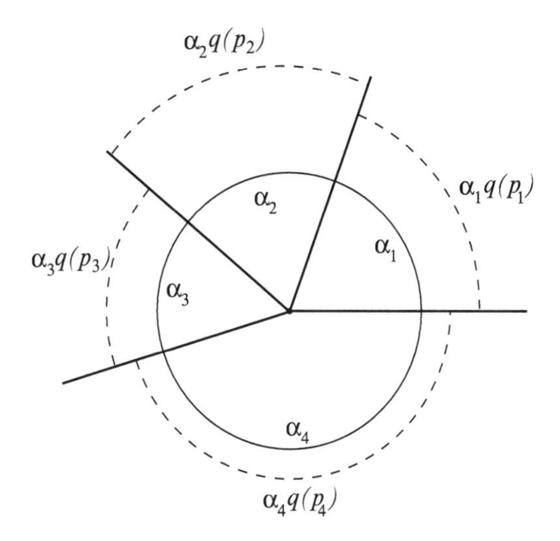

Fig. 5.1 Illustration of market shares and call volumes with four VoIP firms

with prices $p_4 > p_3 > p_1 > p_2$ and market shares $\alpha_4 > \alpha_1 = \alpha_2 > \alpha_3$. The solid circle is normalized to a radius of $\pi^{-\frac{1}{2}}$ to give rise to four slices marked α_i, $i \in N$, whose area corresponds to the market share of each firm. The area of the segments delineated by dashed lines represents the actual call volumes. Three points are to be remarked: First, even though firms 1 and 2 have equal market shares, their call volumes differ due to distinct prices. Likewise it is possible to exhibit different call volumes, even if prices coincide, based on distinct market shares. Second, firm 4, with the largest market share has the lowest average call volume (given by the "radius" of its dashed segment), due to its relatively high price, especially in comparison to firm 1. Third, comparing firms 2 and 3 reveals that the lower market share and the higher price of firm 3 consequently yield a lower call volume in comparison with firm 2, both absolute and relative.

Up to now, we only looked at where calls originate. Where they might terminate, i.e. with customers of the same or another firm, is under consideration next. As the re-routing problem only applies to calls among VoIP firms, we can limit ourselves to calls that find a destination within the given measure of customers and hence to the n firms under consideration. Concerning this flow of calls, we employ the *balanced calling pattern assumption* (BCPA). It appeared first in a two player setting in Laffont et al. (1998a, p. 3) and Laffont and Tirole (2000, p. 189) but is readily extended to the n-player case. In short, the distribution of market shares determines where (on average) the calls terminate.

Take a firm i with market share α_i and call volume $\alpha_i q(p_i)$. Of this call volume, i.e., all calls originated by customers of i, exactly the share α_i also terminates with customers of i. The complementary share $1 - \alpha_i$ terminates with customers of the other VoIP firms $j \in N \setminus \{i\}$. More precisely, this complementary share can be decomposed,

$$1-\alpha_i=\sum_{j\in N\setminus\{i\}}\alpha_j,$$

where each summand α_j equals the share of firm i's call volume terminating with customers of firm j.

Because this holds true for all $i\in N$ firms, we can conclude that the share α_i of the global call volume (which includes firm i) terminates with subscribers of firm i:

$$\sum_{j\in N}\alpha_i\alpha_j q(p_j)=\alpha_i\sum_{j\in N}\alpha_j q(p_j)=\alpha_i Q. \tag{5.2}$$

In (5.2), Q denotes the aggregate call volume of all VoIP firms,

$$Q:=\sum_{i\in N}\alpha_i\cdot q(p_i). \tag{5.3}$$

Its graphical representation in Fig. 5.1 is the sum of the area of all dashed-lined segments.

Alternatively, a given distribution of market shares $\alpha=(\alpha_1,\alpha_2,\dots,\alpha_n)$ can be interpreted as a probability distribution, where α_i, $i\in N$, is the probability for any call to terminate with a customer of firm i. As it turns out, not only is the overall probability for a call to terminate at firm i equal to α_i, the same holds for all conditional probabilities of a call terminating at firm i, given its origin with some firm j customer:

$$P(\text{call to } i|\text{origin in } j)=\frac{P(\text{call from } j \text{ to } i)}{P(\text{origin in } j)}=\frac{\frac{1}{Q}\alpha_j q(p_j)\alpha_i}{\frac{1}{Q}\alpha_j q(p_j)}=\alpha_i \quad \text{for all } \ i,j\in N. \tag{5.4}$$

Therefore, the expression

$$\widetilde{\alpha}_i:=\frac{\alpha_i q(p_i)}{Q} \tag{5.5}$$

is the global probability of a call being initiated with a customer of firm i. Or, the share of firm i's call volume with regards to overall call volume Q.

The balanced calling pattern assumption is illustrated in Fig. 5.2 with $n=3$ firms. For simplicity, firms have equal prices, $p_i=p$ for all $i=1,2,3$, where $q(p)=1$. Given a market share of α_1 for firm 1, the same fraction of its calls, α_1, is directed towards its own customers. The remaining share of calls, $1-\alpha_1=\alpha_2+\alpha_3$, is directed to customers of firm 2 and 3, respectively, to each in the proportion of their market shares. As there are only three firms in this example, this covers firm 1. The situation is analogous for firms 2 and 3, and with respect to call shares certain symmetries arise in this setting: Given the aggregate call volume $Q=1$, the share (and here also the volume) of calls directed from firm 1 to firm 3 is the same as vice versa: $\alpha_1\alpha_3=\alpha_3\alpha_1$. Corresponding results apply for the other two possible pairs of VoIP firms.

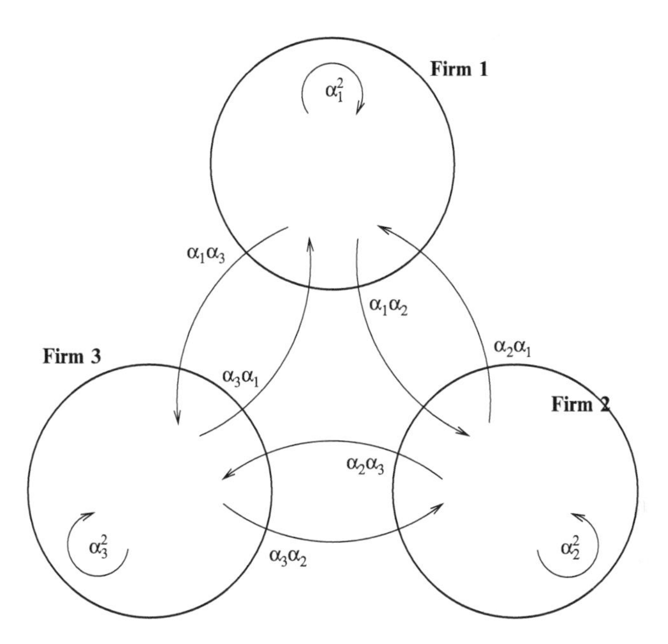

Fig. 5.2 Illustration of the BCPA with $N = \{1,2,3\}$

It is very important to note that these volume symmetries arise only on the basis of equal prices and cannot be generalized. In a general setting, the expression $\alpha_i\alpha_j$ is ambivalent: It could either be the share of calls demanded from firm i at price p_i, which terminate at firm j, or vice versa. But, unless demand is equal at both firms, the actual volumes of calls exchanged between i and j will be different. Here, it is crucial to distinguish between α_i and $\widetilde{\alpha}_i$.

The reasoning for BCPA as such is quite intuitive: On the one hand, callers can usually not distinguish to which VoIP firm the called party subscribes, at least not ex ante on the basis of the phone number. This backs the assumption that calls are uniformly distributed and each subscriber is as likely to be called as any other. Additionally, even if there were a way to know a priori to which firm a customer belongs, prices are generally differentiated rather with respect to geographical distance than with respect to network or VoIP firm affiliation. In our model, where there is but one price for each customer to consider, prices have no influence on the destination of a call either.

Also, we can see the BCPA as some kind of regularity condition: It allows us to drop firms with zero market share from the analysis and at the same time ensures that there actually is call traffic in between any given pair of VoIP firms, $i, j \in N$.

A similar notion of market share based proportionality, albeit in a very different context, is brought forward by Epstein and Rubinfeld (2001, p. 891). There, it is assumed that, following a price increase, the migration of customers of a given firm to its competitors takes place in proportion to the market shares of the competing firms.

Having described the market environment, we now set the stage for the "per-unit" profit a VoIP firm generates purely from its VoIP-to-VoIP operations, on the basis of which we build our model. It is important to note here that we only consider the elements that might contribute to profits from actual calls that both originate and terminate in a VoIP environment. Calls originating or terminating on the PSTN do not enter the analysis, as they do not give rise to gains from peering. As such we also ignore fixed costs that usually arise irrespective of usage or location of a customer.[3] As these omitted elements, too, have an impact on the overall profits of a VoIP firm we slightly abuse the term when subsequently referring to "profits" even though we only consider an extract. Starting with the (operating) costs, we work our way through long distance fees and hybrid-like access charges before finishing with the revenue collected by each firm.

We begin with the assumption of constant marginal costs for initiated calls. Depending on their destination, these costs need not be the same. Here, destination does not refer to (geo)physical location of the two parties involved in a call, but to their affiliation to a VoIP firm. If a customer places a call to another subscriber of the same company, we refer to it as *on-net*; if the called subscriber belongs to another VoIP firm we say *off-net*. This terminology is adopted from Laffont et al. (1998a), yet slightly abused: We completely disregard the physical structure of the networks involved. As soon as caller and called party belong to different VoIP firms, we treat this as an off-net connection, even though they might be part of the same physical network. Consequently, calls in between two customers from the same company are always on-net, even when they originate in different networks.[4,5]

In the first case (on-net), the marginal cost for calls placed is $c > 0$. In the second case (off-net) the additional fee $a > 0$ is levied by and paid to the receiving VoIP firm as an access charge.[6] Note that each firm not only pays access charges for its outgoing calls, but also collects them for incoming connections. This makes access charges a priori an ambiguous flow of money. Only a posteriori will it become clear whether the net payment amounts to a gain or a loss. The level of these access charges is set by a regulator and is equal for all VoIP firms. These access

---

[3]Such costs could arise for billing, servicing, infrastructure maintenance and upgrades, or various other administrative tasks related to customer service.

[4]This is possible, because VoIP services can be nomadic: The only prerequisite for use is broadband internet access, which is provided all over the world by different companies.

[5]Synonymously, *outgoing* calls with respect to a firm (and not a subscriber) refer to calls terminating with another VoIP firm and *incoming* calls are connections originating at another VoIP firm. Technically this should also include calls originating or terminating on PSTN infrastructure, but these are excluded from our analysis.

[6]We deviate from Laffont et al. (1998a) by setting the costs c_0 for call origination and termination equal to zero. As opposed to legacy circuit-switched telephone networks, IP-networks are "always-on" with idle connections and the operating cost for the marginal data-transfer on a given bandwidth is negligible. This is one of the origins of the arbitrage-potential of PSTN-based access fees in a VoIP environment, where such fees as set by the regulator exceed the actual costs they are meant to compensate. (See also European Regulators Group 2009.) We also use the expressions "access" and "termination" charges (or fees) interchangeably, even though technically speaking they are not the same.

charges originated from and hence are intended for interconnecting classic PSTN with a different (read: more expensive) cost structure. Lacking viable alternatives and universal modes of defining the real access cost in a VoIP-setting, the same access charges apply also to VoIP (services) interconnection. In principle, this is not too surprising, since a VoIP firm does the same as a network operator in a PSTN: They provide their customers telephone services, even though they might not own or operate any network infrastructure themselves.

Transit fees, denoted l_i, make up the next component. These fees are charged for the transit of off-net calls from firm i's PSTN-gateway to the gateway of the receiving VoIP firm, say j. Because these transit services are not priced according to the geographical distance they cover – at least not on national markets – but rather according to the number of network hierarchies traversed, we simplify their cost structure. By assumption, each firm $i \in N$ pays $l_i \geq 0$. Consequently, the fee does not depend on (the location of) the destination gateway, reflecting what is known as *element-based charging* in telecommunications networks. That transit prices are not the same for all VoIP firms is reasonable too. They might resort to different suppliers of transit services, be in different brackets of volume discounts, or just operate their own networks to different extents. We limit ourselves to the two polar cases of VoIP firms: Network operators with a full-coverage network on the one side, and service providers without network infrastructure at all, on the other. The former have a PSTN-gateway in all possible locations and therefore incur no transit fees whatsoever. The latter have only one gateway location each, which – without loss of generality – are all in different places and give rise to nonnegative transit fees.[7] These transit fees are an extension to the work by Laffont et al. (1998a,b), and Laffont and Tirole (2000). In their models, because there are only two (homogeneous) networks under investigation, which are furthermore interconnected, the inclusion of transit fees would make little sense.

As far as the pricing of calls is concerned, we assume *nondiscriminatory* pricing, where a firm i charges its customers the same prices p_i for on-net and off-net calls. This assumption is again defended by the fact that phone numbers are allocated either by region or by service, but not by affiliation to networks or in our case VoIP firms. So ex ante, a subscriber is not able to identify at which VoIP firm his call is going to terminate and could therefore not determine the price for a call, if they were to be differentiated in an on-net/off-net fashion. As a matter of fact, a subscriber will not even know ex ante, whether his call terminates with another VoIP firm or a PSTN-based company. The VoIP-related operating revenue of a given firm $i \in N$ is therefore calculated as

$$p_i \alpha_i q(p_i),$$

the product of price and call volume.

[7] Naturally, a network operator could accept incoming calls from service providers to be handed over at the location where the latter have their gateways. This would save the call originating firm the long distance transit fee. But why should operators accept additional traffic on their networks without being compensated for it? This is known in the literature as the *hot-potato routing problem*.

Before deriving the actual profit function, we repeat quickly the assumptions we have introduced so far:

1. Finite set of VoIP firms, $N = \{1, 2, \dots, n\}$, on a national market
2. Unit mass of customers without overlapping subscribers
3. Continuously differentiable demand, based on quasilinear preferences
4. Nondiscriminatory pricing for on-net and off-net calls
5. Balanced calling pattern assumption
6. Only network operators and service providers, nothing in between
7. Nonnegative long distance transit fees l_i
8. Uniform access fee a for all VoIP firms

Based on the preceding, the (operating) profits a VoIP firm i generates from calls originating and terminating on IP-based infrastructure is obtained:

$$\Pi_i(\mathbf{p}) = \alpha_i \left[(p_i - c)q(p_i) - (a + l_i)q(p_i) \sum_{j \neq i} \alpha_j \right] + \alpha_i a \sum_{j \neq i} \alpha_j q(p_j). \tag{5.6}$$

The vector $\mathbf{p} = (p_1, p_2, \dots, p_n) \in \mathbb{R}^n_+$ is the menu of prevalent prices on the VoIP market. It is again to be noted that (5.6) only refers to the operating profits firm i realizes from VoIP calls, but for simplicity it is referred to as "profit" in subsequent parts. The term in square brackets is the firm's core operating profit according to its market share α_i: From the profit of all originating calls (on-net and off-net), $(p_i - c)q(p_i)$, we subtract (for all off-net calls) the access charges and transit fees, a and l_i, in proportion to the market share of all other $n-1$ VoIP firms, α_{-i}.[8] The second term is the revenue from collecting access charges of all incoming calls. The sum of the call volumes of all firms $j \neq i$, i.e. the total incoming call volume, is multiplied with the access charge a. Both terms are weighted with the market share of firm i, α_i.

Because our analysis is focused on the components that allow gains from cooperation expression (5.6) can be rearranged to highlight the total access charges and to separate transit fees:

$$\Pi_i(\mathbf{p}) = \alpha_i \left[(p_i - c)q(p_i) \right] + \alpha_i a \sum_{j \neq i} \alpha_j \left[q(p_j) - q(p_i) \right] - \alpha_i l_i q(p_i) \sum_{j \neq i} \alpha_j. \tag{5.7}$$

Now the first term is firm i's retail profit. The second term depicts the access revenue. It is the difference between the access charges received from and paid to the other $n-1$ VoIP firms. If this difference is negative, i.e. the outgoing call volume is higher than the incoming, we refer to it as an *access deficit*; if it is positive, it is called an *access surplus*. The last term contains the long distance transit fees that accrue for i's outgoing call volume.

[8] We switch between α_{-i} and the summation over all α_j, $j \neq i$, depending on context, but do not mix the expressions within equations or paragraphs.

Because the model contains only a finite number of firms, n, and is restricted to call traffic among these, we operate in a closed environment where all relevant calls can be accounted for. From the profit function it can again be deduced that only firms i with $\alpha_i > 0$ have a meaning in this model.

We now take a closer look at the access revenues of a given firm. Let A_i^j denote the partial access revenue of firm i with firm j, where $i \neq j$:

$$A_i^j = a\,\alpha_i\alpha_j\left[q(p_j) - q(p_i)\right]. \tag{5.8}$$

The following is immediate:

Proposition 5.21. *If the reciprocal partial access revenues of any two distinct firms $i, j \in N$ are nonzero, they are equal except for their sign:*

$$A_i^j = -A_j^i. \tag{5.9}$$

Proof. Adapt (5.8) to firm j and rearrange to factor out -1. ■

The (aggregate) access revenue of firm i, denoted A_i, can be calculated from summing over all its partial access revenues with firms $j \neq i$:

$$A_i = \sum_{j\neq i} A_i^j = a\alpha_i \sum_{j\neq i} \alpha_j\left[q(p_j) - q(p_i)\right]. \tag{5.10}$$

Because for all distinct $i, j \in N$ we know $A_i^j = -A_j^i$ to hold (see Proposition 5.21), the (individual) access revenues cancel out in the aggregate.

Corollary 5.6. *For a finite player set, the sum of all access revenues is equal to 0:*

$$\sum_{i\in N} A_i = 0. \tag{5.11}$$

Proof. Expanding and rearranging the summation,

$$\begin{aligned}
\sum_{i\in N} A_i &= a\sum_{i\in N}\alpha_i\sum_{j\neq i}\alpha_j\left[q(p_j) - q(p_i)\right] \\
&= a\left[\sum_{i\in N}\alpha_i\sum_{j\neq i}\alpha_j q(p_j) - \sum_{i\in N}\alpha_i q(p_i)\sum_{j\neq i}\alpha_j\right] \\
&= a\left[\sum_{i\in N}\alpha_i\sum_{j\neq i}\alpha_j q(p_j) - \sum_{j\in N}\alpha_j\sum_{i\neq j}\alpha_i q(p_i)\right] \\
&= 0,
\end{aligned}$$

it follows that all summands appear twice, albeit with different sign. Because we sum over the partial access revenues of all n players, the term in square brackets must equal 0. ■

Table 5.1 Access surplus comparison for $p_i < p_j$

Σ_i	Σ_j	$\alpha_i \leq \alpha_j$	$\alpha_i > \alpha_j$
≥ 0	> 0	$A_i < A_j$	$A_i \gtreqless A_j$
< 0	≥ 0	$A_i < A_j$	$A_i < A_j$
< 0	< 0	$A_i \gtreqless A_j$	$A_i < A_j$

This result is also quite intuitive. Because partial access revenues cancel out reciprocally, the finite summation over all possible $\binom{n}{2}$ pairs of players equals 0, too.

We also want to stress that due to our static approach, there is no imminent influence of the price p_i on the market share α_i of firm i. This in turn implies that an a priori comparison of access revenues, based only on individual prices or market shares is not possible. This is illustrated by rewriting expression (5.10),

$$A_i = a\alpha_i\Sigma_i \text{ with } \Sigma_i := \sum_{j \neq i} \alpha_j \left[q(p_j) - q(p_i)\right]. \tag{5.12}$$

If we assume firm i to charge its customers a lower price than, say, firm j, we know from the strict monotonicity of the demand function q that $q(p_i) > q(p_j)$; the call volume per customer is greater at firm i than at firm j. Therefore, the sum in (5.12) yields a strictly lower value than a corresponding sum for firm j. One is tempted to assume that when comparing the two firms i and j, certain (qualitative) combinations of market shares and prices allow for such statements as to which firm has the higher or lower access revenue. As it turns out, this need not be the case at all, which we highlight in Table 5.1 below.

Given our initial assumption $p_i < p_j$, three cases arise from the combination of signs of each sum according to (5.12). Both sums can be either nonnegative or strictly negative, or i's sum can be negative and j's nonnegative.[9] These are depicted by rows 1, 3, and 2, of Table 5.1. The two columns represent the relation of the market shares of firms i and j. This leaves us with six possibilities, and as we can see, from market shares and prices of firms i and j alone we cannot infer which one has the larger access surplus.

We have now introduced the players and their environment. We showed how customer demand and market share of the firms determine the volume of outgoing and incoming calls and the resulting access charges for interconnection between firms. From this we modelled the VoIP-related part of a profit function. In the next section, we derive the characteristic function of the coalitional game or communication situation, with the help of network theory. There, we explain how access and transit fees will serve as incentives for VoIP firms to realize gains from cooperation.

[9]The opposite mixed-signs case can only occur when $p_i > p_j$.

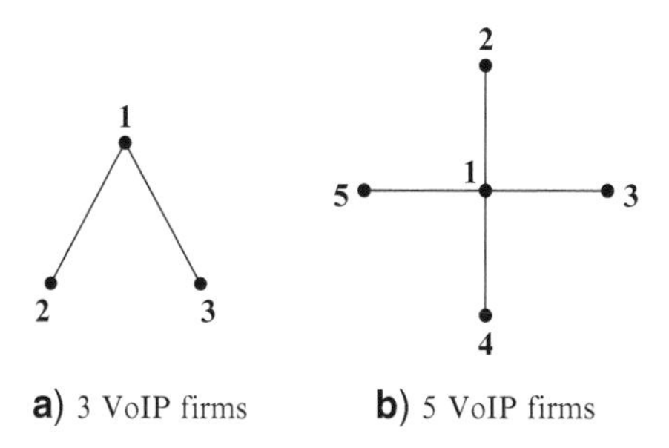

Fig. 5.3 Two exemplary peering networks

5.2.2 *The Characteristic Function of the Game*

For the representation of a coalitional game in characteristic function form, we not only need the set of players $N = \{1,2,\ldots,n\}$, but also a characteristic function $v : 2^N \rightarrow \mathbb{R}$. The latter attaches a real number to each coalition $S \subseteq N$ to describe the gains that can be achieved. In light of the re-routing problem, cooperation leads to gains only between those firms who have a peering agreement in place. To model the peering agreements, we use tools from network theory. The goal of our analysis is to derive a network structure which represents the mesh of peering agreements among the VoIP firms.

The finite set of nodes $N = \{1,2,\ldots,n\}$ of such a network then coincides with the set of players. The links that constitute the network g represent peering agreements. As introduced in Sect. 3.2, links are indexed by the pair of nodes (or players) to which they refer and assume binary values: $g_{ij} \in \{0,1\}$ for all $i,j \in N$. If a link exists between players i and j, we have $g_{ij} = 1$, and $g_{ij} = 0$ otherwise. By convention, $g_{ii} = 0$ for all $i \in N$. Because peering agreements work in both directions, the networks are undirected and all links symmetric, $g_{ij} = g_{ji}$ for all $i,j \in N$. The network g consists of all links with value one: $g = \{g_{ij} | g_{ij} = 1, \text{ with } i,j \in N\}$. The set of peering partners of a player i is therefore identical to his *set of neighbors*, or *neighborhood*. These are the players to which i is linked in network g, denoted $N_i(g) = \{j \in N | g_{ij} = 1\}$. The set of all networks on the player set N is defined as $\mathscr{G}_N := \{g | g \subseteq g^c\}$, where g^c is the complete network, i.e one in which all players are linked, $g_{ij} = 1$ for all distinct $i,j \in N$.

According to this set-up, a peering agreement between two firms is in place, if and only if these firms are linked. Thus, VoIP firm i cooperates exclusively with the members of its neighborhood $N_i(g)$, and there is no peering with another player j to which i is merely connected via some path of length two or more.

Take, e.g., the network (N,g), where $N = \{1,2,3\}$ and $g = \{g_{12}, g_{13}\}$. In this case, firms 1 and 2, as well as 1 and 3 have a peering agreement in place. And even though firms 2 and 3 are connected via firm 1, they cannot realize any gains with one another, because there is no agreement. A further example would be a network (N,g) with $N = \{1,\ldots,5\}$ and $g = \{g_{1j} | j \in N \setminus \{1\}\}$. In this network, VoIP firm

1 peers with all other firms, but each of these other firms peers exclusively with firm 1. This star network as well as the previous example are shown in Fig. 5.3. Through its peering agreements, a VoIP firm i makes available to its neighbors $j \in N_i(g)$ a mapping that assigns to each of its customers' phone numbers the corresponding IP-address. Now as we explained in Sect. 4.4.3, off-net calls from one neighbor to another will not be routed through the PSTN, but rather remain on IP-based infrastructure. Consequently, neither transit fees l_i nor access charges a are incurred by firm i, where the call was initiated. Analogously, the call-receiving firm j will lose the amount it had previously collected in access charges from i. Whereas the distribution of access charges among peering firms is a zero-sum game, the nonnegative transit fee gives rise to a general-sum game. Because the transit fee is charged by a third party and not by the players to one another, a margin is created which opens the door for peering agreements.

What we do not include into our analysis of gains from peering are possible indirect effects on the overall profits of a VoIP firm. By indirect effects, we mean such effects that might result from a possible change in the optimal prize as a consequence of reduced marginal costs through peering. This approach has a number of reasons, which we elaborate on briefly: To begin with, even though the profit-maximizing price of a VoIP firm under peering is always below that without peering, the effects on the profit are ambiguous, since a possible access surplus before peering might outweigh the gains. Also, as noted before, expressions (5.6) and (5.7) only cover the part of VoIP-based operations of firm i and do not take into account calls that originate or terminate on the PSTN, for which the same nondiscriminatory prices are charged, and which might account for a large portion of the total operations of a firm. In addition, the above expressions merely relate to the common monthly period, over which fees are paid and charges netted among firms. Further complications in determining an optimal price arise from the commonly used two-part tariffs, as well as bundled offers more and more common for telecommunications services. Moreover, several studies[10] have shown that the price elasticity of demand for fixed-line telephone calls is sufficiently low to allow us to exclude such indirect effects from our analysis. The fact that despite declining prices for domestic fixed-line calls (see OFCOM 2009, pp. 83 ff.), the volume of the latter has been decreasing or stagnant at best since 2004 in various developed countries (see Ypsilanti and Sarrocco 2009, p. 89) supports these findings. This very price decrease also leads to the our main argument, the demand situation the VoIP firms face: We believe declining prices to be attributable predominantly to the increasing competition on the telecommunications market, as the number of participants offering fixed-line services is still growing in most countries. Technological shocks however, i.e. innovation that diminishes marginal costs, are more rigid in terms of their effect on prices. We base this argumentation on the theory of kinked demand curves and the asymmetric pricing behavior it induces.

[10]See, e.g., the overview in Wheatley (2006) who also suggests that elasticities decrease with the relative price of calls.

The prevailing market price is sustained by all firms, because each firm expects the others to follow, once it lowers its own price. Raising one's price though is expected not to induce any of the competitors to follow suit.[11] The result of these expectations is a kink in the demand curve at the prevailing price and more importantly, a gap in marginal revenues. Due to the latter, the prevailing price might continue to be optimal despite changes in marginal costs. We refer the reader to Maskin and Tirole (1988) and Kreps (1990, pp. 335 ff.) for kinked demand models with homogeneous goods which allow for ranges of possible equilibrium prices. In short, even though it might be the reduced marginal costs resulting from technological shocks that allow for a lower price, we argue that the cause of downward price movements should be attributed to the prevailing competitive development in telecommunications.

Taking the preceding into account, the network restricted value function v^g bears the following characteristics for a given network $g \in \mathcal{G}_N$: By convention, the empty coalition is worthless, $v^g(\emptyset) = 0$. Also, no individual player can realize any gains, which renders all trivial coalitions worthless, too: $v^g(\{i\}) = 0$ for all $i \in N$. Finally, no gains can be realized among firms where no peering agreement has been reached: $v^g(S \cup T) = v^g(S) + v^g(T)$ for all $S, T \subseteq N$ with $S \cap T = \emptyset$ and $N_i(g) \cap T = \emptyset$ for all $i \in S$. So for nonempty, nontrivial coalitions $S \subseteq N$, $|S| =: s \geq 2$, the gains from peering are calculated by summing the *link restricted interconnection payments* of all members $i \in S$. For a given VoIP firm i, the interconnection payment reflects the costs of interconnection with the other $n-1$ players in the absence of peering. These result from costs of transit, corrected by the access revenue. Because we want to highlight the part of these costs that are saved through peering agreements, we account only for firms to which i is linked, hence "link restricted". And within a coalition $S \subseteq N$, the interconnection payment for some $i \in S$ is the difference of his transit fees and the partial access surpluses, both of which arise from business with his neighbors j who are also members of the coalition: $j \in N_i(g) \cap S$. We define this link restricted interconnection payment for player $i \in S$ as follows:

$$\begin{aligned}\widetilde{A}_i(g,S) &:= \alpha_i l_i q(p_i) \sum_{j \in N_i(g) \cap S} \alpha_j - \sum_{j \in N_i(g) \cap S} A_i^j \\ &= \sum_{j \in N_i(g) \cap S} \left[\alpha_j \alpha_i l_i q(p_i) - A_i^j\right], \quad \text{for all } i \in S, S \subseteq N, g \in \mathcal{G}_N. \end{aligned} \tag{5.13}$$

As the summation includes just the firms $j \in N_i(g) \cap S$ that are both directly linked to i and a member of coalition S, savings can only be realized with peering partners. Even though every player potentially contributes to the savings in any coalition $S \subseteq N$ that is connected in g, these contributions arise only between linked players.

---

[11] The phenomenon of kinked demand was first addressed by Hall and Hitch (1939) and Sweezy (1939) based on empirical observations. More recently, theoretical foundations were added to the literature in Maskin and Tirole (1988), Kreps (1990), and Tirole (1997).

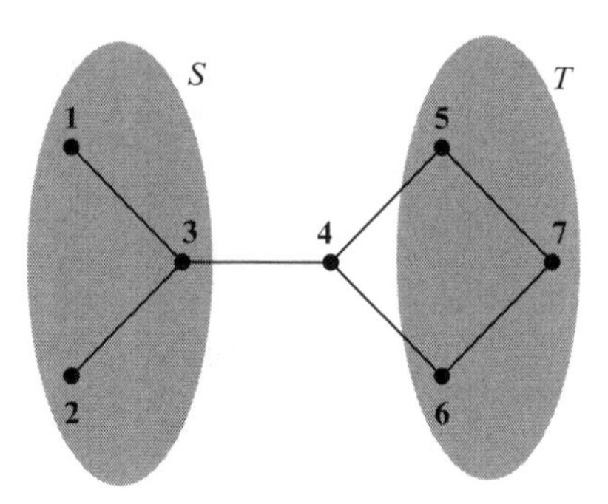

Fig. 5.4 A 7-player network to illustrate features of v^g

Therefore, a characteristic function derived from (5.13) must be a mapping from the product space of all subsets of n players and all networks to the nonnegative real numbers, $v : 2^N \times \mathscr{G}_N \to \mathbb{R}_+$. The functional form is given in the following expression:

$$v(S,g) = \sum_{i \in S} \widetilde{A}_i(g,S) = \sum_{i \in S} \sum_{j \in N_i(g) \cap S} \alpha_j \alpha_i l_i q(p_i) \quad \text{for all } S \subseteq N, g \in \mathscr{G}_N. \quad (5.14)$$

The second equality follows from Proposition 5.21, p. 187, because the access revenues cancel out pairwise within any coalition $S \subseteq N$. The savings that emerge are equal to the sum of the transit fees which, without peering, the members of coalition S would have paid in order to have their calls routed to other firms in the intersection of their respective neighborhood and S. Clearly, the value function only assigns nonnegative values, as all elements are nonnegative themselves. From now on, whenever we speak of "the value function", or v^g, we refer to (5.14). We also use the terms "savings" and "gains" from peering interchangeably.

A graphical illustration of a 7-player situation is depicted in Fig. 5.4 with the underlying network $g = \{g_{13}, g_{23}, g_{34}, g_{45}, g_{46}, g_{57}, g_{67}\}$. For coalition $S = \{1,2,3\}$, the savings according to $v^g(S)$ originate only from members of S that are linked with one another: $v^g(S) = v^g(\{1,3\}) + v^g(\{2,3\})$, because $g_{12} \notin g$. Also, the link g_{34} has no influence on $v^g(S)$, because player 4 does not belong to S.[12] Extending coalition S by player 7 to $S \cup \{7\}$ does not affect the gains: $v^g(S \cup \{7\}) = v^g(S) + v^g(\{7\}) = v^g(S)$, because player 7 is not linked to any member of coalition S, $N_7(g) \cap S = \emptyset$ and because trivial coalitions have a value of 0. Likewise, with coalition $T = \{5,6,7\}$, the value of S and T is given by the sum of both coalitions, because S and T are only connected, but not linked, as is the requirement for additional gains: $v^g(S \cup T) = v^g(S) + v^g(T)$, since no players $i \in S$ and $j \in T$ exist for which $g_{ij} \in g$. Now, with the characteristic function of the peering game established, we give it a more thorough

[12] But there exist nonnegative gains $v(\{3,4\})$, as $g_{34} \in g$.

examination in the next section. We first strip it down to gain additional insights on its explanatory power and then check for some important properties of coalitional games and communication situations.

5.2.3 *The Peering Game and Its Properties*

The peering game whose characteristic function v^g we defined in the previous section can either be given as a coalitional game (N, v^g) for a given network, or as a communication situation (N, v, g). Either way, it is worthwhile to reexamine the construction of v^g and to formalize that it is actually the sum of two mappings, one of them representing a zero-sum game. Thereafter, we revisit selected properties of coalitional games and communication situations to check whether they are satisfied by the peering game.

To gain more detailed insights on the value function v^g, we should substitute the first line of (5.13) into (5.14) to yield

$$\begin{aligned} v(S,g) &= \sum_{i\in S} \widetilde{A}_i(g,S) \\ &= \sum_{i\in S} \left[\alpha_i l_i q(p_i) \sum_{j\in N_i(g)\cap S} \alpha_j - \sum_{j\in N_i(g)\cap S} A_i^j \right] \quad \text{for all } S\subseteq N,\ g\in \mathscr{G}_N. \end{aligned} \tag{5.15}$$

It is now recognizable that v^g is the sum of two different functions that can be interpreted as coalitional games in their own right: First, we have general-sum game regarding the gains from long distance transit. This is actually all that enters the function v^g, because secondly, we have a zero-sum game included in v^g, which covers the access revenues. The latter cancel out pairwise and hence in all coalitions $S \subseteq N$. As a formalization we obtain

$$v(S,g) = v^l(S,g) + v^a(S,g), \tag{5.16}$$

where

$$v^l(S,g) = \sum_{i\in S} \sum_{j\in N_i(g)\cap S} \alpha_j \alpha_i l_i q(p_i) \geq 0 \quad \text{for all } S\subseteq N, g\in \mathscr{G}_N, \tag{5.17}$$

and

$$v^a(S,g) = -\sum_{i\in S} \sum_{j\in N_i(g)\cap S} A_i^j \equiv 0 \quad \text{for all } S\subseteq N, g\in \mathscr{G}_N. \tag{5.18}$$

Function v^l from (5.17) assumes only nonnegative values, because all elements of each summand are nonnegative themselves. Regarding v^a, which is the sum of mutual pairs A_i^j and A_j^i, the equivalence in (5.18) follows from Proposition 5.21.

Even though v^a is negligible in terms of its value and hence impact on v^g, its significance will become apparent when we look at the allocation of gains produced by v^g, especially in the light of network formation in Chap. 6.

As v^g reduces to v^l for our purposes we can see that (additional) gains are realized pairwise, via links between members of a coalition $S \subseteq N$. The impact of an additional link on the value of v^g is independent of coalition S and how many other links there exist: A new link g_{ij} between players i and j, both members of S, will change the value of v^g by $\alpha_i \alpha_j [l_i q(p_i) + l_j q(p_j)]$. As we noted before, v^a is completely unaffected. The gains of an additional link g_{ij} are unconditional on the remainder of both, the network g and coalition S. This fact is summarized in the following proposition:

Proposition 5.22. *The gains of an additional link g_{ij} under value function v^g are independent of g and S:*

$$v(S, g + g_{ij}) - v(S, g) \equiv \alpha_i \alpha_j \left[l_i q(p_i) + l_j q(p_j) \right], \quad \text{for all } g \in \mathcal{G}_N, \; g_{ij} \notin g, \quad \text{for all } S \subseteq N, \; i, j \in S. \tag{5.19}$$

Proof. Follows directly from plugging value function v^g into (5.19). ■

It is important to note that the above is not meant to deny the convexity of v^g, which we establish below: Proposition 5.22 covers the gains of an additional link g_{ij} within some coalition $S \subseteq N$, not the marginal value of an additional member to a coalition $S \subseteq N$. The size of the coalition has no influence on the change in value when linking players i and j, whenever $i, j \in S$. However, adding players to some coalition $S \subseteq N$ can affect the value of v^g, depending on the network g. The following corollary arises from Proposition 5.22:

Corollary 5.7. *In any communication situation (N, v, g) based on the value function v^g, the sets of superfluous and strongly superfluous links coincide. Furthermore, such links exist only between players $i, j \in N$ with $l_i = l_j = 0$.*

Proof. To begin with, Proposition 5.22 holds for all types of links, and hence also for links between those players i and j for which $l_i = l_j = 0$. Consequently $v(S, g) - v(S, g - g_{ij}) = 0$. According to (5.19), added links have no additional value if and only if they are formed between players with zero long distance fees. A superfluous link is one whose addition or removal from any network $g' \subseteq g$ does not affect the value of the grand coalition. A strongly superfluous link is one whose removal in a given network g would not affect the value of any coalition $S \subseteq N$. In both cases, this is fulfilled by (and only by) links between players with zero long distance fees. ■

As it turns out, another consequence of Proposition 5.22 is very interesting. If a link is superfluous for some reference network g, it is so for the complete network g^c and consequently for all networks $g' \subseteq g^c$:

Corollary 5.8. *Under value function v^g, any link g_{ij} being superfluous for network g in communication situation (N,v,g), is also superfluous for network g^c in the otherwise identical communication situation (N,v,g^c).*

Proof. This is a direct consequence of Proposition 5.22. ■

From now on, we will only use the expression *superfluous link*, when speaking of either concept in the context of v^g.

The last insights simplify matters significantly, as the impact (and so also the consideration) of forming a link does not extend beyond the pair of players involved. However, it will affect the outcomes of the various solution concepts employed that in the end determine how the gains from peering will be allocated. This is shown in the next section.

Until then, we examine which types and properties of coalitional games and communication situations (as treated in Sects. 2.3.3, 2.3.4 and 3.4) apply to the game based on v^g. We begin with properties defined explicitly on coalitional games (N,v), where we incorporate the effects of the network g on the value function v but keep g fixed. We then turn to more network-specific properties, where changes in g and their effect on v^g are considered.

5.2.3.1 Null-Players

As a special case of dummy players, null-players, as introduced in Definition 2.18, are players with marginal values of zero for any coalition they might join:[13] $d_i(S) \equiv 0$ for all $S \subseteq N$. In the peering game with v^g, the value-added of some player i generally depends on his "position" in the network g, more specifically on $N_i(g)$, his neighborhood. The most extreme instance of this would be a player without any links at all: Every player i with $N_i(g) = \emptyset$ is a null-player by definition of the value function. Another possibility arises through the long distance fees l_i incurred by the players. Take a player i with $l_i = 0$ and assume that $l_j = 0$ for all players $j \in N_i(g)$ who have peering agreements with i. No savings can be realized among these players, and so i is a null-player in g, regardless of the coalition he joins. Hence, if for all $j \in N_i(g) \cup \{i\}$ we have $l_j = 0$, then i is a null-player.

An extreme case of this situation arises when we consider some component $C \in N/g$, where $l_i = 0$ for all players $i \in C$. Here, peering will not lead to any gains. In the context of communication situations, null-players are referred to as superfluous players (see Definition 3.79 or Slikker and van den Nouweland 2001a, p. 40).

We will see below that the case with the empty neighborhood of a player, as well as the case with the valueless component can be discarded, as either irrelevant or more or less pathological. In contrast, the situation given by the player whose neighbors feature exclusively zero long distance fees is more realistic, especially

[13]Because the stand-alone value of each player under v^g is zero by construction, dummy players are automatically null-players.

for players with very few links. Null-players can therefore not be ruled out a priori and will be included in the treatment, wherever necessary. A slight, yet meaningful distinction is now introduced:

5.2.3.2 Players Without Contribution

With regard to the players' attributes, we believe it is necessary to make another distinction. For this we introduce the notion of *contribution* of a player towards (his marginal value to) a given coalition. This concept is quite elusive to define in general or formal terms, as a characteristic function maps from the set of all coalitions 2^N to the real numbers $\mathbb{R}$. This being the striking feature of cooperative game theory, the part on how exactly the value is created within a coalition $S \subseteq N$ is skipped. The individual contributions of the members of S to $v(S)$ are not revealed, only their marginal values. The idea of a contribution is based on considerations that the marginal value of some player i to coalition S need not arise from merely joining it and "cooperating" (whatever that means). The original members of S might as well have to readjust their cooperation to accommodate the new entrant to coalition $S \cup \{i\}$.

In our design of the value function v^g these contributions are particularly evident which allows for an exact decomposition among the players of a coalition. Suppose two players, i and j, enter a peering agreement and create savings amounting to

$$v^g(\{i,j\}) = \alpha_i \alpha_j \left(l_i \cdot q(p_i) + l_j \cdot q(p_j)\right). \tag{5.20}$$

At the same time (5.20) is the marginal value of player i upon joining coalition $\{j\}$ and vice versa. Disassembling this, the contribution of player i equals $\alpha_i \alpha_j l_i q(p_i)$ and player j's is equal to $\alpha_i \alpha_j l_j q(p_j)$. More generally, under v^g, the contribution of player i to coalition $S \subseteq N$ in network g is defined as

$$c_i^g(S) := \sum_{j \in N_i(g) \cap S} \alpha_i \alpha_j l_i q(p_i) \text{ for all } S \subseteq N,\ i \in S. \tag{5.21}$$

It is easy to verify by comparison with (5.14) that

$$\sum_{i \in S} c_i^g(S) = v(S,g) \text{ for all } S \subseteq N,\ i \in S, \tag{5.22}$$

and so the contributions of the players in a coalition S add up to the value of this coalition under v^g. Therefore, the notions of "contribution towards the marginal value upon joining a coalition" and simply "contribution to a coalition" coincide. Consider the following definition:

Definition 5.103. A *zero-contribution player* in the game defined by v^g is a player i for whom

$$c_i^g(S) \equiv 0 \text{ for all } S \subseteq N,\ i \in S, \tag{5.23}$$

holds.

It follows from (5.21) that the design of the peering game allows for such zero-contribution players: First of all, the pathological case of players without any links:

$$\text{If} \quad N_i(g) = \emptyset, \ \text{ then } c_i^g(S) \equiv 0 \ \text{ for all } \ S \subseteq N, \ i \in S.$$

On this we will not dwell any further. More interesting is the set of players with zero long distance fees, usually represented by network operators:

$$\mathbf{0} := \{i \in N |\ l_i = 0\}. \tag{5.24}$$

Even though they have strictly positive marginal values when entering cooperation with players whose long distance fees are not equal to zero, their contribution in the above sense is nil. From hereon, we will refer to $\mathbf{0}$ as the set of *zero-players*. Note that these players are (as opposed to the pathological kind above) independent of the network.[14] As we shall see, these zero-players will play an important role in our subsequent analysis.

5.2.3.3 Monotonicity

The characteristic function v^g is *monotonic*, if the addition of new members to a coalition cannot result in a lower value. For any two coalitions with $S \supseteq T$, we have $v^g(S) \geq v^g(T)$. Put differently, players must never exhibit negative marginal values: $d_i(S) \geq 0$ for all $S \subseteq N \setminus \{i\}$ and all $i \in N$.

The value function v^g as given in (5.14) fulfills this requirement. All elements are nonnegative and summing over additional members of S can thus not decrease the value of v^g, for any coalition $S \subseteq N$ on any given network g. This can be illustrated more detailed by means of four possible cases of combinations of coalitions and networks:

Case 5.1. $T \setminus S = \{i\}$ and $g_{ij} = 0$ for all $j \in S$
When adding a single member, here i, to S and there exists no link between i and a member of S, nothing changes: $v^g(T) = v^g(S \cup \{i\}) = v^g(S)$.

Case 5.2. $T \setminus S = \{i\}$ and there exists a player $j \in S$ such that $g_{ij} = 1$
When adding a single member, here i, to S and there exists at least one link between i and members of S, the sum in (5.14) will be extended by at least one pair of nonnegative summands: $v^g(T) = v^g(S \cup \{i\}) \geq v^g(S)$.

Case 5.3. $|T \setminus S| > 1$ and $g_{ij} = 0$ for all players $i \in S$, $j \in T \setminus S$
When adding multiple members to S and there exists no links between the new members and some member of S, we face the following situation: $v^g(T) = v^g(S) +$

[14] We are aware of the abuse of notation, but see no potential for confusing $\mathbf{0}$ with a vector of zeroes in our context.

$v^g(T \setminus S) \geq v^g(S)$. If there exist no links among the new members, then $v^g(T \setminus S) = 0$ and $v^g(T) = v^g(S)$. Otherwise, the sum in (5.14) will be extended by at least one pair of nonnegative summands and $v^g(T) \geq v^g(S)$.

Case 5.4. $|T \setminus S| > 1$ and there exist players $i \in S$ and $j \in T \setminus S$ such that $g_{ij} = 1$ When adding multiple members to S and there exists at least one link between one of the new members and some member of S, the sum in (5.14) is extended by at least one pair of nonnegative summands: $v^g(T) \geq v^g(S)$.

Again, it is important to stress that the network g affecting v^g does not change when players are added to S. No links will be formed or removed. A link g_{ij} from some member $i \in S$ to a player $j \notin S$, which is irrelevant at first, might become of interest, whenever this neighbor j is a member of $T \setminus S$.

5.2.3.4 Superadditivity

The characteristic function v^g is *superadditive*, if for any pair of coalitions $S, T \subseteq N$ with empty intersection, $S \cap T = \emptyset$, the following holds:

$$v^g(S \cup T) \geq v^g(S) + v^g(T). \tag{5.25}$$

The merger of any two disjoint coalitions can never diminish the overall gains. Clearly, superadditivity holds for the game based on v^g as defined by (5.14). Whenever there exists no link g_{ij} with $i \in S$ and $j \in T$ (or vice versa), (5.25) holds with equality. Since in this case no neighborhood of a player $i \in S$ has a nonempty intersection with coalition T (and vice versa), the inner sum in (5.14) is not extended by any additional summands through the union of coalitions S and T. Because the inner sum is not affected, the outer sum can be separated into sums over $i \in S$ and $i \in T$, equal to $v^g(S)$ and $v^g(T)$, respectively:

$$\begin{aligned}
v^g(S \cup T) &= \sum_{i \in S \cup T} \sum_{j \in N_i(g) \cap (S \cup T)} \alpha_j \alpha_i l_i q(p_i) \\
&= \sum_{i \in S \cup T} \left[\sum_{j \in N_i(g) \cap S} \alpha_j \alpha_i l_i q(p_i) + \sum_{j \in N_i(g) \cap T} \alpha_j \alpha_i l_i q(p_i) \right] \\
&= \sum_{i \in S} \sum_{j \in N_i(g) \cap S} \alpha_j \alpha_i l_i q(p_i) + \sum_{i \in T} \sum_{j \in N_i(g) \cap T} \alpha_j \alpha_i l_i q(p_i) \\
&= v^g(S) + v^g(T).
\end{aligned}$$

From the disjointness of S and T and *De Morgan's Law*, we obtain

$$N_i(g) \cap (S \cup T) = (N_i(g) \cap S) \cup (N_i(g) \cap T).$$

The two sets in parentheses on the right-hand-side are again disjoint and so the outer summation over $S \cup T$ can be broken up. As soon as links between members of S and T exist, the inner sum in (5.14) changes, because for some players $i \in S$, the intersection of their neighborhoods $N_i(g)$ with T is nonempty (and vice versa). This gives rise to additional pairs of nonnegative summands. A strict inequality could be the result, depending on the long distance fees associated with the links between S and T. Therefore the marginal value of players or coalitions is never strictly negative and the peering game is superadditive.

5.2.3.5 Convexity

The characteristic function v^g is *convex*, if the marginal value of any given player is nondecreasing with respect to the size of a coalition he is to join. As introduced in Definition 2.25, p. 38, we write

$$d_i(S) \leq d_i(T) \quad \text{for all} \quad S \subseteq T \subseteq N, \ i \in N, \tag{5.26}$$

where d_i is now calculated from the network restricted function v^g.

We now check that v^g satisfies this notion of "nondecreasing returns". Within a given network g player i has at least as many links to players in coalition T as he has to players in coalition $S \subseteq T$. Adding him to the larger coalition T cannot yield a lower sum in (5.14) than when adding him to coalition $S \subseteq T$. As the summands are nonnegative, (5.26) must hold with equality at least. This establishes convexity.

We now turn to the more network specific properties of the peering game v^g. These are concerned with changes to the underlying network g rather than with players and coalitions. We begin with a hybrid, decomposability, which was first defined for standard coalitional games but has a very intuitive connection to our peering game.

5.2.3.6 Decomposability

Another interesting feature of the peering game is its decomposability which depends very much (but not exclusively) on the underlying network g. According to Definition 2.31, p. 43, a game is decomposable with respect to a partition $\mathsf{P} = \{P_1, P_2, \ldots, P_m\}$ on the player set N, if

$$v(S) = v(S \cap P_1) + v(S \cap P_2) + \cdots + v(S \cap P_m) \tag{5.27}$$

is fulfilled for all $S \subseteq N$. Each of the $m \geq 2$ *sub*games $(P_j, v|_{P_j})$ is called a component of (N, v). For the peering game v^g, the origin of decomposability is twofold. The network g plays a major role, but the long distance fees l_i can have some influence, too. We begin with a straightforward statement:

Proposition 5.23. *The peering game (N, v^g) is decomposable with respect to the partition N/g induced by the network g on the player set N.*

Proof. It follows from (5.14) that only linked pairs of players can create value under v^g. If g is connected and induces the trivial partition $N/g = \{N\}$, there is nothing to prove. But if we have more than one component, the inner sum of v^g will be zero for all unlinked pairs of players i and j that do not belong to the same component $C \in N/g$. The outer sum in (5.14) can therefore be broken up into the (nonempty) intersections of S with the components $C \in N/g$:

$$\begin{aligned} v^g(S) &= \sum_{i\in S}\sum_{j\in N_i(g)\cap S} \alpha_j\alpha_i l_i q(p_i) \\ &= \sum_{C\in N/g}\sum_{i\in S\cap C}\sum_{j\in N_i(g)\cap S} \alpha_j\alpha_i l_i q(p_i) \\ &= \sum_{C\in N/g}\sum_{i\in S\cap C}\sum_{j\in N_i(g)\cap (S\cap C)} \alpha_j\alpha_i l_i q(p_i) \\ &= \sum_{C\in N/g} v^g(S\cap C) \qquad \text{for all } S\subseteq N. \end{aligned} \tag{5.28}$$

As in the proof on superadditivity above, we can break up the summations over disjoint sets conveniently. In the last step there are by definition no links between members of different components $C \in N/g$, and so $N_i(g) \subseteq C$ whenever $i \in C$. Consequently, for an arbitrary player $i \in C$, the sets

$$\{j \in N | j \in N_i(g) \cap (S \cap C)\} \text{ and } \{j \in N | j \in N_i(g) \cap S\}$$

coincide. ■

In the context of value functions defined on networks, Jackson and Wolinsky (1996, p. 59) refer to the above property as *component additivity*, based on the components N/g induced by network g. Also, decomposability as shown above is not limited to the whole network g and the entire player set N. We can restrict the network to any $g|_S$ with corresponding player set $S \subseteq N$:

Corollary 5.9. *An analogue of Proposition 5.23 holds for every network $g|_S$ with player set $S \subseteq N$ and the corresponding partitions induced by S/g.*

Proof. Replace the set of players N by $S \subseteq N$. ■

So far, the player sets corresponding to the components from decomposition, $P_j \in \mathsf{P}$, and the sets of nodes induced by the network, $C \in N/g$ coincided. But, in addition to the partition induced by the network, N/g, another possibility arises for v^g to be decomposable. It stems from allowing for zero-players with $l_i = 0$. The following stylized example shows this.

Example 5.1. Suppose two sets S and T partition N: $S \cup T = N$ and $S \cap T = \emptyset$. Furthermore, suppose that every possible path between any two players $l \in S$ and $k \in T$ includes the link g_{ij}. That is, if we removed the link g_{ij}, the network

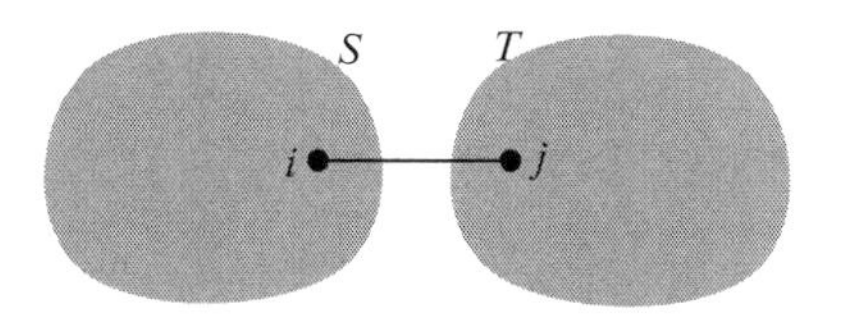

Fig. 5.5 Every path between players in sets S and T includes the link g_{ij}

$g - g_{ij} =: g'$ would have two components, S and T: $N/g' = \{S, T\}$, whereas before $N/g = \{N\}$. The situation under g is illustrated in Fig. 5.5.

Now assume that players i and j are both zero-players with $l_i = l_j = 0$. Then,

$$v^g(N) = v^g(S) + v^g(T) = v^{g'}(S) + v^{g'}(T) = v^{g'}(N), \tag{5.29}$$

and the communication situation (N, v, g) is decomposable, even though network g has only one component. The situation illustrated in Fig. 5.5 can be generalized to multiple links between members of the sets S and T, as long as these links are formed by zero-players on each side.

With the preceding in mind, we can now formulate a sufficient condition to ensure the peering game is not decomposable:

Proposition 5.24. *The peering game (N, v^g) is not decomposable, if all n players are connected and no set $S \subseteq N$ exists, for which every path from an element in S to an element in $N \setminus S$ includes a link g_{ij} with $i \in S$, $j \in N \setminus S$ between two zero-players.*

$$\begin{aligned}
&\text{If } N/g = \{N\} \text{ and } \nexists\, S \subseteq N \text{ such that} \\
&\text{for all } h \in S,\ k \in N \setminus S \text{ and for all } p_{hk} \in P_g(h,k) \\
&\exists\, g_{ij} \in p_{hk} \text{ with } l_i = l_j = 0,\ i \in S,\ j \in N \setminus S \\
&\text{then } v^g \text{ is not decomposable.}
\end{aligned}$$

$P_g(h,k)$ is the correspondence $P : \mathcal{G} \times N \times N \rightrightarrows 2^g$ assigning to every pair of players all possible paths that connect them via g. It was introduced in expression (3.10), p. 122.

Proof. As $N/g = \{N\}$, there is no nontrivial partition induced by the network g whose component values add up to the value of the grand coalition $v^g(N)$. If all n players have nonzero long distance fees, no decomposition is possible according to (5.14). This also holds if there are players i with $l_i = 0$, as long as such players are not linked to one another, i.e. there exist no links g_{ij} with $l_i = l_j = 0$. Now suppose such links g_{ij} exist in g. Collect them all in the set denoted $g_{\mathbf{0}}$ as defined by

$$g_{\mathbf{0}} := \left\{ g_{ij} \middle|\ l_i = l_j = 0,\ i, j \in N \right\}.$$

If we remove them from network g to obtain $g \setminus g_{\mathbf{0}} =: g'$, and network g' is still connected, i.e. $|N/g'| = 1$, there exists no set $S \subseteq N$ as described above, because all n players are connected via links that generate strictly positive gains through v. But, if $|N/g'| \geq 2$, there exist sets of players, the members of which were connected under g, albeit through a path including a link $g_{ij} \in g_{\mathbf{0}}$. This is a contradiction to our initial assumption. ■

Corollary 5.10. *An analogue of Proposition 5.24 holds if the characteristic function v^g is restricted via $v^g|_C$ on the players in components $C \in N/g$.*

Proof. Replace the set of players N by some $C \in N/g$ and take a coalition S satisfying $S \subseteq C$. Other than that, the proof is identical. ■

Naturally, the finest decomposition achievable via v^g is the partition of N into singleton sets. But this corresponds to the pathological cases, where either $g = g^{\emptyset}$ is the empty network or $l_i = 0$ for all $i \in N$. Even though formally possible, when we speak of "components" we refer to sets with existing links between players within. This comes naturally in our context, as unlinked coalitions cannot produce any savings. From this follows another proposition:

Proposition 5.25. *A game (N, v^g) is not decomposable, if the underlying network g is the complete network g^c, and there exists a player $i \in N$ with strictly positive long distance fees, $l_i > 0$.*

Proof. Since exactly $n-1$ players are linked to each player, hence also to a player with $l_i > 0$, no proper subset $S \subsetneq N$ exists for which all paths between members in S and $N \setminus S$ include a link with zero-players on both ends. If S is such that $l_j = 0$ for all $j \in S$, then also for each $j \in S$ there must be a link to the one player i with the strictly positive long distance fee. ■

5.2.3.7 Critical Link Monotonicity

A communication situation is said satisfy this property (see Definition 3.81, p. 128), if the removal of a crucial link increases the average value of either new component above the average value of the initial component. In other words, one component has a high value (relative to the number of its members) while the other one has a lower value. The peering game based on v^g does not generally fulfill this critical link monotonicity, which the following example will illustrate:

Example 5.14. We consider a peering game with the characteristic function v^g. The player set is given by $N = \{1,2,3,4\}$, and the network is a line network, $g = \{g_{12}, g_{23}, g_{34}\}$. Let the parameters α_i, p_i, and l_i be such that the gains created among the links of g are the following: $v^g(\{1,2\}) = 5$, $v^g(\{2,3\}) = v^g(\{3,4\}) = 1$, and consequently, $v^g(N) = 7$. If we remove the critical link g_{23} from g, it holds for the new network $\tilde{g}$ that

$$v^g(N) \geq v^{\tilde{g}}(\{1,2\}) + v^{\tilde{g}}(\{2,3\}).$$

Nevertheless,

$$\frac{v^g(C)}{|C|} \geq \max\left\{\frac{v^{\widetilde{g}}(C_i)}{|C_i|};\frac{v^{\widetilde{g}}(C_j)}{|C_j|}\right\},$$

with $N/g = C$ and $N/\widetilde{g} = \{C_i, C_j\}$ is violated, because

$$\frac{v^g(C)}{|C|} = \frac{7}{4} < \frac{5}{2} = \frac{v^{\widetilde{g}}(\{1,2\})}{|\{1,2\}|}.$$

5.2.3.8 Network Monotonicity

A communication situation is monotonic on g, if none of the links $g_{ij} \in g$ affect the value of $v(S,g)$ negatively, regardless of coalition $S \subseteq N$. This is satisfied by v^g, because any added link leads to two nonnegative additional (inner) summands in (5.14) whenever $i, j \in S$ and leaves v^g unaffected otherwise.

5.2.3.9 Link Anonymity

While the peering game v^g is anonymous regarding the players, it is not so in terms of links. The marginal value of a link is individually constructed from parameters linked to the players at both ends. Hence, the value of a coalition clearly does not depend solely on the number of links between its members, but also on the specific members which are being linked.

This concludes our analysis of the peering game v^g itself. We have established that it does fulfill a number of convenient properties. Having to deal only with nonnegative marginal values simplifies things a lot. In particular, convexity ensures nonemptiness of the core and has further implications on allocation rules. Decomposability, when applicable, lets us focus on the components one by one, without distorting overall results. And the fact that v^g is not link anonymous supports our view to base a peering situation on players after all, and not on links between them.

How these properties influence the outcomes of the various solution concepts will be subject of the following section. There, we take a look at some possible allocation methods and determine their advantages and disadvantages with respect to our application, i.e. the gains from peering through cooperation.

5.3 Allocation of Gains from Peering

5.3.1 Allocate What?

This section revolves around the allocation of gains from peering in the VoIP peering game. Before we really go into the matter of allocation, we have to ask ourselves the question what exactly we are allocating, or more precisely, what our point of origin is. Implicitly, we answered this already by including the access fees into our value function v^g. It is constructed to incorporate all costs a VoIP firm can save by peering, which, in our model, includes long distance transit fees and access charges. But, by expression (5.14), the cooperative gains of any coalition of firms result from savings on long distance fees alone. The change in access revenue drops out completely, as we have shown in (5.16) ff.; even though it arises on the same basis as gains from long distance fees, i.e. pairs of players, the access charges are paid by the VoIP firms to each other and not to some external third party like a long distance provider. Because the change in access revenues through peering constitutes a zero-sum game, coalitions of players are unaffected by this. It is the individual player that is likely to be affected from access fees in his payoffs. While being able to sideline the readjustment of access revenues for now, it will be pivotal in Chap. 6, when it comes to network formation and the problem with whom to peer is considered. For the subsequent analysis of allocation we can ignore access charges entirely and focus on the gains from transit.

As far as allocation methods are concerned, we again work our way from more general concepts to the specific ones: We begin with set-valued solutions, most notably the core, and investigate whether the least-core can yield a point-valued reduction of the set of outcomes generally produced by the former. Then we turn to the point-valued solution concepts, where at first the Shapley and Myerson values are applied. The second approach are weighted solution concepts, i.e. the Weighted Shapley value and the weighted bargaining solutions by Nash and Kalai–Smorodinsky. We close with a comparison and discussion of the results.

5.3.2 The Core of the Peering Game

In this section, we take a closer look at the core of the VoIP peering game and try to find intuitive characteristics that allow us to describe it more easily and possibly in more detail. These will have to result from the network g and the influence it has on v^g. As we have checked, the peering game is convex, so we benefit from some results derived earlier. Most importantly, the stable set solution and the core coincide, so we can restrict ourselves here to the latter concept.[15] Equally important, the core of any convex game is nonempty, which means that a solution of this type always exists.[16]

As we noted earlier, the core is a solution concept requiring coalitional rationality: Any allocation $\mathbf{x}$ in the core must meet the demands of all coalitions $S \subseteq N$: $x(S) \geq v(S)$. Naturally, this includes the polar cases of singleton coalitions if $S = \{i\}$, $i \in N$, as well as the grand coalition when $S = N$. The first is referred to as individual rationality, the second implies an efficient, non-wasteful, allocation.

[15] For this result, see Theorem 2.5, p. 81, or Shapley (1971).

[16] We refer the reader to either Theorem 2.3, p. 74, or Shapley (1971).

If we now take these conditions to the peering game, the singleton case needs little further mention: A single VoIP firm cannot create any savings on its own and $v^g(\{i\}) = 0$ for all $i \in N$. Hence, any nonnegative core allocation, $x(\{i\}) \geq 0$, satisfies the individual player i by definition of the core, as he is at least as well off as without peering at all.

For the grand coalition N, efficiency dictates that in any core allocation $\mathbf{x} \in C(N, v^g)$ all savings are to be exhaustively distributed among the n players of the game. Other than a global lower and upper bound, the network g does not seem to give us any useful information as to how core allocations can be characterized more thoroughly for these polar cases. However, this changes when we look at intermediate coalitions. Within those coalitions, we focus first on allocations to individual players, as the network g sets exact limits how much they could obtain without violating the core requirements. From there, we extend the notion to general coalitions.

The smallest value-creating coalition under v^g, a pair of players, with no other links than to one another, is our starting point. We can do so without loss of generality, because the value some players i and j create is independent of the remaining network $g - g_{ij}$, according to Proposition 5.22.

Therefore, consider a simple network $g = \{g_{ij}\}$ consisting of only one link between distinct players i and j. Any allocation in the core needs to be no less than what they could realize jointly, $x(\{i,j\}) \geq v(\{i,j\})$, and must hold with equality because of efficiency. Even though gains from peering can be attributed directly to their originators, the pairs of players, the core is silent about the distribution within such pairs, other than $x(\{i\}) \geq v(\{i\}) = 0$ to hold for all $i \in N$ as well. Thereby, any such situation is characterized by the following three conditions:

$$\begin{aligned} x_i + x_j &= v(\{i,j\}) \\ 0 \leq x_i &\leq v(\{i,j\}) \\ 0 \leq x_j &\leq v(\{i,j\}) \end{aligned}$$

The upper bound for the payoff of player i is the total gain achieved through the link g_{ij}, which would consequently leave player j with nothing, and vice versa.

For players with more that just one neighbor $j \in N_i(g)$, these conditions are still valid for each link by itself. Within a general network g, the maximal payoff of a single player $i \in N$ without violating the core requirements is given on the right-hand side of

$$x_i \leq \sum_{j \in N_i(g)} v^g(\{i,j\}) \quad \text{for all} \quad i \in N,\ \mathbf{x} \in C(N, v^g),\ g \in \mathcal{G}_N.$$

When met with equality, player i is allocated the total value created from links to all his neighbors $j \in N_i(g)$. This does not necessarily mean that these neighbors $j \in N_i(g)$ are allocated nothing, as they might have other neighbors than just player i with whom they realize gains from peering.

In terms of networks, the most a player i can be allocated is the value created within the maximal star network embedded in g, for which he serves as central node.[17] Because savings are generated analogously for all players, if i was to be allocated strictly more, some coalition S, not necessarily intersecting $N_i(g) \cup \{i\}$, would have to be allocated strictly less than what it could achieve on its own. These findings can be summarized:

Theorem 5.9. *In the peering game v^g, the following must hold for all core allocations $\mathbf{x} \in C(N, v^g)$:*

$$x_i \leq \sum_{j \in N_i(g)} v^g(\{i, j\}) \text{ for all } i \in N. \tag{5.30}$$

Proof. We prove by contradiction. Suppose there is an allocation $\mathbf{x} \in C(N, v^g)$ for which some player i is allocated an amount strictly larger than the right-hand side of (5.30). Now, according to (5.14), the value of the grand coalition, $v^g(N)$, in network g is given by $v^g(N)$. To highlight the sum of the marginal value created by some player i when joining the grand coalition, we write

$$v^g(N) = v^g(N \setminus \{i\}) + \sum_{j \in N_i(g)} v^g(\{i, j\}). \tag{5.31}$$

The second term in (5.31) is exactly the aforementioned value of the maximal star network with center i, corresponding to $d_i(N)$. Note that the gains possibly created between players $j \in N_i(g)$ are still included in the term $v^g(N \setminus \{i\})$. Because $\mathbf{x}$ is assumed to be in the core, $x(N) = v^g(N)$ holds. Now, if $x_i > \sum_{j \in N_i(g)} v^g(\{i, j\})$, it follows that $x(N \setminus \{i\}) < v^g(N \setminus \{i\})$ in order for (5.31) to be satisfied. This violates the conditions for the allocation $\mathbf{x}$ to be in the core. ■

If the left-hand side of (5.30) is pushed to where it just holds with equality, another interesting observation can be made:

Corollary 5.11. *Any core allocation $\mathbf{x} \in C(N, v^g)$ for which (5.30) holds with equality for some $i \in N$ must be located on the boundary of the core:*

$$\text{For all } \mathbf{x} \in C(N, v^g) \,\middle|\, \exists\, i \in N \text{ with } x_i = \sum_{j \in N_i(g)} v^g(\{i, j\}) \Rightarrow \mathbf{x} \in bd\, C(N, v^g). \tag{5.32}$$

Proof. Suppose there exists some allocation $\mathbf{x}' \in int\, C(N, v^g)$ for which (5.30) holds with equality. Then, there must also exist an $\varepsilon \in \mathbb{R}_{++}$ such that an open ball around $\mathbf{x}'$ with radius ε is included in the core: $B(\mathbf{x}', \varepsilon) \subsetneq C(N, v^g)$. But then we could increase the allocation to player i at the expense of some other player by the amount ε to $x_i = x'_i + \varepsilon$ and $\mathbf{x}'$ is still in the core. Deducing from (5.31) and the fact that $x'(N) = v^g(N)$, this leads to $x'(N \setminus \{i\}) < v^g(N \setminus \{i\})$, a violation of the core requirements. This is a contradiction to our initial assumption $\mathbf{x}' \in int\, C(N, v^g)$. ■

[17] This amount is limited to $d_i(N)$, the maximal marginal value of player i for a given network g.

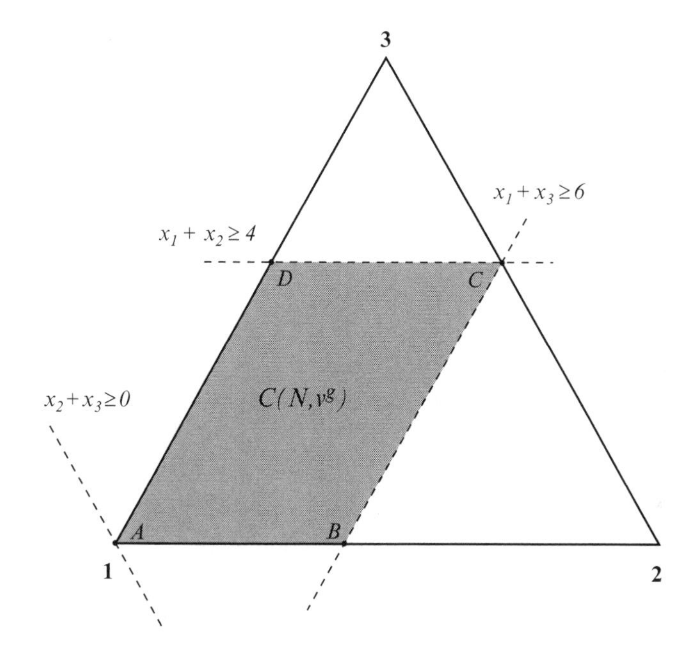

Fig. 5.6 Peering game with core $C(N, v^g)$

The notion of upper bounds on allocations is readily extended to general coalitions $S \subseteq N$: Intuitively, this upper bound is the sum of two parts: First, the value created among its members $i \in S$. Secondly, in accordance with previous considerations, each $i \in S$ could also be allocated the value created with neighbors not in S. Formally, we write

$$x(S) \leq v^g(S) + \sum_{i \in S} \sum_{j \in N_i(g) \setminus S} v^g(\{i, j\}) \quad \text{for all } S \subseteq N. \tag{5.33}$$

For $|S| = 1$ this expression collapses to (5.30). The proof is analogous to the one presented for the single player case, and hence omitted. An illustration is given in the two examples below.

Example 5.15. In a peering game v^g with $N = \{1, 2, 3\}$, situated on a line network $g = \{g_{12}, g_{13}\}$, the characteristic function v^g shall assume the following values: $v^g(\emptyset) = 0$ and $v^g(\{i\}) = 0$ for all $i \in N$. Furthermore, $v^g(\{1, 2\}) = 4$, $v^g(\{1, 3\}) = 6$, $v^g(\{2, 3\}) = 0$. Consequently, $v^g(N) = 10$. The core $C(N, v^g)$ is depicted in Fig. 5.6 as the shaded area with the corner points A,B,C, and D. Their coordinates are given by

$$A = (10, 0, 0),$$

$$B = (6, 4, 0),$$

$$C = (0,4,6),$$
$$D = (4,0,6),$$

where each entry corresponds to the allocation of the respective player. In this example the number of vertices of the core (four) is less than the number of possible permutations on the player set (six). For permutations $(1,2,3)$ and $(1,3,2)$ the marginal vector is identical, given by C, because it does not make a difference in which order players 2 and 3 join player 1 to form the grand coalition. Gains are only realized with player 1 but not between players 2 and 3. The same holds for permutations $(3,2,1)$ and $(2,3,1)$, corresponding to marginal vector A. Permutations $(3,1,2)$ and $(2,1,3)$ are represented by B and D, respectively. Which marginal vectors coincide in terms of payoffs depends on the underlying network structure g. For a comparison, see Example 5.16 below.

In Example 5.15, the points A,B,C, and D are allocations in which (5.30) holds just with equality for some player who receives his maximally possible amount for any core allocation. The corresponding allocations are located on the boundary of the core. Also, allocations for which (5.30) is violated are outside the core, possibly even outside the simplex. With three players this is straightforward to see: Whenever some player i receives more than his "maximum", the other $N \setminus \{i\}$ players are allocated less than they could realize independently. Such allocations are then outside $C(N, v^g)$. Unfortunately, in the three player case, the upper bounds of a given player or coalition always coincide with a boundary of the simplex and do not lend themselves to a distinct graphical representation.

Example 5.16. If we modify the peering game from Example 5.15 by adding link g_{23} to obtain the complete network g^c and set the gains for this pair of players to $v^g(\{2,3\}) = 2$, the value of the grand coalition also rises: $v^g(N) = 12$.

As before, the core $C(N, v^g)$ is depicted as the shaded area in Fig. 5.7. But this time there are more extreme points of the core: A,B,C,D,E, and F with coordinates

$$A = (10,0,2),$$
$$B = (10,2,0),$$
$$C = (6,6,0),$$
$$D = (0,6,6),$$
$$E = (0,4,8),$$
$$F = (4,0,8).$$

These vertices are represented by the following permutations: $(2,3,1)$, $(3,2,1)$, $(3,1,2)$, $(1,3,2)$, $(1,2,3)$, and $(2,1,3)$, in alphabetical order. Now, all marginal vectors are distinct.

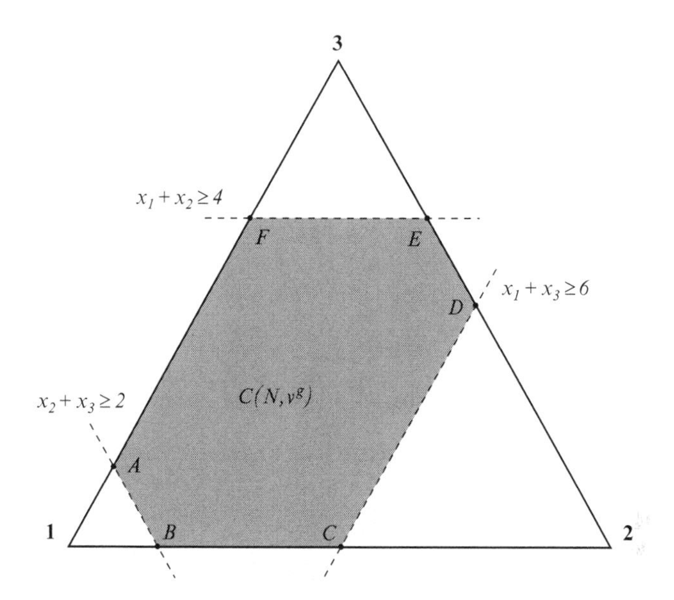

Fig. 5.7 Peering game with core $C(N, v^g)$

Interlude: *The attentive reader will have noticed the connection to marginal worth vectors in convex games. The upper bound for x_i in (5.30) is exactly the entry of player i in all marginal worth vectors $\mathbf{x}^{\pi}$ whose permutations specify $\pi(i) = n$, i.e. with player i in the last spot.*[18] *In convex games, this coincides with the largest possible marginal worth player i can realize.*[19] *But even though marginal worth vectors and the vertices of the core in a convex game coincide, the allocations for which (5.30) holds with equality comprise a much larger set than the possibly n! vertices. They must all lie on the boundary of the core, but are not restricted to the extreme points of the latter. To see this, consider the relationship between neighboring vertices of the core:*[20] *For each vertex $\mathbf{x}^{\pi}$, there are as many as $n-1$ neighboring vertices, but there is only one, where the last player under π*

[18] This fact is not restricted to convex games. In our context it arises as property of the peering game v^g.

[19] This is not the whole truth, especially not in convex games like the peering game: Expression (5.30) holds with equality in all marginal worth vectors where i is in the position allowing for his maximal marginal value, which need not necessarily be the last spot in a permutation π. In the peering game for example, all marginal vectors qualify in which only players not linked to i come after him in π.

[20] Neighboring vertices are such marginal vectors x^{π} and $x^{\pi'}$ whose underlying permutation π and π' differ only in two consecutive slots of the ordering: For some distinct players $i, j \in N$, for which $\pi(j) = \pi(i) + 1$, to yield a neighboring marginal vector, the permutation π' must fulfill $\pi'(j) = \pi(i)$, $\pi'(i) = \pi(j)$, and $\pi'(k) = \pi'(k)$ for all $k \in N \setminus \{i, j\}$. This is a special case of a *transposition* of π into π', as two consecutive (and not any two) elements are switched.

is moved to the penultimate spot in the new ordering.[21] *So for all other neighboring vertices, the allocation to the last player i in the marginal worth vector will remain constant. Only the allocations to some players earlier in the order will change, as we move along the edge from $\mathbf{x}^{\pi}$ to, say, $\mathbf{x}^{\pi'}$. Then, for each allocation $\mathbf{y} \in [\mathbf{x}^{\pi}, \mathbf{x}^{\pi'}] \subseteq C(N,v)$ on this edge, it holds that $x_i = \sum_{j \in N_i(g)} v^g(\{i,j\})$ and we are on the boundary of the core. So the set of allocations where some player i is allocated $x_i = \sum_{j \in N_i(g)} v^g(\{i,j\})$ is a proper superset (uncountable at that) of the marginal vectors, or vertices of the core. However, it does not cover the whole boundary (or surface) of the core. To see this, just consider two neighboring vertices for which the last player i with $d_i(N) > 0$ changes places with the second to last one. The edge connecting them need not contain an allocation $\mathbf{x}$ with some player j whose allocation is at the described limit.*[22] *This is also quite intuitive, since there are more ways of violating one of the core inequalities than by just exceeding the limit allocation for some single player. This interlude can be illustrated by means of Examples 5.15 and 5.16, in particular their corresponding figures. They allow to visualize the content of this interlude.*

These considerations lead to a more specific characterization of the core when dealing with the peering game as defined above: Not only is there the usual minimum (total) allocation for every coalition $S \subseteq N$, and the global upper bound embodied in the efficiency criterion $x(N) = v^g(N)$. Now we can even be more specific about every possible coalition.[23] Formally, the more precise definition for the core looks like this:

$$C(N,v^g) = \left\{ \mathbf{x} \in \mathbb{R}^n_+ \mid v^g(S) + \sum_{i \in S} \sum_{j \in N_i(g) \setminus S} v^g(\{i,j\}) \geq x(S) \geq v^g(S),\ S \subseteq N \right\}. \tag{5.34}$$

This does not make the set of core allocations any smaller than under the common definition. But now, for every one of the $2^n - 1$ half-spaces, there exists one half-space parallel to it, yet facing in the opposite direction. This, in general, is a closer bound on the core conditions than the one given by the boundary of the simplex. As we have seen, players can potentially draw all gains which they create with their neighbors. Even though this still qualifies for a core allocation, it is hardly enforceable, as each player at the end of a given link is equally necessary to realize the gains from it. Why, after all, should any firm pass on those gains and not the firm on the other end of the link? This is especially true in our application to VoIP peering: One firm would not only forgo those gains passively, but it would actively have to pay the other the amount it previously paid for long distance transit.

---

[21] See Example 5.15 for a convex game with less core vertices than permutations on N.

[22] If no player i with $d_i(N) > 0$ for some permutation exists, the game is inessential.

[23] In geometrical terms, the limit before was given by the set of imputations, i.e. the simplex. Now, every coalition's allocation $x(S)$ is bounded by the intersection of two "opposing" half-spaces. They are both within the simplex insofar as the new upper bound is not slack.

In summary, the underlying network does allow for some intuitional restrictions on the core as to which allocations are admitted and which are not. These restrictions are only player- or coalition-specific and by no means global and reveal little information on what the payoffs to the complementary players are to look like without violating any core requirements. But at least, by means of the network g, the upper bound of the payoffs of a coalition can be used as a quick method to sort out allocations $\mathbf{x}$ that are not in the core. Taking the core allocations as a starting point in the search for an applicable allocation is very useful, as no VoIP firm is made worse off than without any cooperation at all.

With intuition-supported boundaries on what a single player or a coalition can expect to be allocated without violating the core requirements, and before turning to truly point-valued solution concepts, we try to narrow down the allocations provided by the core in another way. We employ the concept called ε-core, but contrary to its common use. As the peering game is convex, emptiness of the core is not a problem to be concerned with. Rather, the core is likely to be an uncountable set of allocations in up to $n-1$ dimensions, which we would like to reduce as far as possible, albeit still in the spirit of the core. Therefore, we go the other direction, and instead of loosening the core inequalities, we will tighten them, to admit as few allocations as possible.

5.3.3 Downsizing with the Least-Core

One way of ensuring the nonemptiness of the core of a general cooperative game without imposing assumptions on the underlying characteristic function v is to loosen the core requirements, given by the $2^n - 1$ inequalities $x(S) \geq v(S)$ for all $S \subseteq N$. In that context, we introduced the ε-core and subsequently the least-core in Sect. 2.4.3.10, p. 83. According to this concept, all half-spaces representing the proper core inequalities for $S \subsetneq N$ are moved by the same amount ε, until their intersection is a nonempty set. In our peering game, we know the core to be nonempty due to the convexity of v^g. Most likely, the core even contains an uncountable number of elements, which makes picking the "right" allocation very difficult, or at least dependent on some other concept. It was already hinted at in Footnote 59, p. 59, that the sign of ε is not predetermined. It is equally possible to choose a negative value for ε, in order to sharpen the core requirements and thereby downsizing the core. For this to make sense, we will have to slightly alter the definition of the ε-core, originally given in (2.149) on p. 83:

$$C_\varepsilon(N, v^g) := \{\mathbf{x} | \mathbf{x} \in \mathbb{R}^n,\ x(N) = v^g(N),\ x(S) \geq v^g(S) - \varepsilon \ \text{ for all } \ S \subsetneq N\}. \quad (5.35)$$

With the inequality for the grand coalition N excluded from modification through ε, the playing field, i.e. the simplex of imputations $I(N, v^g)$, will not expand as the absolute value of ε increases. Consequently, within a given simplex, the half-spaces representing the core inequalities will move closer together as $|\varepsilon|$ goes

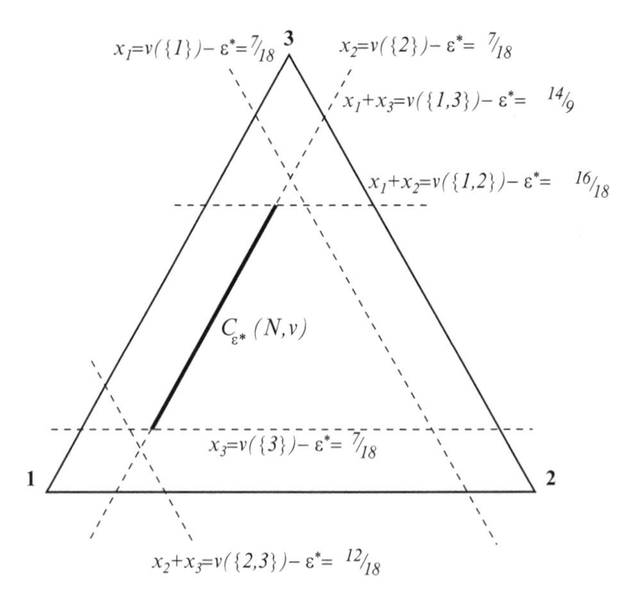

Fig. 5.8 $C_{\varepsilon^*}(N, v^g)$ in a communication situation with $\varepsilon^* = -\frac{7}{18}$

up. Unfortunately, there is no guarantee that the least-core will turn out to be a singleton, not even in convex games. This is illustrated by Shapley and Shubik (1969, pp. 54 f.). Below, we provide an example for a specific peering game v^g:

Example 5.17. In a peering game v^g, let the player set be given by $N = \{1,2,3\}$, the network be complete, so that $g = g^c$, and the characteristic function assume the following values: $v^g(\emptyset) = 0$, $v^g(\{i\}) = 0$ for $i = 1,2,3$, $v^g(\{1,2\}) = \frac{1}{2}$, $v^g(\{1,3\}) = \frac{7}{6}$, $v^g(\{2,3\}) = \frac{5}{18}$, and $v^g(N) = \frac{35}{18}$.[24] This situation is depicted in Fig. 5.8 (which is not to scale), the ε-core itself being the thick black line.

The calculation of the minimal ε, i.e. one where the core is just not empty, is fairly straightforward in this 3-player example. Because core allocations are efficient, it holds that $x_i + x_j = v^g(N) - x_k$ for any $i \neq j \neq k$. Rearranging the core conditions for our game,

$$
\begin{aligned}
x_1 + x_2 &\geq v^g(\{1,2\}) - \varepsilon^*, \\
x_3 &\geq v^g(\{3\}) - \varepsilon^*, \\
x_1 + x_3 &\geq v^g(\{1,3\}) - \varepsilon^*, \\
x_2 &\geq v^g(\{2\}) - \varepsilon^*, \\
x_2 + x_3 &\geq v^g(\{2,3\}) - \varepsilon^*, \\
x_1 &\geq v^g(\{1\}) - \varepsilon^*,
\end{aligned}
$$

[24] These slightly odd values originate from parameters set in Example 5.18, p. 216. They are calculated through the characteristic function v^g.

we can now check the above relation for every pair of players to find the minimal ε. In doing so, we assume for each such coalition that the core condition holds just with equality. For the first pair, players 1 and 2, we obtain

$$\frac{1}{2} - \varepsilon = \frac{35}{18} - (0 - \varepsilon), \tag{5.36}$$

which results in $\varepsilon = -\frac{13}{18}$. For the coalition of players 1 and 3, we get $\varepsilon = -\frac{7}{18}$, and for players 2 and 3 the value is given by $\varepsilon = -\frac{15}{18}$. Hence, the only value that satisfies all equations and for which the core is nonempty is $\varepsilon^* = -\frac{7}{18}$. For any smaller ε, the core vanishes completely. As can be seen from Fig. 5.8, the least-core is not a singleton but a line: The allocation to player 2 is fixed at $x_2 = \frac{7}{18}$, but players 1 and 3 can distribute the remaining amount of $\frac{28}{18}$ among them, as long as $x_1 \geq \frac{9}{18}$ and $x_3 \geq \frac{7}{18}$.

Again, we are left with an uncountable set of allocations from which to choose. Especially surprising in this setting is the fact that the range of allocations to choose from only affects players 1 and 3, as player 2's allocation is fixed, notwithstanding that the peering game is based on a complete network and hence the players are in a symmetrical position in the network.

Example 5.17 shows that narrowing down the core of a peering game to its least-core can potentially yield a solution as unsatisfying as the original core or even more so. In the first place, because we are still not certain to arrive at a point-valued solution in form of a unique allocation. In addition, within the resulting set of allocations, some players might have a constant payoff, but no intuitive support from the network structure g is provided as to which players are or can be affected.[25]

In order to avoid such problems, we now turn to point-valued solution concepts. As we know, these do not recommend a range of acceptable solutions, but specify one precise allocation. It will then also be possible for us to compare such outcomes with respect to a pre-set playing field, given by the core of the peering game, $C(N, v^g)$, or even the least-core $C_{\varepsilon^*}(N, v^g)$.

5.3.4 *The Shapley Vector in the Peering Game*

The first allocation rule we apply to the peering game v^g is the Shapley value, introduced in Shapley (1953b). It is without doubt the most influential allocation rule and should not be left unconsidered in any application such as ours. The resulting allocation is consistent with the marginality principle on which the operator is based. Unfortunately though, differences between VoIP firms are not accounted for as detailed as one would wish for in our context.

[25] If we modify Example 5.17 with the link g_{23} removed from $g = g^c$ and the value function v^g updated accordingly, the qualitative outcome is identical: Player 2 receives a fixed allocation in the resulting set-valued least-core and players 1 and 3 have an uncountable menu to "choose" from.

Due to the dependence of the characteristic function on the network structure, the computation of the Shapley value in our peering game v^g can be simplified considerably. Also, the reduced formula for computation can readily be interpreted. The following theorem captures this:

Theorem 5.10. *In the peering game v^g, the Shapley value ϕ_i of any player $i \in N$ is the total of half the gains realized with each neighbor $j \in N_i(g)$:*

$$\phi_i(N, v^g) = \frac{1}{2} \sum_{j \in N_i(g)} \alpha_i \alpha_j \left(l_i q(p_i) + l_j q(p_j) \right) \text{ for all } i \in N,\, g \in \mathcal{G}_N. \tag{5.37}$$

Proof. The above result can already be obtained by reasoning along the axioms on which the Shapley operator is based.[26] Evidently, the savings generated by each pair of players depend equally on the willingness to cooperation of both players. The marginal value of either player – upon joining the other to form the coalition $\{i, j\}$ – is the same, $v^g(\{i, j\})$, because $v^g(\{i\}) = v^g(\{j\}) = 0$. By the axiom of symmetry, linked pairs of players must share savings from cooperation in equal parts. Then, each player i is assigned half of the gains created through a link to another player j, and in the aggregate half of the gains that he allows for by cooperation with all players in his neighborhood, $j \in N_i(g)$.

For a more formal proof recall the standard equation of the Shapley value of a given player i in the peering game corresponding to expression (2.202), p. 99:

$$\phi_i(N, v^g) := \sum_{\substack{S \subseteq N \\ i \in S}} \frac{(s-1)!\,(n-s)!}{n!} \left[v^g(S) - v^g(S \setminus \{i\}) \right] \text{ for all } i \in N,\, g \in \mathcal{G}_N. \tag{5.38}$$

Substituting for the value function v^g from (5.14), the expression in square brackets can be reworked,

$$\begin{aligned} v^g(S) - v^g(S \setminus \{i\}) &= \sum_{i \in S} \widetilde{A}_i(S, g) - \sum_{k \in S \setminus \{i\}} \widetilde{A}_k(S \setminus \{i\}, g) \\ &= \sum_{i \in S} \sum_{j \in N_i(g) \cap S} \alpha_i \alpha_j l_i q(p_i) \\ &\quad - \sum_{k \in S \setminus \{i\}} \sum_{m \in N_k(g) \cap (S \setminus \{i\})} \alpha_k \alpha_m l_k q(p_k) \\ &= \sum_{j \in N_i(g) \cap S} \alpha_i \alpha_j \left(l_i q(p_i) + l_j q(p_j) \right). \end{aligned} \tag{5.39}$$

This simplification, especially the last step, is possible, because in the sums over players in S and $S \setminus \{i\}$, respectively, all elements cancel out, but those

[26] A similar proof can be found in Brown and Housman (1988, p. 5). It is given in another context for the class of so-called weighted graph games.

corresponding to player i, i.e. all gains with other players who are both in S and i's neighborhood. We can now rewrite expression (5.38):

$$\phi_i(N, v^g) = \sum_{\substack{S \subseteq N \\ i \in S}} \frac{(s-1)!\,(n-s)!}{n!} \left[\sum_{j \in N_i(g) \cap S} \alpha_i \alpha_j \left(l_i q(p_i) + l_j q(p_j) \right) \right]. \tag{5.40}$$

Summing over the cardinality of the coalitions S instead of summing over all coalitions that include player i, we obtain

$$\phi_i(N, v^g) = \sum_{s=1}^{n} \binom{n-1}{s-1} \frac{(s-1)!(n-s)!}{n!} \cdot \sum_{j \in N_i(g)} \frac{\binom{n-2}{s-2}}{\binom{n-1}{s-1}} \rho_{ij}, \tag{5.41}$$

where for simplicity,

$$\rho_{ij} := \alpha_i \alpha_j \left(l_i q(p_i) + l_j q(p_j) \right).$$

The change from sets to their cardinalities in the summations is explained as follows. The expression $\binom{n-1}{s-1}$ is the possible number of coalitions of size s which also contain player i. The second sum, previously in square brackets, is now independent of the intersection of the coalition S with the neighborhood of player i, $N_i(g) \cap S$. To adjust for the expression $\binom{n-1}{s-1}$ up front, for each neighbor $j \in N_i(g)$ we now must include the expression

$$\frac{\binom{n-2}{s-2}}{\binom{n-1}{s-1}},$$

which is equal to the probability of player j being a member of a coalition of size s when i is also a member. This probability is calculated by dividing the number of coalitions of size s that contain both i and j by the number of coalitions of size s to which at least player i belongs. For any given cardinality s, these probabilities are identical for all $j \in N_i(g)$ neighbors of player i. Spelling out the binomial coefficients to the form where

$$\binom{n}{k} = \frac{n!}{k!(n-k)!}$$

allows us to simplify matters further:

$$\begin{aligned}\phi_i(N, v^g) &= \sum_{s=1}^{n} \left[\frac{(n-1)!}{(s-1)!(n-s)!} \frac{(s-1)!(n-s)!}{n!} \cdot \right. \\ &\qquad \left. \sum_{j \in N_i(g)} \frac{(n-2)!}{(s-2)!(n-s)!} \frac{(s-1)!(n-s)!}{(n-1)!} \rho_{ij} \right] \\ &= \frac{1}{n(n-1)} \sum_{s=1}^{n} (s-1) \sum_{j \in N_i(g)} \rho_{ij}\end{aligned}$$

$$= \left[\frac{n(n+1)}{2n(n-1)} - \frac{n}{n(n-1)} \right] \sum_{j \in N_i(g)} \rho_{ij}$$

$$= \frac{1}{2} \sum_{j \in N_i(g)} \rho_{ij}. \tag{5.42}$$

Then, substituting for ρ_{ij} yields the desired solution,

$$\phi_i(N, v^g) = \frac{1}{2} \sum_{j \in N_i(g)} \alpha_i \alpha_j \left(l_i q(p_i) + l_j q(p_j) \right) \text{ for all } i \in N,\ g \in \mathcal{G}_N. \tag{5.43}$$

This concludes the proof of Theorem 5.10. ■

What we just established is in a sense quite curious for its intended application. Gains originating on some link g_{ij} are split right down the middle between players i and j under the Shapley operator. This is perfectly in line with the axioms on which the operator is based but might give rise to some reasonable objections a VoIP firm i could have if subject to such an allocation. Of course, without a peering partner no gains can be realized at all by any given firm $i \in N$. However, the Shapley allocation does not coincide with the contributions which the firms make to the value created under v^g. Because why, after all, should one firm give up some of its very own contribution to the gains, given by $\alpha_i \alpha_j l_i q(p_i)$,[27] and not the other? In this regard it is crucial to consider that the amount of the contribution one firm makes is specific to its peering partner, but essentially independent of the actions of the counter-party in peering agreement g_{ij}. As neither firm is more crucial or involved when peering is agreed upon, this indeed sounds like a strange proposal. The contributions[28] of players i and j to $v(\{i,j\})$ are possibly different, but the Shapley operator splits up the total equally among both players. This effect might be distorted in the aggregate when a firm i takes into considerations all its peering partners $j \in N_i(g)$. But, unless these effects happen to cancel out exactly, there will always be at least one firm which loses out on the Shapley allocation. In Example 5.18 we can see how firms with relatively higher contributions are bound to lose out under Φ[29]:

Example 5.18. Consider a peering game where $N = \{1,2,3\}$, $g = g^c$, and v^g is set up according to (5.14), with parameters given on the left side of Table 5.2. Each row represents one player and contains his market share, demand, and long distance fees. For simplicity, we use uniform long distance fees normalized to one: $l_i = l = 1$ for all $i = 1,2,3$. Calculated from this, the right side of Table 5.2 contains the origin

[27] Unless $l_i q(p_i) = l_j q(p_j)$ holds, one firm will always have to "subsidize" the other.

[28] Again, by a contribution of player i, we mean the amount $\alpha_i \alpha_j l_i q(p_i)$, and not the marginal value of player i upon joining j to form coalition $\{i,j\}$.

[29] According to the balanced calling pattern assumption, "relatively higher contributions" for some firm i are the result of demand $q(p_i)$ and long distance fees l_i, or rather their product, only. The call distribution via market shares, $\alpha_i \alpha_j$, is taken into consideration equally on both ends of a link g_{ij}.

Table 5.2 Players' parameters and corresponding contributions under v^g

	α_i	$q(p_i)$	l_i
1	$\frac{1}{2}$	4	1
2	$\frac{1}{6}$	2	1
3	$\frac{1}{3}$	3	1

	1	2	3	Σ
1	0	$\frac{1}{3}$	$\frac{2}{3}$	1
2	$\frac{1}{6}$	0	$\frac{1}{9}$	$\frac{5}{18}$
3	$\frac{1}{2}$	$\frac{1}{6}$	0	$\frac{2}{3}$

of the gains realized, that is, the contributions of the players. Here, an entry in row i, column j, is the amount of savings player i contributes to $v^g(\{i,j\})$ when peering with player j. The last entry of each row are the aggregate contributions of the respective player. These sum up to $v^g(N) = 1\frac{17}{18}$. The Shapley value is readily calculated by adding diagonally opposite entries, dividing them by half, and assigning these halves to the players involved. The result is $\Phi(v^g) = (\frac{5}{6}, \frac{7}{18}, \frac{13}{18})$. Clearly, player 1 is allocated less than the sum of his contributions. Players 2 and 3, however, benefit from the Shapley operator. The former from peering with both, players 1 and 3, while the latter benefits only from peering with player 1. His "losses" from peering with player 2 are more than offset.

This example demonstrates how the allocation under the Shapley operator differs from what players contribute to the gains from peering. According to Theorem 5.10, these differences originate pairwise on each link g_{ij}.

Generally, the degree of distortion of some firm i under the Shapley operator does not depend on the value of the product $l_i q(p_i)$ alone. It depends also on the position firm i has within the network g, that is, on the set of its neighbors $j \in N_i(g)$, and for which neighbors $l_i q(p_i) \neq l_j q(p_j)$.

Consequently, this weakness of the Shapley operator is most significant when it comes to VoIP firms with zero long distance fees. Such a firm i with $l_i = 0$ is entitled to a nonzero allocation under Φ, whenever it has a neighbor $j \in N_i(g)$ with $l_j > 0$. This is a rather unpleasant side-effect, as firm i's contribution to $v^g(\{i,j\})$ is exactly zero.[30]

As we have seen, in a peering game based on v^g the Shapley value is particularly easy to calculate. Also, the simplified expression (5.37) lends itself to a straightforward interpretation of how the gains are allocated among the VoIP firms. Nevertheless, we believe the Shapley operator reveals a major flaw for its use in the VoIP peering game: Some firms are allocated more and some less than their actual contributions to the gains from peering agreements. In other words, the average marginal values calculated by the Shapley operator are misleading in this case, because they do not in general coincide with the contributions of the

[30] An interesting approach for further research arises in this setting: What is the influence of the distortion through the Shapley vector on the VoIP firms' incentive to lower their costs for long distance traffic l_i, be it through negotiating better deals with their long distance carriers or through extending their own physical network? How does this incentive change with respect to a firm's neighbors and the relative levels of their contributions?

respective firms. As these contributions arise from individual long distance fees (and not from any actual effort within the coalition), which are paid – in the absence of peering – to third parties, it is hard to motivate, why one VoIP firm should continuously subsidize another under the scheme suggested by the Shapley operator. This is especially striking for VoIP firms that are zero-players at the same time.

Before we turn our attention to weighted solution concepts to possibly alleviate this problem, we consider the Myerson value and show why the resulting allocation will not differ from the one suggested by the Shapley operator. More precisely, we show that we already applied the Myerson value, albeit in disguise.

5.3.5 *The Myerson Value*

In this section we apply to our peering game an allocation rule derived from the Shapley value, which by construction accounts for properties regarding the network g of a communication situation (N, v, g). It became known as the *Myerson value* and was presented in detail in Sect. 3.5, p. 134. Originally introduced in Myerson (1977), the author proposed a value γ that would satisfy *component efficiency* and *fairness* of an allocation.

What the Myerson value does, in essence, is to refine the characteristic function v of any given cooperative game to incorporate an underlying network g:

$$v^m(S) := \sum_{C \in S/g} v(C) \quad \text{for all } S \subseteq N,\ g \in \mathscr{G}_N. \tag{5.44}$$

In (5.44) the network g determines that only players can achieve gains from cooperating with one another if they are connected to one another, i.e. in the same component. In addition, the network also highlights the players' relative importance within some coalition $S \subseteq N$, when it comes to the distribution of gains.

As by definition, the Myerson value is the Shapley value using a refined characteristic function, we will see that applying the Shapley operator to the network refined value function v^g, as we did in Sect. 5.3.4, must produce no other allocation than the Myerson value. This is due to the fact that our characteristic function v^g is even more refined than v^m as used in Myerson (1977); the former constrains gains through cooperation not only to components of the network g, but even exclusively to linked pairs of players. We point out the relation between the two value functions in the following proposition:

Proposition 5.26. *The value function $v(S, g)$, as defined in (5.14), captures the effects of the network restricted value function from Myerson (1977), v^m, from (5.44). It is also decomposable with respect to the components $C \in N/g$, induced by network g.*

Proof. Merely rearranging (5.14) leads to the desired result:

$$\begin{aligned} v(S,g) &= \sum_{i\in S}\sum_{j\in N_i(g)\cap S} \alpha_j\alpha_i l_i q(p_i) \\ &= \sum_{C\in S/g}\sum_{i\in C}\sum_{j\in N_i(g)\cap C} \alpha_j\alpha_i l_i q(p_i) \\ &= \sum_{C\in S/g} v(C,g). \end{aligned} \tag{5.45}$$

■

From this another important corollary arises:

Corollary 5.12. *The Myerson value* γ *and the Shapley value* Φ *result in the same allocation when based on* v^g *as defined in (5.14):*

$$\gamma(v^g) \equiv \Phi(v^g) \text{ for all } v^g. \tag{5.46}$$

Proof. Follows directly from (5.45). ■

Because the characteristic function v^g from (5.14) is at the heart of our application, the Myerson value does not refine the results of the Shapley value. The same problems occur under the Myerson value that we encountered before with the Shapley operator: Individual contributions are neglected in favor of marginal values.

We continue now with weighted allocation concepts, hoping to find a plausible and intuitive solution that allows for allocations corresponding to the players' contributions rather than to their marginal values. Our first approach is the Weighted Shapley value, followed by the weighted versions of solutions to the bargaining problem.

5.3.6 The Weighted Shapley Value

In Sect. 5.3.4 we applied the Shapley operator to a peering game based on v^g. As we saw, the resulting allocation need not coincide with the contributions that the VoIP firms make towards the value of v^g, possibly leading to difficulties in terms of implementation. In an attempt to avoid this problem, we turn to the Weighted Shapley value as introduced in Shapley (1953a). It is an allocation method which takes into account characteristics of players not expressed directly through the value function, in our case the divergence of the contribution and the marginal value of a player. We investigate if, and under which conditions, it is possible to find a weight system that allows for the an allocation corresponding to the players' contributions. We find that such an allocation is extremely appealing in terms of implementation, as it would minimize the necessary administrative interaction between VoIP firms, once peering is in place. Also, any disagreements regarding the disclosure of sensitive information among the (competing) firms could be avoided. Ideally, this weight system would also exhibit an intuitive connection to the players' parameters determining their contributions. Instead of restating the concept entirely, we will only list the pieces we need in this context. For a more comprehensive treatment, we refer the reader to Sect. 2.4.6, p. 102, and the original literature stated there.

Our aim is to find appropriate weights, or rather, a weight system satisfying the problem addressed above. That is, we look for an allocation $\mathbf{x}^*$ which assigns to all VoIP firms the sum of their contributions within a given peering network:

$$\mathbf{x}^* \in \mathbb{R}^n_+ \text{ with } x_i = \sum_{j \in N_i(g)} \alpha_j \alpha_i l_i q(p_i) \text{ for all } i \in N. \tag{5.47}$$

All x_i^*s are nonnegative, equal to zero for all players $i \in \mathbf{0}$ and strictly positive for all others with nonempty neighborhood. For two reasons we should now check whether the allocation $\mathbf{x}^*$ is an element of the core: For one, the requirements for a core allocation are appealing and satisfying them can only strengthen the case for an allocation $\mathbf{x}^*$. Second, as we know from Proposition 2.16, p. 107, the set of all Weighted Shapley values and the core coincide in convex games and hence also in the peering game v^g. Consequently, if $\mathbf{x}^* \in C(N, v^g)$, then $\mathbf{x}^*$ can be reproduced by a weight system ω under the Weighted Shapley operator.

Claim. In the peering game v^g, the allocation $\mathbf{x}^*$ as specified in (5.47) is an element of the core:

$$\mathbf{x}^* \in C(N, v^g). \tag{5.48}$$

Proof. To check whether $x^*(S) \geq v^g(S)$ for all $S \subseteq N$ we begin by with the individual rationality conditions assuming $|S| = 1$. The lowest value a player i can attain in the allocation given by (5.47) is zero. This is the case, whenever $l_i = 0$ or $N_i(g) = \emptyset$.[31] As by (5.14) it holds that $v^g(\{i\}) = 0$ for all $i \in N$, regardless of long distance fees, the constraint of individual rationality is fulfilled.

Next, we check for larger coalitions with $|S| > 1$. Here, again, we can see that if we add up the contributions as given in (5.47) of all players linked within S, we receive, according to (5.14), an amount equal to or larger than the value $v^g(S)$:

$$x^*(S) = \sum_{i \in S} \sum_{j \in N_i(g)} \alpha_j \alpha_i l_i q(p_i) \geq \sum_{i \in S} \sum_{j \in N_i(g) \cap S} \alpha_j \alpha_i l_i q(p_i) = v(S, g). \tag{5.49}$$

This does not discriminate between players with zero and nonzero long distance fees and also holds for $S = N$, the grand coalition, in which case also $N_i(g) = N_i(g) \cap N$ for all $i \in N$ and so (5.49) holds with equality. Therefore, the allocation $\mathbf{x}^*$ is in the core of the game. ■

[31] We again ignore the pathological cases of zero demand or zero market share.

With $\mathbf{x}^* \in C(N, v^g)$ we proceed to find a weight system ω that allows us to recreate $\mathbf{x}^*$. This lexicographic weight system $\omega = (\lambda, \Sigma)$ consists of two parts. The first element is an n-dimensional vector $\lambda \in \mathbb{R}^n_{++}$, the second an ordered partition on the set of players $\Sigma = (S_1, S_2, \ldots, S_m)$. To convey as much intuition as possible, we try to find ω via the marginalistic expression of the operator, which was given in (2.214), p. 105:

$$\phi_i^{\omega}(v^g) = \sum_{\pi \in \Pi_\Sigma} p_\omega(\pi) \cdot \left[v^g(S_{\pi,\pi(i)}) - v^g(S_{\pi,\pi(i)} \setminus \{i\})\right] \text{ for all } i \in N. \tag{5.50}$$

In a brute force approach one would (try to) solve for every possible partition of the player set N a system of n equations and n unknowns:

$$\phi_i^{\omega}(v^g) = \sum_{\pi \in \Pi_\Sigma} p_\omega(\pi) \cdot \left[v^g(S_{\pi,\pi(i)}) - v^g(S_{\pi,\pi(i)} \setminus \{i\})\right] \stackrel{!}{=} \sum_{j \in N_i(g)} \alpha_j \alpha_i l_i q(p_i), \tag{5.51}$$

where the unknowns for which to solve are the strictly positive weights $\lambda \in \mathbb{R}^n_{++}$ that constitute the probabilities $p_\omega(\pi)$. These probabilities are described through expressions (2.212) and (2.213) on p. 105. As the number of possible partitions literally explodes when the player set N grows,[32] we search for an approach to limit the number of partitions to be investigated.

Under $\mathbf{x}^*$, zero-players are assigned a value of zero. Because $\lambda \in \mathbb{R}^n_{++}$ is strictly positive in all its entries, as soon as a zero-player i yields a nonzero marginal value under some ordering π_Σ, he also receives a nonzero allocation: $\phi_i^{\omega}(v^g) > 0$. We therefore need to restrict the possible permutations on N to ensure that no zero-player can have a nonzero marginal value. In short, we have to make sure no permutation π_Σ exists allowing a zero-player to succeed (directly or indirectly) some "nonzero" player j with $l_j > 0$ to whom he is linked. This, of course, has to be accomplished via the ordered partition Σ and how we group the players across its elements.

One possible approach would set the following conditions on Σ: No zero-player i in some $S_k \in \Sigma$ can be linked to any nonzero player j in S_k. Additionally, none of the zero-players in S_k must be linked to a nonzero player in some element $S_h \in \Sigma$ preceded by S_k in the order specified by Σ. The ordered partition must be such that

$$\text{for all } k = 1, \ldots, m, \text{ and } i \in S_k,\ l_i = 0,\ \nexists g_{ij} \in g \,|\, j \in S_h,\ h \geq k,\ l_j > l_i = 0. \tag{5.52}$$

As this expression hinges almost solely on specific links, changes to the network g might render a qualifying partition unapt to generate $\mathbf{x}^*$. More importantly, these conditions need not be satisfiable in the first place, given certain networks. Think

[32] The number of possible partitions of a set with n elements is given by the nth *Bell Number*, denoted B_n. For the empty set and singletons we have $B_0 = B_1 = 1$, all following values can be calculated with the recursive formula $B_{n+1} = \sum_{k=0}^{n} \binom{n}{k} B_k$. For more details, see Rota (1964, p. 500).

only of a single zero-player linked to all other players, or at least to all other nonzero players. He can not only not be element of any $S_k \in \Sigma$ which also contains regular players. He also must be member of an element of Σ preceding all those elements containing regular players, in order to avoid a nonzero marginal value. These considerations lead us to a much more restrictive condition to impose on the structure of Σ. It is, at the same time, much simpler to describe and supported by intuition:

Proposition 5.27. *If the ordered partition Σ of the weight system ω satisfies*

$$\Sigma = (S_1, S_2)$$
$$\text{with } S_1 = \{i \in N \mid l_i = 0\}$$
$$\text{and } S_2 = N \setminus S_1,$$

then there exists weights $\lambda \in \mathbb{R}^n_{++}$ for which $\Phi^\omega(v^g) = \mathbf{x}^*$.

Proof. Because expression (5.50) calculates the weighted marginal value of some player i, we look for permutations of the player set that exhibit the lowest possible marginal value for zero-players. This, for example, is always the case when one such player follows another, because zero long distance fees give rise to zero gains. Hence, in any order π_{S_1}, there is no marginal value assigned to any of the zero-players $i \in S_1$ under the operator $\Phi^\omega(v^g)$. This is true regardless of the weights λ_i of players $i \in S_1$.

We can now turn our attention to the nonzero players $j \in S_2$ and the task of assigning them a value equal to their contributions according to $\mathbf{x}^*$. Grouping all these and only these players in S_2 results in a total number of $|S_1|! \cdot |S_2|!$ permutations of the player set as restricted by Σ. Nevertheless, as all permutations π_{S_1} on S_1 produce identical marginal values (of zero) for all $i \in S_1$, there are only $|S_2|!$ distinct marginal vectors $\mathbf{x}^\pi$ entering expression (5.51).

The closed convex hull of these marginal vectors, $co\{(\mathbf{x}^\pi)_{\pi \in \Pi_\Sigma}\}$ is a subset of the core. More precisely, it is the only subset, in which $x_i = 0$ for all $i \in S_1$ and $x_j \geq 0$ for all $j \in S_2$. From our previous characterization of the core (see Chap. 2.4.3.2) and its comparison with the set of all Weighted Shapley values on convex games (see Chap. 2.4.6), we know that $\mathbf{x}^*$ must lie in this subset. Invoking Minkowski's Theorem (see Footnote 91, p. 109) allows us to conclude that some convex combination of (or probability measure on) these extreme points $\mathbf{x}^\pi$ (or possibly of a subset of them) exists, for which $\mathbf{x}^*$ is the outcome. Hence, weights λ_j with $j \in S_2$ can be found that produce an appropriate p_λ on Π_{S_2}. ■

As mentioned above, Proposition 5.27 provides a fairly strong sufficient condition and is not necessarily the only way to find an ω which satisfies $\Phi^\omega(v^g) = \mathbf{x}^*$. One big advantage is certainly that its ordered partition of the player set is "network-proof": Changes to the underlying network g, even though they have a potential influence on v^g, create no need to adjust the partition Σ in any way. Not even (nonzero) players with empty neighborhoods pose a problem. This is certainly not

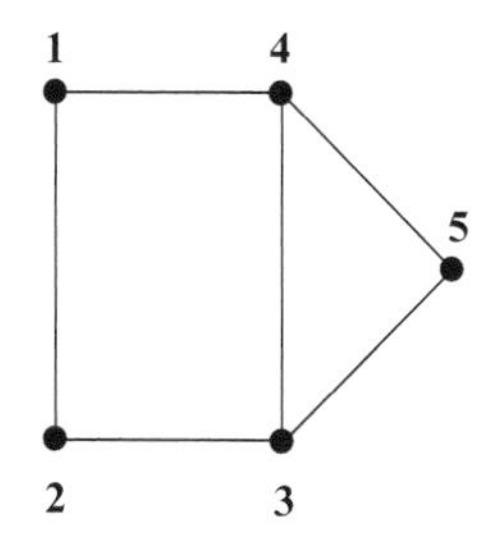

Fig. 5.9 Network represented by g

Table 5.3 Players' parameters and corresponding contributions under v^g

	α_i	$q(p_i)$	l_i
1	$\frac{1}{3}$	2	0
2	$\frac{1}{3}$	2	0
3	$\frac{1}{9}$	3	1
4	$\frac{1}{9}$	4	1
5	$\frac{1}{9}$	5	1

	1	2	3	4	5	Σ
1	0	0	0	0	0	0
2	0	0	0	0	0	0
3	0	$\frac{1}{9}$	0	$\frac{1}{27}$	$\frac{1}{27}$	$\frac{15}{81}$
4	$\frac{4}{27}$	0	$\frac{4}{81}$	0	$\frac{4}{81}$	$\frac{20}{81}$
5	0	0	$\frac{5}{81}$	$\frac{5}{81}$	0	$\frac{10}{81}$

the case for the more lax conditions imposed by (5.52), because the prescribed ordering in the latter depends on how nonzero and zero-players are linked across elements of Σ.

We now provide an illustrating example of finding an ω for which $\Phi^{\omega}(v^g) = \mathbf{x}^*$ in the 5-player case, followed by some generalizations to both, the n-player case and Proposition 5.27.

Example 5.19. Consider a peering game v^g with the player set $N = \{1,2,3,4,5\}$ and the network $g = \{g_{12}, g_{14}, g_{23}, g_{34}, g_{35}, g_{45}\}$, as depicted in Fig. 5.9. Following Example 5.18 we list in Table 5.3 the parameters of the players (left) and the corresponding contributions under v^g (right). As before, we try to keep matters as simple as possible in terms of market shares, demand, and long distance fees. Players 1 and 2 are assigned zero long distance fees, $l_1 = l_2 = 0$, and hence they all have entries of zero in the contributions table on the right. We save the hassle of spelling out the value function and proceed to the marginal vectors instead. With an ordered partition $\Sigma = (\{1,2\},\{3,4,5\})$, there are $2! \cdot 3! = 12$ possible orderings, but only 6 distinct marginal vectors, as players 1 and 2 always have a marginal value of zero under any permutation $\pi \in \Pi_{\Sigma}$.

$$\mathbf{x}^{345} = \begin{pmatrix} 0 \\ 0 \\ \frac{9}{81} \\ \frac{19}{81} \\ \frac{17}{81} \end{pmatrix}, \mathbf{x}^{435} = \begin{pmatrix} 0 \\ 0 \\ \frac{16}{81} \\ \frac{12}{81} \\ \frac{17}{81} \end{pmatrix}, \mathbf{x}^{453} = \begin{pmatrix} 0 \\ 0 \\ \frac{24}{81} \\ \frac{12}{81} \\ \frac{9}{81} \end{pmatrix},$$

$$\mathbf{x}^{543} = \begin{pmatrix} 0 \\ 0 \\ \frac{24}{81} \\ \frac{21}{81} \\ 0 \end{pmatrix}, \mathbf{x}^{534} = \begin{pmatrix} 0 \\ 0 \\ \frac{17}{81} \\ \frac{28}{81} \\ 0 \end{pmatrix}, \mathbf{x}^{354} = \begin{pmatrix} 0 \\ 0 \\ \frac{9}{81} \\ \frac{28}{81} \\ \frac{8}{81} \end{pmatrix}.$$

Because we can ignore the ordering within S_1, we write, in a slight abuse of notation, for example $\mathbf{x}^{345}$ for the marginal worth vector, where $\pi \in \Pi_\Sigma$ assigns the ordering $(3,4,5)$ to S_2.[33]

The probability $p_\omega(\pi_\Sigma)$ for a permutation π_Σ on Σ is, according to (2.212) and (2.213), the product of $p_\lambda(\pi_{\{1,2\}})$ and $p_\lambda(\pi_{\{3,4,5\}})$. Inserting this into the marginalistic expression (2.214) for the Weighted Shapley operator yields

$$\begin{aligned} \Phi^\omega(v^g) = {} & p(12) \cdot \Big[p(345) \cdot \mathbf{x}^{345} + p(435) \cdot \mathbf{x}^{435} + p(453) \cdot \mathbf{x}^{453} \\ & + p(543) \cdot \mathbf{x}^{543} + p(534) \cdot \mathbf{x}^{534} + p(354) \cdot \mathbf{x}^{354}\Big] \\ & + p(21) \cdot \Big[p(345) \cdot \mathbf{x}^{345} + p(435) \cdot \mathbf{x}^{435} + p(453) \cdot \mathbf{x}^{453} \\ & + p(543) \cdot \mathbf{x}^{543} + p(534) \cdot \mathbf{x}^{534} + p(354) \cdot \mathbf{x}^{354}\Big] \\ = {} & \underbrace{(p(12) + p(21))}_{\equiv 1} \cdot \Big[p(345) \cdot \mathbf{x}^{345} + p(435) \cdot \mathbf{x}^{435} + p(453) \cdot \mathbf{x}^{453} \\ & + p(543) \cdot \mathbf{x}^{543} + p(534) \cdot \mathbf{x}^{534} + p(354) \cdot \mathbf{x}^{354}\Big] \\ \stackrel{!}{=} {} & \mathbf{x}^*. \end{aligned} \quad (5.53)$$

[33] The entries of $\mathbf{x}^{345}$ are explained as follows: The first two, for players 1 and 2 are equal to zero, because Σ only admits permutations on N in which the first two players also occupy the first two spots and hence their marginal values are zero. Player 3's entry of $\frac{9}{81}$ is value of the marginal gains he realizes with all players previous in the permutation to which he is linked: In our case only the (zero-)player 1, and so player 3's marginal value is $\frac{9}{81}$. Player 4's marginal value is the sum of all contributions with player 1 and player 3, i.e. $\frac{4}{27} + \frac{4}{81} + \frac{1}{27} = \frac{19}{81}$. Player 5's marginal value in this permutation is given by the sum of all contributions with players 3 and 4, $\frac{5}{81} + \frac{4}{81} + \frac{5}{81} + \frac{1}{27} = \frac{17}{81}$. The remaining marginal vectors $\mathbf{x}^\pi$ are calculated analogously.

Abbreviating notation, expression $p(345)$, for example, stands for the probability $p_\lambda(\pi)$, where π specifies the order $(3,4,5)$ on S_2. The desired allocation $\mathbf{x}^* = (0,0,\frac{15}{81},\frac{20}{81},\frac{10}{81})$, which assigns every player the sum of his contributions, is calculated in the very right column of Table 5.3.

As the first two entries of all vectors in (5.53) are zeroes, and the remaining probabilities of orders on S_2 are given solely in terms of λ_3, λ_4, and λ_5, we are left to solve a system of three equalities and three unknowns.[34] In our example, a desired weight system $\omega = (\lambda, \Sigma)$ to yield $\mathbf{x}^*$ as the value computed by the Weighted Shapley operator is

$$\omega = ((a, b, 0.0082, 0.0110, 0.0137), (\{1,2\}\{3,4,5\})),$$

where a and b are any strictly positive real numbers. The somewhat cumbersome check of this solution is left to the reader.

Even though Example 5.19 is limited to 5 players, some generalizations can be made regarding the n-player case. The most important one concerns the number of equations and unknowns to solve. Whenever we assign players to Σ as specified in Proposition 5.27, we can rearrange the probabilities in expression (5.51). Because every probability on π_Σ can be decomposed into $p_\omega(\pi_\Sigma) = p_\lambda(\pi_{S_1}) \cdot p_\lambda(\pi_{S_2})$, it is possible to factor out the sum $\sum_{\pi \in \Pi_{S_1}} p_\lambda(\pi_{S_1})$ which by construction is equal to one.[35] Because the entries of all players $i \in S_1$ in the marginal vectors are zeroes, as well as their allocation x_i^*, we can also reduce our system of equations to $|S_2|$ equations with as many unknowns. This is quite intuitive, because the permutations on the set of zero-players within Σ are irrelevant to us. For the same reasons the weights λ_i for players $i \in S_1$ are irrelevant.

One way to relax Proposition 5.27 would be to refine the ordered partition Σ by splitting up S_1 into multiple sets, as long as these refined elements appear before S_2 in the ordering. The probabilities on permutations on the refined elements can be factored out analogously and, again, sum to one by construction. The resulting set of marginal vectors is exactly the same as before. Crucial is merely that all the zero-players precede the nonzero ones in any ordering π_Σ admissible by Σ.

Depending on the dimension of the core and on where in the convex hull $co\{(x^\pi)_{\pi \in \Pi_\Sigma}\}$ the allocation $\mathbf{x}^*$ lies, it might also be possible to represent $\mathbf{x}^*$ by a convex combination of less than $|S_2|$ elements, allowing for a possible refinement on S_2 within Σ. Hence, aside from the arbitrary weights for the zero-players, there may exist multiple solutions ω satisfying $\Phi^\omega(v^g) = \mathbf{x}^*$. But as Proposition 5.27 provides a fairly straightforward way of achieving allocation $\mathbf{x}^*$, we see no need to investigate this matter any further.

---

[34] Take for example the permutation π_{S_2} with ordering $(3,4,5)$. The value of the corresponding probability is $p_\lambda(\pi_{S_2}) = \frac{\lambda_4 \cdot \lambda_5}{(\lambda_3+\lambda_4)\cdot(\lambda_3+\lambda_4+\lambda_5)}$.

[35] See the second to last line of (5.53) in Example 5.19 with the only 2 possible permutations on players 1 and 2.

We have shown how the Weighted Shapley value, using an appropriate weight system ω, can produce a solution $\mathbf{x}^*$ which assigns to every player i his aggregate contributions arising from peering with his neighbors $j \in N_i(g)$. Unfortunately, aside from the partition of the player set into zero- and nonzero players, no intuitive characterization can be made regarding the weights λ. They do not in general coincide with market share, neither global, nor relative within coalitions. Given the balanced calling pattern assumption, this might not be too surprising, as it is not merely the market shares, but rather the call volumes that give rise to gains under v^g. But even for the fee-valued call volumes $\alpha_i l_i q(p_i)$ produced by a VoIP firm i, no intuitive relation to ω seems to exist. This makes not only the calculation of the weights a rather tedious business, but also raises the question on which basis such a weight system should be motivated for the allocation. Further research should be put into this, especially when using the slightly different weight systems as proposed by Monderer et al. (1992, p. 31).

Our final means of allocation in this application will be the concept of a bargaining solution in the context of the bargaining problem. This approach is not directly comparable to the previously studied solution concepts, because it is not based strictly on cooperative games. Nevertheless, we can deduce the necessary domain from the value function v^g without much trouble. As we will see, weighted solutions reflecting the players' contributions are much better supported by intuition in the bargaining environment.

5.3.7 *Bargaining Solutions*

This section is devoted to a slightly different approach to peering among VoIP firms. From the peering game v^g we extract the relevant information to obtain a bargaining problem as introduced in Sect. 2.4.7. In such a setting, bargaining leads either to an allocation decision taken unanimously by all players or to the reversion to the disagreement point, i.e. the status quo ante. This is referred to as "pure bargaining", which we will uphold, albeit in a pairwise manner rather than on the whole player set N at once.

We begin by defining the bargaining problem derived from the peering game v^g and subsequently turn to the bargaining solutions. We touch only very briefly on the regular variants of both the Nash solution and the Kalai–Smorodinsky solution (as both yield allocations identical to the Shapley value) and concentrate rather on their weighted counterparts.

5.3.7.1 The Bargaining Problem in the Peering Game

As we have hinted at above, we split up the n-player bargaining problem into a 2-player variant, based on the links in network g. This allows for a more intuitive approach, especially with respect to illustrations. Nothing is lost on the results, as the overall allocation of any player is the sum of his allocations in the $n-1$ 2-player problems to which he belongs.

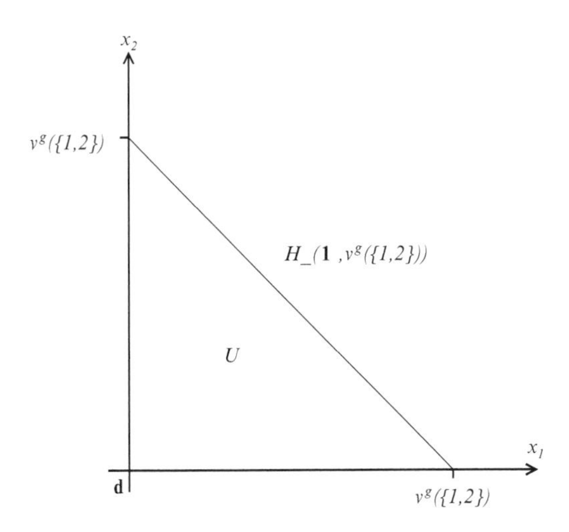

Fig. 5.10 The feasible set U for a 2-player situation in the peering game

In a first step, we want to derive the feasible set U from the value function v^g. As we are concerned only with 2 players at a time, the surplus from cooperation among linked players is given by

$$v^g(\{i,j\}) = \alpha_i \alpha_j (l_i q(p_i) + l_j q(p_j)), \quad \text{for all } i,j \in N \text{ with } g_{ij} \in g, \tag{5.54}$$

and zero whenever $g_{ij} \notin g$. Labelling the players in question as player 1 and 2, respectively, the feasible set is given by the intersection of half-space $H_-(\mathbf{1}, v^g(\{1,2\}))$ with the positive quadrant. This is depicted in Fig. 5.10. Because the absence of peering describes the status quo ante, the disagreement point $\mathbf{d}$ coincides conveniently with the origin. The feasible set U is not only convex, but also a symmetric, compact subset of $\mathbb{R}^2$, and it is comprehensive. Most importantly, it is non-degenerate, unless $v^g(\{i,j\}) = 0$, which will only happen if $l_i = l_j = 0$ for distinct players $i,j \in N$, and leads to a situation where there is no surplus to allocate in the first place. Note that the feasible set U may change across pairs of linked players according to $v^g(\{i,j\})$.

Alternatively, one could also define the feasible set as to incorporate the drop of access fees when peering. Then, instead of a symmetric maximal allocation of $v^g(\{i,j\})$, we now get for one player, say i, the maximal allocation $v^g(\{i,j\}) + A^i_j$, because without access fees a, player i's access charges, which amounted to a deficit before peering, no longer apply.[36] A correction for the player with the pre-peering surplus is not needed, as the most he can achieve is $v^g(\{i,j\})$ if the payment

[36] We remember that $A^j_i = a\alpha_i\alpha_j(q(p_j) - q(p_i))$ are the access revenues with firm j in the absence of peering. As $A^j_i = -A^i_j$, we assume without loss of generality that it is player i who has the pre-peering deficit.

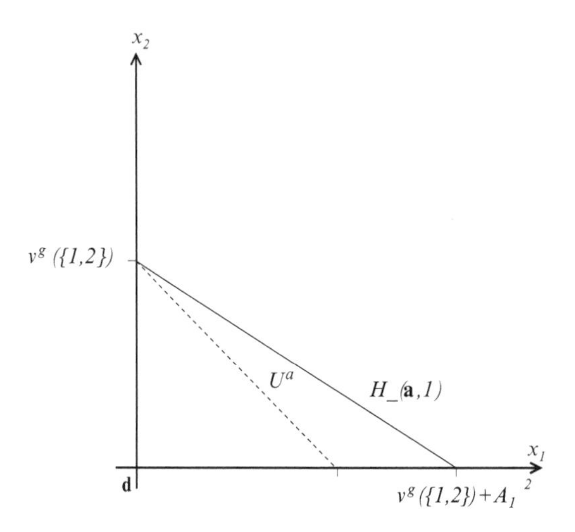

Fig. 5.11 The feasible set U^a for a 2-player situation in the peering game

from the access charges is upheld regardless of peering. We depict this no longer symmetric feasible set U^a in Fig. 5.11. The corresponding half-space is $H_{-}(\mathbf{a},1)$, where $\mathbf{a}' = (v+a,a)$. The reformulation of the bargaining problem to include access fees has to be handled with caution though. In case of two zero-players the problem becomes a degenerate one, because there exists no allocation $\mathbf{x} \in U^a$ for which $x_i > d_i = 0$ for both $i = 1,2$. Also, as we will see below, it is no longer possible in certain situations to assign the access deficit to either player without loss of generality. We can now proceed to the solution concepts themselves.

5.3.7.2 The Nash Solution

In this section we only briefly concern ourselves with the standard Nash solution, in favour of its weighted counterpart. Adapting expression (2.227) from p. 114 to the 2-player case with the disagreement point normalized to the origin yields

$$f^N(U) = \arg\max_{x \in U} x_i x_j \text{ for all distinct } i,j \in N. \tag{5.55}$$

Being related to the geometric average, expression (5.55) assumes a maximum whenever $x_i = x_j$ for all pairs of players with $i \neq j$. As proposed by the Shapley value, this solution again splits the gains of any given link g_{ij} equally among both players. Things change, when we switch to the feasible set U^a and consider access fee correction. As soon as the normal vector of the hyper-space bounding U^a differs from $\mathbf{1}$, the arg max of $f^N(U^a)$ is no longer an equal split of the surplus, but rather one in favor of the player with the previous access deficit. Now, a problem can arise, as this bias does not take into account which of the players has the larger

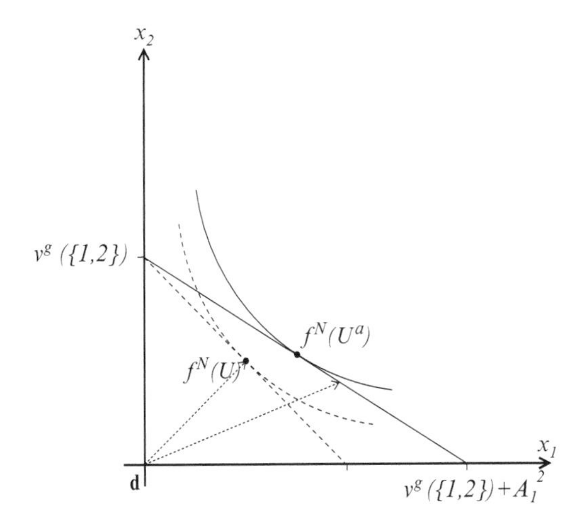

Fig. 5.12 Nash solutions for feasible sets U and U^a

contribution to $v^g(\{i,j\})$. This is because the player with the access deficit need not also be the one with the lower contribution. In the extreme case, one of the players is a zero-player. For an illustration of the Nash solution on both domains, see Fig. 5.12, where it is player 1 with the pre-peering access deficit. (For the moment, we ask the reader to ignore the dotted vectors.)

To account for the different contributions that players can make to any cooperation, we apply the Weighted Nash solution to the bargaining problem, as introduced in Definition 2.74, p. 115. Adapted to a 2-player setting for the peering game, it is given by the following expression:

$$f^{N_\gamma}(U) = \arg\max_{\mathbf{x}\in U} x_i^{\gamma_i} x_j^{\gamma_j} \quad \text{for all } i \neq j \in N,\ \gamma_i, \gamma_j \in \mathbb{R}_+. \tag{5.56}$$

In order to solve for the maximizer, we set up the Lagrangian with the corresponding first-order conditions, given the weights γ_i and γ_j:

$$\mathscr{L}(x_i, x_j, \lambda) := x_i^{\gamma_i} x_j^{\gamma_j} + \lambda\,(v^g(\{i,j\}) - x_i - x_j) \tag{5.57}$$

$$\frac{\partial \mathscr{L}}{\partial x_i} = \gamma_i x_i^{\gamma_i - 1} x_j^{\gamma_j} - \lambda \stackrel{!}{=} 0 \tag{5.58}$$

$$\frac{\partial \mathscr{L}}{\partial x_j} = \gamma_j x_i^{\gamma_i} x_j^{\gamma_j - 1} - \lambda \stackrel{!}{=} 0 \tag{5.59}$$

$$\frac{\partial \mathscr{L}}{\partial \lambda} = v^g(\{i,j\}) - x_i - x_j \stackrel{!}{=} 0. \tag{5.60}$$

Solving for x_i and x_j yields

$$x_i^* = \frac{v^g(\{i,j\})}{1+\frac{\gamma_j}{\gamma_i}} \quad \text{and} \quad x_j^* = \frac{v^g(\{i,j\})}{1+\frac{\gamma_i}{\gamma_j}}, \tag{5.61}$$

which add up to $v^g(\{i,j\})$, hence satisfying (5.60). Proceeding in two steps, we first split up $v^g(\{i,j\})$ according to (5.21) and (5.22) (see p. 196) into $c_i^g(\{i,j\})$ and $c_j^g(\{i,j\})$, substituting into (5.61):

$$x_i^* = \frac{\overbrace{c_i^g(\{i,j\}) + c_j^g(\{i,j\})}^{=v^g(\{i,j\})}}{1+\frac{\gamma_j}{\gamma_i}} \text{ and } x_i^* = \frac{c_i^g(\{i,j\}) + c_j^g(\{i,j\})}{1+\frac{\gamma_i}{\gamma_j}}. \tag{5.62}$$

In the second step, we set $c_i^g(\{i,j\}) \stackrel{!}{=} x_i^*$ and $c_j^g(\{i,j\}) \stackrel{!}{=} x_j^*$ in (5.62) to solve for a pair of relative weights γ_i^* and γ_j^* needed to obtain the allocation that corresponds to the players' individual contributions:

$$\gamma_i^* = \mu \cdot c_i^g(\{i,j\}) \quad \text{and} \quad \gamma_j^* = \mu \cdot c_j^g(\{i,j\}) \text{ for all } \mu > 0. \tag{5.63}$$

With $\mu = 1$, a pair of weights can be chosen that are actually equal to the players' contributions, which the reader is welcome to check. In all, the Weighted Nash solution allows us to find weights that assign to each player his contributions, and, being equal to these contributions when $\mu = 1$, they are also very intuitive. It has to be noted though that the weights are not "global", i.e. a weight γ_i^* is only applicable in the 2-player bargaining problem with a specific counterpart j. To obtain the overall allocation of any given player, his allocations under this pairwise bargaining have to be added with respect to all players in his neighborhood. The last bargaining solution we consider is the Kalai–Smorodinsky solution below.

5.3.7.3 The Kalai–Smorodinsky Solution

We again focus mostly on the weighted counterpart of the Kalai–Smorodinsky solution to bargaining problems in this section. As opposed to the Nash solution, the Kalai–Smorodinsky solution directly incorporates the possible maximal allocations to players within the feasible set U, as defined above. Based on the peering game, its 2-player variant for all linked pairs of players is given by

$$f^{KS}(U) = \max_{k \in (0,1]} k \cdot \mathbf{a}(U) \text{ subject to } f^{KS}(U) \in U, \tag{5.64}$$

where $\mathbf{a}'(U) = (v^g(\{i,j\}), v^g(\{i,j\}))$, because U is symmetric. The Kalai–Smorodinsky solution also proposes an equal split of the gains realized between VoIP firms i and j. And, the consideration of the access revenues via U^a yields a

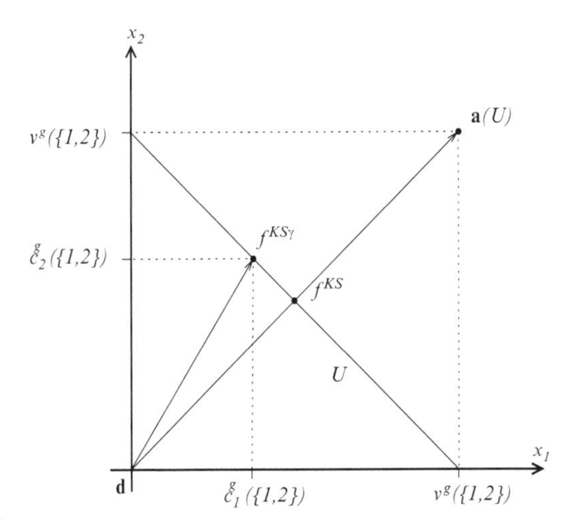

Fig. 5.13 (Weighted) Kalai–Smorodinsky solutions for feasible set U

solution similar to the one above: Because U^a is no longer symmetric, the vector $\mathbf{a}'(U^a) = (v^g(\{i,j\}) + A^i_j, v^g(\{i,j\}))$ is now skewed in direction to the player (here i) who had an access deficit before peering. We illustrate this, not to scale, in Fig. 5.12. The dotted vectors' intersections with the boundaries of the respective feasible sets are the allocations given by the Kalai–Smorodinsky solution. With feasible set U, the solution happens to coincide with the Nash solution in the peering game, which need not be the case for the asymmetrical U^a. Again, the Kalai–Smorodinsky solution can not account for the different cases that might arise under an access-fee-corrected feasible set, i.e. whether the VoIP firm with the pre-peering access deficit is the one with higher contributions or not, and possibly even a zero-player.

Advancing to the Weighted Kalai–Smorodinsky solution, we adapt the vector $\mathbf{a}(U)$, where $a_i(U) = \max\{x_i | \mathbf{x} \in U\}$ for all $i = 1,2$ to

$$\mathbf{a}^\gamma(U) = \Big(\gamma_i \cdot a_i(U), \gamma_j \cdot a_j(U)\Big)', \tag{5.65}$$

for all distinct pairs of players $i, j \in N$. Both the regular and the Weighted Kalai–Smorodinsky solutions are illustrated in Fig. 5.13, denoted by f^{KS} and f^{KS_γ}, respectively. They are located on the border of the feasible set U, the allocation f^{KS_γ} assigning each player exactly its contribution $c_i^g(\{1,2\})$, $i = 1,2$. In order to obtain this allocation the following weights, derived from the relative contributions to $v^g(\{i,j\})$, are applied:

$$\gamma_1 = \frac{l_1 q(p_1)}{l_1 q(p_1) + l_2 q(p_2)} \quad \text{and} \quad \gamma_2 = \frac{l_2 q(p_2)}{l_1 q(p_1) + l_2 q(p_2)}. \tag{5.66}$$

The intuitional link is more obvious now, as the weights conform to the relative shares in call volume within a pair of players. Again it should be noted that the weights are pair-specific, i.e. that the weight γ_i for some player $i \in N$ is in general not meaningful in conjunction with any other player. The Weighted Kalai–Smorodinsky solution is applicable to any meaningful pairs of players $i, j \in N$, but caution is advised when pairing up zero-players $i, j \in \mathbf{0}$: To begin with, they have no gains to realize, and, more formally, the weights from expression (5.66) are not defined whenever $l_i = l_j = 0$. The "global" allocation is again calculated for each player by adding up all respective allocations with his neighbors.

Due to the symmetry of the bargaining problem derived from the peering game v^g, and hence ultimately due to the equal marginal values of players i and j to $v^g(\{i, j\})$, the standard bargaining solutions of Nash and Kalai–Smorodinsky yield solutions that split the gains equally between linked players. In effect, this is the same solution as provided by the Shapley and the Myerson values. Only when we resort to the weighted variants of these bargaining solutions are we again able to find a solution which reflects the contributions of the players. In a pairwise setting, the weights assigned to each of the players have very intuitive connections to the contributions and the relative call volume of the players. The next section recapitulates the solutions produced so far, providing a comparison and a brief discussion of the latter.

5.3.8 *For Comparison Only*

Up to now this chapter produced the peering game v^g, the properties of which were discussed, and to which we applied several application rules as well as bargaining solutions. As the reader has certainly noticed, we ended up with two different allocations throughout. The first we will refer to as the *standard allocation*, as it resulted from all the allocation rules in their standard variant. The second we shall call the *contribution allocation*, which came forth after applying appropriate weights in the weighted variants of the solution concepts employed. We now discuss each allocation in turn.

The standard allocation represents a result which is in line with economic theory and hence not surprising at all: The gains from peering are split in half between the pairs of players among which they arise, i.e. among those players that are linked to one another. This allocation reflects that no individual player can, by himself, realize any gains, but rather depends on cooperation. It also shows that less "significant" players with smaller call volumes and hence lower contributions have some leverage at their disposal when it comes to distributing the gains: They can reap more under the standard allocation than they contribute to the overall gains, because without them, there would be no surplus at all. In the extreme case, even network operators with no gains from long distance fees are entitled to half of the surplus to which they allow for with their cooperation.

The contribution allocation, in contrast, specifically (and exclusively) takes into account the contributions. Players are assigned exactly the amount they contribute to the overall surplus, i.e. the savings they realize on their part through cooperation.

This depends directly on their own call volume, as well as on their long distance fees. Any potential leverage is excluded in this consideration. As we will argue now, we consider the contribution allocation the more apt solution for an implementation in a real-life setting.

Our main objection regarding the standard allocation concerns the equal treatment of firms within the pairs where the gains are realized. Even though VoIP firms generally exhibit different characteristics with respect to size, call volume, and long distance fees, each is allocated half its pairwise gains. In other words, the exact share of the pie which is contributed on the basis of a firms characteristics is not taken into account. All that is needed to establish cooperation in terms of peering is the agreement and subsequent provision of the routing information corresponding to the subscribers' telephone numbers. This (negligible) effort is identical for all firms regardless of their characteristics. Even more strikingly, from a technical point of view, the required effort does not even increase with the number of peering partners. This is where we see the advantage of the contribution allocation: It relies on how much of the gains do originate with a given VoIP firm. The allocation explicitly considers the individual characteristics in terms of size, call volume, and long distance fees, and not only the pairwise aggregate.

No doubt, in a real life setting, the "smaller" firms (those with lower contributions within a pair) would exercise some of their leverage in order to benefit on the dependence of their peering partner on them to realize any gains at all. But whether an equal split can generally be seen as a realistic solution is doubtful. We believe the contribution allocation to be more realistic in implementation, especially if one sees peering as a repeated game. It does not require any verification of the peering firms' parameters (which might change over time) to ensure a correct allocation, since every firm keeps its "own" gains. This reduces the need for administration on the side of the firms as well as for a regulatory body to check for compliance. Especially in a setting of central peering, which seeks to avoid direct interaction between peering VoIP firms, the contribution allocation is very appealing.

To motivate the contribution allocation economically, one could envision an initial one-shot deal, followed by the repeated peering game over an infinite horizon. In the one-shot deal the "smaller" firms are compensated to some extent through a one-off payment, as they are necessary to reap any gains at all. The subsequent infinitely repeated peering game then distributes the gains according to the contributions of the players. Such an (individually rational) allocation can be upheld arguing along the lines of the classical *Folk Theorem*, see Fudenberg and Maskin (1986, pp. 536 f.), where the so-called *minimax* payoff corresponds to the payoff without peering.[37] The theorem's requirement of players to discount future payoffs sufficiently little also allows us to treat the preceding one-off payment as negligible.

[37] We refer the reader to expression (2.9), p. 28. It reflects the idea of the minimax payoff, being the amount a coalition, or player, can secure for himself, when the remaining players pick strategies in order to minimize his payoff. In our case, it is the payoff without peering, i.e. zero.

In this chapter we have first derived the peering game based on v^g. This characteristic function captures the gains VoIP firms can realize by entering peering agreements with one another. On the one hand, gains arise from longs distance fees which no longer accrue once peering is in place. On the other hand, the access revenues, i.e. the charges exchanged between VoIP firms for terminating calls, influence the individual gains. Access revenues appear as surplus or deficit, depending on whether a firms has higher incoming or outgoing traffic, respectively.

The crucial point in the peering game is not only that value is created on the basis of pairs of linked players. More importantly, when looking at these smallest coalitions with nonzero marginal value, the latter can be taken apart even further into contributions of their players. The contributions are the share of the nonnegative savings of each VoIP firm to the gains from peering. In going through the different solution concepts for the peering game, we look for a result that would take into account exactly these contributions rather than the marginal values of players or the overall values of coalitions only. This approach also factors in VoIP firms with zero long distance fees: Their marginal value can be strictly positive, even though no savings (and hence contributions) originate from them.

As we established via convexity of the peering game, the nonempty core does indeed contain such an allocation. But as the core generally results in an uncountable set and not only in the desired allocation, there is no further advance on the basis of the core alone. Also the least-core can result in an uncountable set of allocations, and as we leave to the reader to check by means of Example 5.17, there is no guarantee that our desired allocation is contained in it.

The Shapley and the Myerson value both operate on a marginalistic principle and do not incorporate the notion of contributions. Only for the Weighted Shapley value can a weight system be found so that its proposed allocation reflects the sum of each player's contributions. Unfortunately, this weight system provides little in terms of intuition as to why it is chosen other than to implement $\mathbf{x}^*$. However, this deficiency can be remedied using the weighted bargaining solutions of Nash and Kalai–Smorodinsky. For these concepts, intuitively plausible weights in terms of (relative) call volumes exist.

In the next chapter, we employ an allocation procedure which by construction assigns the contribution allocation to the VoIP firms. On its basis, we investigate different means of network formation, i.e. such represented by games in strategic and extensive form. Distinguishing further, we look at two scenarios, one in which firms peer on a pairwise basis and one where a central instance for peering exists. Issues of network stability are also included.

Chapter 6
Network Formation in Peering

6.1 Overview

This chapter revolves around the process of network formation among VoIP firms. Under which conditions will the firms enter peering agreements and what will the resulting cooperation network look like? We apply a number of different approaches, not only with respect to the modus operandi of network formation, but also with regard to the general settings in which the firms operate. Each such situation is analyzed from two polar viewpoints regarding the access fees. In one scenario, we ignore any changes to access revenues, regardless of peering, while in the other we presume that access fees are dropped completely, once peering is agreed upon. Our goal is to find a simple, transparent and realistic access charge regime under which all VoIP firms have the incentive to peer with as many of others as possible. This is also in the best interest of a regulator, as it promotes the transition to comprehensive NGN infrastructure, leaving behind the need to consider multiple types of networks (see European Regulators Group 2009).

Before we get into the details of network formation, we introduce the allocation procedure according to which the gains are distributed among the players. It is one that explicitly considers the contributions of the individual players to any value realized and assigns what we called the *contribution allocation* in the preceding chapter. As we argued there, this allocation seems to us the most practical to implement in a real-life setting and could follow a one-off payment between VoIP firms to initiate peering. We relate to the properties previously established for allocation rules.

Our first formation scenario is one of bilateral peering agreements where firms enter such agreements pairwise, if to their advantage. The scenario is modelled on a graph with undirected links and the consent of both players is required to form a link, yet only one is needed to quit such an agreement. We begin with the strategic form model of network formation and then focus on some stability issues. The sequential form model, as we will see, is less appealing in this context.

P. Servatius, *Network Economics and the Allocation of Savings*, Lecture Notes in Economics and Mathematical Systems 653, DOI 10.1007/978-3-642-21096-9_6,

Another possible setting would be for the national regulator to establish a central peering instance: Here any VoIP firm can become a member, and, by doing so automatically establishes peering agreements with all other members. This approach is more along the lines of unilateral link formation, at least with respect to the existing members. Every VoIP firm can decide by itself whether or not to join the peering instance or to quit its membership. Here, the sequential formation process is the more appealing one, as we assume that it is known to everybody who already is a member and who is not. Nevertheless, stability notions and strategic form are also considered. The chapter is concluded with considerations on how a regulator could influence the VoIP environment to promote peering among the firms.

6.2 A Contribution-Based Allocation Procedure

As network formation takes place in light of the payoffs players can expect in a given network, we now introduce the allocation procedure which we apply in the rest of this chapter. As we will see below, our *allocation procedure* is not an allocation rule in the sense of Definition 2.50, p. 86. Even though it is specifically tailored to the peering game v^g from the previous chapter (cf. p. 192) it is defined on the latter's components in terms of contributions, rather than on the values v^g itself assumes. For reference, we restate the characteristic function here,

$$v(S,g) = \sum_{i \in S} \sum_{j \in N_i(g) \cap S} \alpha_j \, \alpha_i \, l_i \, q(p_i) \quad \text{for all } S \subseteq N, g \in \mathscr{G}_N,$$

where already by construction the value zero is assumed for $S = \emptyset$, all coalitions with $|S| = 1$ and the empty network $g = g^c$. As originally shown in on p. 196 in Sect. 5.2.2, our characteristic function can further be decomposed:

$$v(S,g) = \sum_{i \in S} c_i^g(S) \ \text{ for all } \ S \subseteq N, \ i \in S, \tag{6.1}$$

where

$$c_i^g(S) := \sum_{j \in N_i(g) \cap S} \alpha_i \alpha_j l_i q(p_i) \ \text{ for all } \ S \subseteq N, \ i \in S. \tag{6.2}$$

To refresh our memories we quickly relist the ingredients of the above expressions. α_i is the strictly positive market share of firm i, with respect to the continuum of customers normalized to one. Market shares sum to one over all $i \in N$. The demand for calls, $q(p_i)$, is based on the price of a call as charged by firm i, and l_i are the long distance fees a firm i incurs for having to route a call through the PSTN to the destination firm. The termination charges a do not explicitly appear but are of major importance later on. These are paid from the call-initiating firm to the firm where the call ends to reimburse for the costs of termination.

We stressed the fact that in the peering game an outcome is desirable which allocates to every player his contributions to the overall gains achieved, i.e. the contribution allocation, and not some average of his marginal values. As we showed, this allocation can be reproduced with the Weighted Shapley value and both weighted bargaining solutions we considered, the one by Nash and the other by Kalai–Smorodinsky. We neutrally denote the allocation procedure γ for this specific purpose of network formation in the peering game. It takes the following form:

$$\gamma_i(\mathbf{c}^g) = \sum_{j \in N_i(g)} c_i^g(\{i,j\}) = \sum_{j \in N_i(g)} \alpha_j \, \alpha_i \, l_i \, q(p_i) \quad \text{for all } i \in N. \tag{6.3}$$

At this point the reader will have noticed the difference to a standard allocation rule. The argument of the function γ in (6.3) is not the coalitional game v^g, but the individual players' contributions. For a given network g, we denoted by $\mathbf{c}^g$ the $(n \times (2^n - 1))$-matrix where each row corresponds to a player $i \in N$, and the elements to the values of $c_i^g(S)$ for the respective $\emptyset \neq S \subseteq N$:

$$\mathbf{c}^g := \begin{pmatrix} c_1^g(S_1) & c_1^g(S_2) & \cdots & c_1^g(S_{2^n-1}) \\ c_2^g(S_1) & c_2^g(S_2) & \cdots & c_2^g(S_{2^n-1}) \\ \vdots & \vdots & \ddots & \vdots \\ c_n^g(S_1) & c_n^g(S_2) & \cdots & c_n^g(S_{2^n-1}) \end{pmatrix}. \tag{6.4}$$

As such, γ does not map from the product space of peering games, i.e. $\mathbf{V}_N$ for a given network g, into $\mathbb{R}^n$, but rather from a subset of the nonnegative orthant of the $\binom{n}{2}$-dimensional real numbers.[1] It is therefore not an allocation rule as previously defined, but in order to maintain the intuition we have established for them, we now illustrate how γ is related. The first link is given by (6.1), which shows how the elements from $\mathbf{c}^g$ are aggregated to yield the characteristic function of the peering game v^g, the argument of a classic allocation rule. Next, we show how through this link the properties defined on cooperative games and allocation rules can be carried over into this setting. In more comprehensive vector notation, expression (6.1) translates into

$$\left(\mathbf{1}' \cdot \mathbf{c}^g\right)' = \mathbf{v}^g, \tag{6.5}$$

from which it can be concluded that any operation on the right-hand side, i.e. to the game, can analogously be performed on the left-hand side. Likewise, any property of v^g, or rather its systematic origin can be analyzed in more detail on the left-hand side. And it is this representation that makes it convenient to actually retrace

[1] Even though the matrix $\mathbf{c}^g$ is an element of the Euclidean space with $n \times (2^n - 1)$ dimensions, $n \cdot (n-1)$ are sufficient to construct a $\mathbf{c}^g$ consistent with the peering game v^g: As gains from cooperation are initially based on links between players, we can construct the entire matrix from knowing the individual contributions to all 2-player coalitions for a given network g.

possible changes to the value function v^g to the contributions of the players. Also, it is especially well suited to illustrate the properties of the peering game and how value arises among cooperating players.

So, in the following, whenever we attach properties to γ that were originally introduced for allocation rules and are furthermore based on coalitional games, we do so on the basis of $\mathbf{c}^g$, as all operations performed thereon conveniently hold true on v^g. So whenever we refer to the game as argument of γ, we actually refer to the elements that make up the game, i.e. those in $\mathbf{c}^g$ as given by (6.4). If the systematic of the latter causes the game v^g to exhibit a certain property, this can be used in conjunction with γ, as if its argument was indeed the game. This now brings us to the properties of the allocation procedure γ, where we cover informally and with the intuition described above the properties of general allocation rules elaborated in Sects. 2.4.4 and 3.4. The first properties stem from standard coalitional games and take the underlying network as given, i.e. as incorporated into the value function. Thereafter we turn to properties that were specific to communication situations in which the effects on payoffs under γ through alterations to network structure are considered. We start with the properties that are satisfied by γ.

Nonnegativity of γ is established without complications. The lowest payoff any player can be assigned is zero. This is the case for all players $i \in N$ that are unlinked, $N_i(g) = \emptyset$, and have hence decided completely against peering, as well as for those with zero long distance fees, the zero-players $i \in \mathbf{0}$ with $l_i = 0$.

Remarkable is also the absence of externalities under γ. Any payoff assigned to a given player i can be attributed directly to the links of which he is a part. Consequently, no player $k \in N \setminus \{i, j,\}$ is ever influenced (in terms of his payoffs under γ) by the formation or deletion of a link g_{ij} in which he does not participate.

The allocation procedure γ is both *additive* and *homogeneous*, and therefore *linear*. This can be deduced from its functional form, see (6.3), as well as from that of v^g given in (6.1). Multiplying the argument $\mathbf{c}^g$ by some constant or adding one is directly carried over to the allocation under γ.

By construction of v^g, any *dummy player* is automatically a *null-player*, too. Under γ, a payoff of zero is associated with such players, because either they are unlinked, or all their links give rise to zero contributions on their part in the marginal value.

There are no negative marginal values for the players in the (monotonic) game v^g and so all summands of equation (6.3) are nonnegative. With this, γ satisfies the property of *positivity*.

Allocation procedure γ also exhibits *coalitional monotonicity*. Whenever the value created through some link $g_{ij} \in g$ increases, players i and j can be no worse off, according to (6.3). This is the case regardless of which coalition i and j belong to. It should also be considered that by construction of v^g, increases in the value of some coalition S must be traceable to an underlying link $g_{ij} \in g$. For a given network g, this increase must therefore be consistent for all subsets of S containing players i and j, as well as all supersets of S.

According to the property of *strong monotonicity*, if the marginal value of some player is no less in game v^g than in game u^g, his payoff under γ must not decrease in u^g. Checking expression (6.3), we see this is true, because whenever the marginal value for a player is (weakly) increased, this results from either his contribution or the contribution to a player he is linked to. Hence, even in the case where one player's contribution remains constant, and the (weak) increase in his marginal value is due solely to some other player's (weakly) increased contribution, the former player's allocation cannot go down.

Taking advantage of our more detailed specification in terms of contributions, we can also introduce a property called *contributional monotonicity*:

Definition 6.104. An allocation procedure γ is said to be *contributionally monotone*, if an weak increase in some players contribution will not lead to a strictly lower allocation under γ:

$$c_i^g(S) \geq \widetilde{c}_i^g(S) \text{ for all } S \subseteq N \Rightarrow \gamma_i(\mathbf{c}^g) \geq \gamma_i(\widetilde{\mathbf{c}}^g), \text{ for all } i \in N,\ g \in \mathscr{G}_N. \quad (6.6)$$

Comparing (6.3) and (6.4) reveals that a higher contribution will necessary lead to a higher payoff under γ.

Symmetry is also fulfilled by allocation procedure γ, because the payoffs assigned to players are based solely on their position within the network g, i.e. on their links with other players. Relabelling the players in g does not influence the allocation.

Efficiency, too, is satisfied by γ, as all gains from cooperation are distributed among the players: $\mathbf{1}' \cdot \gamma(\mathbf{c}^g) = v^g(N)$. The relation is immediate, given the fact that $N_i(g) = N_i(g) \cap N$:

$$\sum_{i \in N} \gamma_i(\mathbf{c}^g) = \sum_{i \in N} \sum_{j \in N_i(g) \cap N} \alpha_j\, \alpha_i\, l_i\, q(p_i) = v^g(N). \quad (6.7)$$

The same reasoning as to why the null-player property is fulfilled by γ holds true for the *carrier axiom*: Only players with strictly positive marginal value to some coalition $S \subseteq N$ are assigned a nonnegative payoff. And, since there can exist players with $l_i = 0$, whose marginal values are strictly positive but their contributions to it equal zero, it is possible that even a strict subset of the carrier already is allocated the value of said carrier under v^g. Nevertheless, even the minimal carrier must include all zero-players.

The allocation procedure γ satisfies the property of *core consistency* in connection with the peering game v^g, the latter exhibiting convexity and hence a nonempty core. We refer the reader to Claim 5.3.6, p. 220, where $\gamma(\mathbf{v}^g) = \mathbf{x}^*$, as (5.47) and (6.3) reveal.

As the notions of a strongly and regularly superfluous link coincide under v^g, so do the strongly and regular *superfluous link properties* on allocation procedure γ. Such a link can only occur between players i and j with zero transit fees, $l_i = l_j = 0$. The link g_{ij} will consequently not lead to additional value under v^g and appear as zero valued summands in the algebraic form of γ given by (6.3).

Given a network g, a superfluous player i under v^g is one with either $l_i = l_j = 0$ for all neighbors $j \in N_i(g)$, or alternatively $N_i(g) = \emptyset$. He is not involved in the creation of any gains, and so his links, if any, can safely be removed from g. This in turn does not affect γ, and so the *superfluous player property* is fulfilled.

Along this line, we also find γ to be *network flexible*: In networks g' giving rise to the same value as the complete network g^c, given by $v^{g'}(N) = v^{g^c}(N)$, this value must originate on the same links and is allocated in the same fashion under γ.

Another very important property is that of *component efficiency*. It is clearly fulfilled by allocation procedure γ, because there are no externalities, i.e. the gains from a link g_{ij} are distributed among players i and j alone. By definition, this cannot be outside the component to which g_{ij} belongs. For the same reason, γ is *component decomposable*.

The *improvement property* is trivially fulfilled, as the situation it applies to cannot occur under γ: No player $k \in N \setminus \{i, j\}$ will ever be affected in his payoffs, when i and j form (or sever) a link between them.

Also very important is the property of *link monotonicity*. It ensures that players involved in forming a link do not end up worse off. As the entries of $\mathbf{c}^g$ and hence the summands of γ are nonnegative, it is satisfied.

Unfortunately, there are also some plausible and convenient properties which allocation procedure γ does not satisfy:

Aggregate monotonicity cannot hold by definition of the underlying value function v^g. As gains from cooperation can be traced back to links between players, it is impossible for the grand coalition N to have a higher value, without some subset $S \subsetneq N$ also exhibiting this increase due to a link.

The often-cited property of *fairness*, which ensures that both players are affected equally from forming or removing a link between them, is intentionally not fulfilled. Allocation procedure γ is constructed as to consider explicitly the contributions of the players which need not be symmetric. This also withdraws any basis for the *balanced contributions property*.

Link anonymity of γ is not fulfilled, as the players who give rise to the value of a link need not have identical market shares, demand, and long distance fees.

Weak link symmetry, even though it seems to be a fairly light assumption is also violated by γ. Take an arbitrary link g_{ij} with $l_i > 0$ and $l_j = 0$. Then, it is easy to see that player j's payoff is not affected by the addition or removal of g_{ij}, while player i's payoff is.

Summing up, allocation procedure γ from (6.3) meets a number of desirable properties such as linearity, different forms of monotonicity, symmetry, (component) efficiency, and last but not least contributional monotonicity. Others properties which seem very plausible at first glance are, however, not satisfied. Among the latter we find fairness and in particular weak link symmetry, which is designed to overcome the indifference of players to agree to the formation of a link between them. As we will see below, this indifference is to play a crucial role.

In the context of network formation, the access revenues are of importance for the first time. In the peering game v^g they disappear, due to their zero-sum nature, and are consequently also swallowed by any allocation rule based on v^g, as well as the contribution based γ. Nevertheless, they do play an important role, as they are being dropped under peering: As soon as two companies enter a peering agreement, their net access revenues are de facto reversed in sign as compared to before. One firm loses its surplus, while the other benefits from a disappeared access deficit. This change, denoted by A_i^g for some player i in network g, is given by

$$A_i^g = \sum_{j \in N_i(g)} A_i^j = \sum_{j \in N_i(g)} a\alpha_i\alpha_j \left[q(p_j) - q(p_i)\right]. \tag{6.8}$$

It is the sum of all the partial access payments to i's peering partners. We denote the vector containing all n VoIP firms entries, by $\mathbf{A}^g \in \mathbb{R}^n$. If its ith entry is negative, player i gains from peering in terms of access revenues. Before peering, he had an access deficit, as his outgoing call volume to the firms $j \in N_i(g)$ was higher (in the aggregate) than the call volume incoming from these firms. If it is positive, player i will lose out from dropped termination fees, which, aggregated, he collected more than paid out before peering. In case he only peers with other VoIP firms whose prices are identical, i.e. $p_i = p_j$ for all $j \in N_i(g)$, its access revenues are equal to zero, $A_i^g = 0$, and are unaffected by peering.

Interpreting this change in access revenue as the costs of forming a link suggests the use of a certain class of network formation games. See, e.g., Slikker and van den Nouweland (2001b, pp. 193 ff.). They present an approach to network formation in which establishing links between players can only be done at a cost. Each link g_{ij} is assigned a nonnegative cost, possibly uniform, that must be split up in some way between the players, usually the initiators i and j. Unfortunately, these approaches are of little help to allocation in the peering game: Based on access revenues, the "aggregate" cost of a link g_{ij} in our context is given by the sum of the partial access payments of players i and j with respect to one another. Even though A_i^j and A_j^i generally assume nonzero values each, their sum is zero, as Proposition 5.21 (see p. 187) shows: $A_i^j + A_j^i \equiv 0$ for all $i, j \in N$.

It can only be concluded that this change in access revenues must be considered in network formation to yield a meaningful result. How the access revenue correction is incorporated into the network formation game will be the subject of the individual formation processes treated below. We begin with pairwise network formation where links are formed only with the consent of both parties involved.

6.3 Bilateral Network Formation Through Pairwise Peering Agreements

6.3.1 *Overview*

In this section we set the environment for the VoIP firms such that peering is done freely on the basis of pairwise agreements. In order for a peering agreement to be established, both firms, say i and j, have to agree to the formation of a link g_{ij} between them. Based upon all these pairwise decisions, a network g will result. This process of network formation can be conducted in two ways: First through a strategic form game, where all peering decisions are modelled to be taken simultaneously. Secondly, by means of an extensive form game, in which players, or rather pairs of such, act sequentially regarding their decisions to peer. We will see that the latter is of little appeal in this framework. Both forms are treated in the order in which they were mentioned above, with considerations on network stability in between.

6.3.2 Network Formation in Strategic Form

In Sect. 3.6.3, p. 147, the network formation game in strategic form, $\Gamma(N,v,\gamma)$, was introduced. Its triplet (N,S,f) is readily adapted to our VoIP peering situation: The player set N naturally corresponds to the set of VoIP firms, and S to the space of pure strategies of the n players. There is no limited or unidirectional peering, either a link exists between players i and j, or it does not. Consequently, an individual strategy s_i is a subset of $N\setminus\{i\}$, containing those and only those VoIP firms, with which i would like to peer. The individual space of pure strategies is given by $\mathsf{S}_i = 2^{N\setminus\{i\}}$. Whether a peering agreement will be formed depends only on the willingness of the potential peering partner to do so.

The payoff function $f: CS_N \times \mathsf{S} \to \mathbb{R}^n$ can be amended to the specific situation, given the allocation procedure γ and the underlying network $g(s)$, the latter formed on the basis of the strategy tuple $s \in \mathsf{S}$. Most generally, we obtain for all players $i \in N$ the following expression,

$$f_i(s) = \gamma_i(\mathbf{c}^{g(s)}), \tag{6.9}$$

where the payoffs are solely based on the γ. This setup can be changed to accommodate situational differences, as we will see below.

We have shown in the previous chapter that the class of games v^g contains only convex games, i.e. the marginal gains of all players $i \in N$ from joining ever larger coalitions do not diminish, regardless of the underlying network g. In the context of v^g, this also means that no player $i \in N$ has a strictly negative marginal value, regardless of the coalition $S \subseteq N\setminus\{i\}$ he might join. Therefore, given a network g, no higher value can be realized than under the grand coalition: $v^g(N) \geq v^g(S)$ for all $S \subseteq N$ and $g \subseteq g^c$. Additionally, v^g is monotonic on g, so adding links to the network g will never result in a lower value for any coalition:

$$v(S,g) \geq v(S,\hat{g}), \quad \text{for all } g \in \mathcal{G},\ \hat{g} \subseteq g,\ S \subseteq N. \tag{6.10}$$

Therefore, an appropriate allocation procedure γ should present the necessary incentives to the players to choose an equilibrium strategy $s^* \in \mathsf{S}$ for which the resulting network $g(s^*)$ is the complete network g^c, or at least one that is payoff equivalent.

We elaborate on what an "appropriate" γ is, and then show that $s^c := (s_i)_{i\in N}$ with $s_i = N \setminus \{i\}$ for all $i \in N$ is indeed an equilibrium strategy. Over the course of our analysis, we will see that the allocation procedure γ as specified in (6.3) satisfies what was termed appropriate above. The first property we require is link monotonicity: No player involved in the formation of a link should be made worse off, else he would have an incentive to deviate from such a strategy by unilaterally severing this link. This is sufficient to establish the strategy s^c as one constituting an undominated Nash equilibrium.

Proposition 6.28. *If γ is an allocation procedure satisfying link monotonicity, the strategy s^c constitutes an undominated Nash equilibrium in the network formation game $\Gamma(N,v,\gamma)$*

Proof. We refer the reader to the identical proof of Proposition 3.18, p. 149. Despite different assumptions on γ in Proposition 3.18, the proof only relies on the property of link monotonicity which is implied by those different assumptions there. ■

Unfortunately, link monotonicity of γ is not sufficient to ensure that all strategies s^* which constitute undominated Nash equilibria also lead to a network $g(s^*)$ which is efficient, so that $v(N,g(s^*)) = v(N,g^c)$. Here, the zero-players for which we allow introduce some difficulties: Because the allocation procedure γ is constructed so it can allocate gains from an additional link g_{ij} unilaterally to either player i or player j, it does not satisfy weak link symmetry, according to which both players, if any, must strictly profit from forming link g_{ij}. A considerably weaker result than the one given by Proposition 3.19, p. 150, is the consequence, due to the lack of weak link symmetry of allocation rule γ.

Proposition 6.29. *Any strategy s^* that constitutes an undominated Nash equilibrium in the network formation game $\Gamma(N,v,\gamma)$, leads to a network $g(s^*)$ in which at least all nonzero players $i \in N \setminus \mathbf{0}$ with $l_i > 0$ are fully connected. More precisely, the resulting network $g(s^*)$ is bounded from below and above,*

$$g^c|_{N\setminus\mathbf{0}} \subseteq g(s^*) \subseteq g^c,$$

where $g^c|_{N\setminus\mathbf{0}}$ is the complete network restricted to the players with $l_i > 0$.

Proof. Suppose this was not the case. Then, a strategy s^*, undominated and constituting a Nash equilibrium, must exist, such that $g_{ij} \notin g(s^*)$, where for players i and j we have $l_i, l_j > 0$. Because s^* is a Nash equilibrium, we conclude that $j \notin s_i^*$ and $i \notin s_j^*$.[2] Now, consider strategies $\widetilde{s}_i^* := s_i^* \cup \{j\}$ and $\widetilde{s}_j^* := s_j^* \cup \{i\}$ and note that $g(\widetilde{s}_i^*, s_{-i}^*) = g(s_i^*, s_{-i}^*) = g(s^*)$. Then

$$f_i(\widetilde{s}_i^*, s_{-i}^*) = \gamma_i(N,v,g(\widetilde{s}_i^*, s_{-i}^*)) \geq \gamma_i(N,v,g(s_i^*, s_{-i}^*)) = f_i(s_i^*, s_{-i}^*) \text{ for all } s_{-i}^* \in \mathsf{S}_{-i},$$

[2] If either $j \in s_i^*$ or $i \in s_j^*$ was the case, the respective other player has an incentive to deviate to increase his payoff by forming the link g_{ij}!

where the inequality ist strict for $S_{-i} = (\widetilde{s}_j^*, s_{-j}^*)$. Therefore, strategy s_i^* is dominated by $\widetilde{s}_i^*$, an argument which holds analogously for player j and poses a contradiction to our assumption of s^* constituting an undominated Nash equilibrium. The upper bound, g^c, can by definition not be broken. ■

Unfortunately, for all zero-players $i \in \mathbf{0}$ with $l_i = 0$, the strategy s_i^* is not dominated by strategy $\widetilde{s}_i^*$ as constructed above. By design, zero-players are allocated no value under γ which renders them indifferent to forming a link albeit their (possibly strictly) positive marginal value. Then, an undominated Nash equilibrium strategy s^* does not guarantee an efficient network $g(s^*)$, where $v(N, g(s^*)) = v(N, g^c)$.

Because the (undominated) Nash equilibrium is a concept that supports only strategies from which individual deviations are not worthwhile, we will also investigate the refinements of the Nash equilibrium concept known as Coalition-Proof and Strong Nash equilibrium. These take into account possible deviations of coalitions with more than just one player, when allowing for equilibria. We begin with the weakest concept, the Coalition-Proof Nash equilibrium, and work our way down the set-inclusion chain to the restrictive version of the Strong Nash equilibrium.

The Coalition-Proof Nash equilibrium is a concept that allows only strategies which are consistently robust to deviations by coalitions of any size. "Consistently" in this context requires not only the equilibrium strategy to be deviation proof, but also the deviation itself in terms of further (counter-)deviations. The concept was originally introduced by (Bernheim et al., 1987) and discussed in more detail in Definition 2.8, p. 18.

But as we will see momentarily, the equilibrium strategies produced by this refinement do not guarantee an efficient network either. More precisely, they yield the same result as stated in Proposition 6.29.

Proposition 6.30. *Any strategy s^* that constitutes a Coalition-Proof Nash equilibrium in the network formation game $\Gamma(N, v, \gamma)$, leads to a network $g(s^*)$ such that*

$$g^c|_{N \setminus \mathbf{0}} \subseteq g(s^*) \subseteq g^c.$$

Proof. As the equilibrium network $g(s^*)$ is bounded above by the complete network g^c, we only need to show that it is not possible that the lower bound is violated and that $g^c|_{N \setminus \mathbf{0}} \subseteq g(s^*)$ holds indeed. Under allocation procedure γ, a deviation from an existing link will never make a player strictly better off, regardless of his long distance fees, i.e. whether he is a zero-player or not. An initial deviation of any coalition $R \subseteq N$ from the equilibrium strategy s^* to form one or more additional links can therefore never be counter-deviated profitably by any subset $T \subseteq R$.

Likewise, if there was a pair of nonzero players for which $g_{ij} \notin g(s^*)$, a deviation would always be profitable: If $j \notin s_i^*$ and $i \notin s_j^*$, then strategies $\widetilde{s}_i^*$ and $\widetilde{s}_j^*$ as defined above would be strictly profitable for players i and j in a deviating coalition. If either $j \notin s_i^*$ or $i \notin s_j^*$, both of the players i and j would profit strictly from either i or j deviating to $\widetilde{s}_i^*$ or $\widetilde{s}_j^*$. Hence the links between all nonzero players must be part of $g(s^*)$. ■

This result is not surprising. In essence, the Coalition-Proof notion merely looks at all possible decompositions of the original game with respect to the player set; in each such decomposition, the strategy choice of the subset of players in response to the equilibrium strategy played by the complementary set of players must not constitute a deviation from the respective entries of s^*. And deviations will only occur, if the players are strictly better off by choosing such a strategy as a response. Under allocation procedure γ this will not be the case for the set of zero-players, when they are faced with a deviation that would establish a link with a nonzero player.

With the Coalition-Proof refinement not being a refinement at all under γ (when compared to the undominated Nash equilibrium) we now turn to an even more rigourous concept, the Strong Nash equilibrium. It was originally introduced by Aumann (1959) and comes in two nuances, one more, and the other less restrictive (see Definition 2.7, p. 17). The more restrictive one rules out as candidates for equilibrium all strategies that lead to a Pareto improvement for a group of deviating players. No coalition can exist in which as little as one member benefits strictly from deviating while the others are held at par at least. The less restrictive one excludes "only" those strategies that allow for weak Pareto improvements through which all members of a deviating coalition will be strictly better off. This difference can be illustrated very well in our context; it will prove crucial when it comes to the formation of efficient networks. We continue the analysis with the less restrictive notion of the Strong Nash equilibrium, beginning with the following statement:

Proposition 6.31. *Any strategy s^* that constitutes a Strong Nash equilibrium (in the less restrictive sense) in the network formation game $\Gamma(N,v,\gamma)$ leads to a network $g(s^*)$ such that*

$$g^c|_{N\backslash\mathbf{0}} \subseteq g(s^*) \subseteq g^c.$$

Proof. Analogous to the proof of Proposition 6.29; suppose this is not the case and there exists a network $g(s^*) \subsetneq g^c|_{N\backslash\mathbf{0}}$. Now, any nonexisting link between nonzero players i and j makes a deviation to form this link strictly profitable to both players under γ. The strategy tuple s^* can therefore not be an equilibrium strategy by the definition of a Strong Nash equilibrium (in its less restrictive sense). What "remains" then are only links between zero and nonzero players or purely between the former. As these do not give rise to mutually (strictly) beneficial deviations, such links may or may not exist in any equilibrium strategy. ■

As it turns out to be, the less restrictive version of the Strong Nash equilibrium will leave us in the same position as before. Since a deviation must strictly benefit all deviating players, the refinement allows for equilibrium strategies whose result is a non-efficient network as characterized above. The desired, efficient, network will at last occur when we apply the restrictive notion of the Strong Nash equilibrium.

Proposition 6.32. *Any strategy s^* that constitutes a Strong Nash equilibrium (in the more restrictive sense) in the network formation game $\Gamma(N,v,\gamma)$ leads to an efficient network $g(s^*)$, where*

$$v(N,g(s^*)) = v(N,g^c).$$

Proof. We again proceed by contradiction, assuming an equilibrium strategy s^* for which $g(s^*)$ is not efficient. In such a case, there must be a pair of players $i, j \in N$ whose strategies are such that $j \notin s_i^*$ and $i \notin s_j^*$, and consequently $g_{ij} \notin g(s^*)$. Now, if these two players deviate to the strategies $\widetilde{s}_i^* := s_i^* \cup \{j\}$ and $\widetilde{s}_j^* := s_j^* \cup \{i\}$, the resulting network will be such that $g_{ij} \in g(\widetilde{s}_i^*, \widetilde{s}_j^*, s_{N\setminus\{i,j\}}^*)$ and possibly efficient. In any case, players i and j are no worse off under γ than before. More precisely, if both players are such that $l_i, l_j > 0$ they will be strictly better off under γ, and if one of them is a zero-player, only the other will benefit strictly from forming link g_{ij}. Either way, the deviation from s^* to $(\widetilde{s}_i^*, \widetilde{s}_j^*, s_{N\setminus\{i,j\}}^*)$ is one which is excluded by the Strong Nash equilibrium. Hence, a strategy s^* cannot constitute an equilibrium, if it does not give rise to an efficient network. ■

So, curiously enough, we obtain the desired network structure (where g is efficient with respect to v^g) only under the most restrictive equilibrium notion, the Strong Nash equilibrium. This concept is usually criticized for being too strong for many games in terms of existence, especially when the cardinality of the player set exceeds three.[3] As far as existence of Strong Nash equilibria in a network formation game $\Gamma(N, v, \gamma)$ is concerned, the strategy s^c as applied in Proposition 6.28 will suffice. And because the restrictive notion of the Strong Nash equilibrium is a subset of its less restrictive cousin in terms of admissible strategies, the latter notion again is a subset of the Coalition-Proof Nash equilibrium, existence is assured for the previously applied concepts as well.[4]

What we have consequently ignored so far are the access fees, or rather that they are dropped under a peering agreement: How will the players' decisions be influenced, given that now the gains from long distance fees that are allocated by γ will have to be corrected by the access revenues. As introduced above, the access revenue of a VoIP firm i, denoted A_i^g (see (6.8)) is the balance from access fees charged to other firms for incoming calls and paid to these firms for outgoing (off-net) calls. Incorporating this access payment into the network formation game $\Gamma(N, v, \gamma)$ leads to an altered payoff function $\widetilde{f}$, which is now based on both γ and $\mathbf{A}^g$:

$$\widetilde{f}_i(s) = \gamma_i(\mathbf{c}^{g(s)}) - A_i^{g(s)}. \tag{6.11}$$

We are now confronted with the network formation game $\widetilde{\Gamma}(N, v, \gamma)$ based on the triplet $(N, \mathsf{S}, \widetilde{f})$. In this game the payoff function loses many of the convenient properties with which it was endowed before. This change is caused by the introduction of the correction factor $\mathbf{A}^g$. Whenever some player i agrees to form a link g_{ij}, and player j does likewise, their payoffs result not only from their contributions to the marginal value of this link, as allocated by γ, but also from the correction A_i^j and A_j^i, respectively. As we know, this correction factor can only

[3] See, e.g. Bernheim, Peleg, and Whinston (1987, p. 3) or Slikker and van den Nouweland (2001b, p. 148).

[4] See Fig. 2.3, p. 20, for this relation of the equilibrium concepts.

be beneficial to one of both players (assuming it differs from zero, $-A_i^j = A_j^i \neq 0$), while it decreases the payoff of the other player by the same amount. Whether it is still worthwhile to agree to form the link for the player whose correction will lower his payoff is a priori not determined. It depends not only on the difference of access charges and long distance fees, but also on the difference in absolute call volume. Formally, when we substitute for γ and $\mathbf{A}^g$, a player i faces the following situation:

$$\begin{aligned} \widetilde{f}_i(s) &= \sum_{j \in N_i(g(s))} \alpha_j\, \alpha_i\, l_i\, q(p_i) - \sum_{j \in N_i(g(s))} a\, \alpha_i \alpha_j\, [q(p_j) - q(p_i)] \\ &= \sum_{j \in N_i(g(s))} \alpha_j \alpha_i\, [q(p_i)(l_i + a) - q(p_j)a]\,. \end{aligned} \tag{6.12}$$

Remember that in this network formation game a link between players i and j is formed, i.e. $j \in N_i(g(s))$ (and vice versa) if and only if $j \in s_i$ and $i \in s_j$. This reflects the bilateral character of link formation, which allows a single player to decide against forming or maintaining a link. Let us now take a closer look at the individual strategies, for example s_i of player i. In (6.12), each summand reflects the net gains player i makes with some peering partner $j \in N_i(g(s))$. Because these are independent of one another, we can look at them separately to determine whether player j is an element of s_i.

In order to engage in bilateral peering , the amount player i saves from long distance fees must be large enough to balance what he might lose from the abolition of termination fees under peering, all with respect to player j. The same must hold for the peering partner, when we look at things from the other side.

Dissecting the payoff function given in (6.12), we can extract i's net gains from peering with j:

$$\alpha_i \alpha_j\, q(p_i) \cdot l_i - \alpha_i \alpha_j\, [q(p_j) - q(p_i)] \cdot a. \tag{6.13}$$

The first term represents what is saved via long distance fees, the second is the partial net access payment of firm i with respect to firm j. If expression (6.13) is positive as a whole, player i gains from a link with j. If it is negative, then peering is not worthwhile for player i, because his partial access surplus (the spread between fees collected from and paid to j) is larger than the cost saved through abolishing long distance transit to j. This applies analogously to firm j, which can be seen by exchanging indices i and j in expression (6.13).

Combining the above considerations for players i and j, we infer that entering a peering agreement, i.e. forming a link, requires both players to be no worse off than before; for this the following inequalities, derived from (6.13), must hold simultaneously:

$$q(p_i) \cdot l_i \geq [q(p_j) - q(p_i)] \cdot a, \tag{6.14}$$

$$q(p_j) \cdot l_j \geq [q(p_i) - q(p_j)] \cdot a. \tag{6.15}$$

Condition (6.14) corresponds to player i and requires that a partial access surplus with j be no larger than the savings from the long distance fees. Only then can i benefit from cooperation. In case of an access deficit of firm i, i.e., for $q(p_i) > q(p_j)$, the right hand side is negative and the condition is trivially fulfilled. Analogue reasoning along (6.15) holds for firm j.

We summarize our findings in the following proposition:

Proposition 6.33. *In the network formation game $\widetilde{\Gamma}(N,v,\gamma)$, consider any pair of players $i,j \in N$ and an undominated Nash equilibrium strategy s:*

- *If a link between players i and j is formed, $g_{ij} \in g(s)$, then conditions (6.14) and (6.15) must be fulfilled.*
- *If (6.14) and (6.15) both hold with* **strict** *inequality, then $g_{ij} \in g(s)$.*

Proof. The proof follows directly from a comparison of (6.14), (6.13), and (6.12): Wether (6.14) is fulfilled or not will determine the sign of (6.13) and hence its influence on the payoffs in (6.12). If the link $g_{ij} \in g(s)$ exists, it must be the case that (6.14) and (6.15) hold with equality at least. If not, and, say (6.14) is violated, then the strategy s_i leading to this outcome is dominated by a strategy $s_i' = s_i \setminus \{j\}$. Player i can strictly increase his payoff by choosing s_i' instead, which would result in $g_{ij} \notin g(s_i', s_{-i})$. By the same logic, if, say, inequality (6.14) holds strictly, strategy s_i' is dominated by s_i and player i would strictly lose from not agreeing to establish link g_{ij}. This is true analogously for player j ■

Again, Proposition 6.33 is silent on links that have no influence on the payoffs of one or both players. But unlike before, we cannot make the easy distinction between zero and nonzero players. In the initial network formation game $\Gamma(N,v,\gamma)$, such a link was only possible if at least one zero-player was involved in its formation. Now, in the game $\widetilde{\Gamma}(N,v,\gamma)$, this can also occur without any involvement of zero-players: It cannot be ruled out that (6.13) is equal to zero because the pre-peering access surplus and the gains from long distance fees cancel each other out exactly. Then, the link in question has no influence on the payoff of player i, even though $l_i > 0$. This implies likewise that the zero-players, even though they do not contribute to gains from long distance fees, will generally experience a change in payoff under $\widetilde{f}$. That is, unless a zero-player i forms a link with some other player j, where both have identical prices $p_i = p_j$. Then (6.13) is trivially equal to zero.

In the light of these findings, we will have to revisit the solution concepts treated in the previous part, to see if we can still make similar predictions about the networks resulting in a given equilibrium of the game $\widetilde{\Gamma}(N,v,\gamma)$.

The modification of the network formation game from Γ to $\widetilde{\Gamma}$ changes the payoff function f in a way where the value of

$$A_i^g = \sum_{j \in N_i(g)} A_i^j$$

is simply subtracted from the value assigned to i by γ. One could interpret this a measure to allocate the change in access payments, which arises from dropping the access fees under peering, even before they cancel out in $\mathbf{c}^g$ and v^g, and "evade" γ.

But now, with the correction factor A_i^g incorporated into $\widetilde{f}_i$, two major changes have to be considered: First, the partition of the player set into zero and nonzero players is no longer of any use in the network formation game, as all players's payoffs can now be affected by forming links. Secondly, the game $\widetilde{\Gamma}(N,v,\gamma)$ is no longer monotonic in terms of linkage. Forming (additional) links might now make some players strictly worse off, when a former access surplus exceeds the gains from dropping long distance fees.

These changes have a profound impact on the equilibrium notions we have studied above, because we can no longer guarantee any specific network structure as the result of some equilibrium strategy s^* when f is replaced with $\widetilde{f}$. The results of Propositions 6.28–6.32 cannot be reproduced in the game $\widetilde{\Gamma}(N,v,\gamma)$. So, instead of analyzing the results of possible equilibrium strategies in the network formation game, we approach from vis-à-vis by searching for network structures which satisfy the stability notions introduced in Sect. 3.6.4, p. 151. In other words, we look for networks, in which, given the payoff function $\widetilde{f}$, no player or group of such has incentives to deviate with the consequence of increasing their payoffs. Can the resulting network $g(s)$ meet the conditions imposed by the notions of stability?

6.3.3 *Stability of Networks*

This section focusses on the stability of networks which resulted from the strategic network formation game $\widetilde{\Gamma}(N,v,\gamma)$, especially from its payoff function $\widetilde{f}$. But instead of analyzing the players' strategies $s \in \mathrm{S}$, we are more interested in the resulting network $g(s)$. Given such a network we check for possible lucrative deviations where players have incentives to add or remove links between them. The first notion we want to study in this context is that of pairwise stability, as introduced in Definition 3.100, p. 152. Informally, a network is pairwise stable, if each pair of players is content with the status quo (in terms of linkage) between them. If there is a link between them, its existence does not make either player strictly worse off; and if there is no link, adding it could not result in a weak Pareto improvement making both players strictly better off.

We deviate from the standard definition insofar as we use the payoffs $\widetilde{f}$ from the network formation game and not the plain allocation procedure γ for our analysis. In other words, we take into account the access charge correction $\mathbf{A}^g$, once peering is in place. Rewriting the definition of pairwise stability with the current notation leads to the following:

Definition 6.105. A network $g(s)$ resulting from the network formation game $\widetilde{\Gamma}(N,v,\gamma)$ is called pairwise stable, if the following conditions are met:

1. For all links $g_{ij} \in g(s)$, we have $q(p_i) \cdot l_i \geq [q(p_j) - q(p_i)] \cdot a$ and $q(p_j) \cdot l_j \geq [q(p_i) - q(p_j)] \cdot a$.

2. For all links $g_{ij} \notin g(s)$, if $q(p_i) \cdot l_i > [q(p_j) - q(p_i)] \cdot a$, then $q(p_j) \cdot l_j < [q(p_i) - q(p_j)] \cdot a$.

The similarity to Proposition 6.33 is immediate. While the first condition is practically identical, the second condition is slightly more comprehensive: Now, not only links g_{ij}, from which both i and j benefit strictly, must be part of the pairwise stable network g, but also those, whose existence leaves one party unchanged in termes of payoffs, while the other benefits strictly. But again, Definition 6.105 remains silent about the presence of links where both players' payoffs are unaffected. The above considerations lead to the following proposition:

Proposition 6.34. *In the network formation game $\widetilde{\Gamma}(N, v, \gamma)$ a strategy vector s constitutes an undominated Nash equilibrium, if the network $g(s)$ is pairwise stable.*

Proof. Compare Proposition 6.33 and Definition 6.105 to see that the former includes the conditions specified in the latter. ■

By definition, pairwise stability provides payoff-conditions according to which a network is considered stable, but without any intuitional reference as to what such networks could look like. In the setting with access fee correction, the partition of the player set into zero and nonzero players according to their long distance fees is generally no longer meaningful. This is because under $\widetilde{f}$, the zero-players' payoffs are unlikely to remain unaffected by the formation of links, at least not in the aggregate. We nevertheless stick to the nomenclature and provide a short example to emphasize this change:

Example 6.20. Suppose the player set numbers 15 VoIP firms, $N = \{1, 2, \ldots, 15\}$, where firms 5, 10, and 15 are zero-players, i.e. $\mathbf{0} = \{5, 10, 15\}$. Now, suppose we have prices

$$p_1 = p_2 = \ldots = p_5 < p_6 = \ldots = p_{10} < p_{11} = \ldots = p_{15},$$

and a network g with three components, each comprised exclusively of players with equal prices:

$$N/g = \{\{1,2,3,4,5\}, \{6,7,8,9,10\}, \{11,12,13,14,15\}\} =: \{C_1, C_2, C_3\}$$

If we analyze one component at a time, pairwise stability of, say, C_1 is satisfied, as soon as the nonzero players are all linked to one another and to all zero-players. Links among the zero-players can but need not be established. Because the prices are symmetric within any component, the access revenues of all players are identical to zero and $\widetilde{f} = f$.

Now, if we take more than just a single such component into consideration, the specified network g need not be pairwise stable at all. Depending on the parameters, now also players from different components might have the incentive to engage in peering and establish additional links. In a pairwise stable network, this will occur as long as the price-induced demand differences (i.e. the access revenues) do not overshadow the gains from dropping long distance fees. This applies likewise to zero and nonzero players, even though the former cannot accrue gains from long distance fees.[5] And even though the prices, and hence the access revenues, are identical for all players within a component, the long distance fees can differ individually, as we have not assumed any symmetry there. Consequently, we can make no general prediction as to whom of the firms is likely to peer across components in a pairwise stable network.

Despite the assumptions placed on the prices charged by the VoIP firms, Example 6.20 still leaves considerable scope in terms of what constitutes a pairwise stable network. This is mostly due to the individual long distance fees.

Generally speaking, other than assuming the existence of network-operating firms i with $l_i = 0$, we refrain from placing an further assumptions on the long distance fees. Even though it might be argued that these fees are often differentiated through volume discounts, we believe that in our setting this should not be incorporated, as high call volumes can be the result of low prices as well as high market shares.

Under conditions where none of the parameters is a priori fixed for individual players, a pairwise stable network need not have any links at all, let alone enough to be efficient. The only general statement we can make in excess of Definition 6.105 concerns the zero-players:

Proposition 6.35. *In a pairwise stable network g, a zero-player $i \in \mathbf{0}$ is part of some link $g_{ij} \in g$ only if his partial access revenue A_i^j with player j is nonpositive:*

$$g_{ij} \in g \;\Rightarrow\; A_i^j \leq 0 \quad \textit{for all } i \in \mathbf{0}, j \in N$$

Proof. To begin with, we want to emphasize the fact that Proposition 6.35 only states a necessary condition, not a sufficient one. The reasoning behind this proposition is quite apparent: As zero-players have nothing to gain from dropping long distance fees, the only way they could be better off under peering is through the access revenues. We now have to distinguish two cases, one where the peering partner is another zero-player, and the other one, where the peering partner is a nonzero player.

In the first case, in order to conform to the conditions of a pairwise stable network, both zero-players must have the same price and hence demands, so that their partial access revenues are identical to zero. Otherwise, one player would gain strictly from peering, while the other would lose strictly, which contradicts pairwise stability.

In the second case, the zero-player's partial access revenues A_i^j can also be strictly negative (a deficit), as long as the gains of the peering partner j are larger than the access surplus he loses under peering. If both have a net gain, pairwise stability is not violated. ■

[5] Note that a zero-player will only agree to peer with out-of-component firms whose prices are higher than his own, i.e. where he has an access deficit which would then turn into a positive payoff.

The same result can be expressed in much simpler terms, i.e. in such of prices instead of access revenues:

Corollary 6.13. *In a pairwise stable network g, there exist no links g_{ij}, where $i \in \mathbf{0}$ is a zero-player and $p_i > p_j$.*

Proof. If a zero-player peers with a firm that has a lower price, his partial access revenues A_i^j are strictly positive, which can be seen from (6.13). This is a contradiction to Proposition 6.35. ■

The payoff function $\widetilde{f}$ is based on two elements: First, an allocation procedure γ which assigns values not according to marginal values of the players, but according to their contributions upon joining a coalition $S \subseteq N$. Secondly, the access revenue correction factor $\mathbf{A}^g$ which results from dropped access fees when peering. Both elements are in turn derived via a value function v^g which is pairwise additive, i.e. value is generated on the basis of pairwise peering agreements, without any externalities beyond the players forming them. Due to the incorporation of access revenues, the zero-players $i \in \mathbf{0}$ generally experience some change in payoff in the aggregate whenever they decide to peer. (This excludes the pathological cases of absolute price symmetry.)

Because the notion of a strongly stable network incorporates the definition of obtainable networks, which relies on link-wise deviations, no further refinement is to be expected from this equilibrium notion.

Proposition 6.36. *For any network g based on the network formation game $\widetilde{\Gamma}$, especially with respect to the value function v^g and the allocation procedure γ, the following holds: The network g is pairwise stable if and only if it is strongly stable.*

Proof. From the definition of obtainability (cf. p. 153), we can see that the path from some network g to g' can be dissected into a series of steps, each involving the formation or deletion of a single link. Because of the characteristics of v^g and γ, the order in which the steps are taken is irrelevant. Each additional or removed link will affect (only) the payoffs of the two players at its end, regardless of the remaining network. Now, by definition, no links can be added to a pairwise stable network that make a player strictly better off, without making the other strictly worse off. Likewise the removal of a link cannot make either player better off, and hence, a pairwise stable network is also strongly stable.

For the other direction, suppose the initial network g is strongly stable. Then, by definition, all obtainable networks g' in which one member of S is strictly better off also lead to some other member of S being strictly worse off. In terms of adding a link, this coincides directly with the definition of pairwise stability, where $|S| = 2$. If reaching g' involves deleting links, we again face the problem that for example the deletion of a single link can be done unilaterally, in which case $|S| = 1$. Then, if player $i \in S$ profits strictly, there is no other player $j \in S$ who has to lose in order for g to be strongly stable (see Corollary 3.5, p. 154). ■

In addition to the coincidence of pairwise and strong stability from the previous proposition, we can only make some fairly general statements about zero-players in such networks and under which conditions they can or cannot have links to others. (See Proposition 6.35 and Corollary 6.13.) To wrap up this section on the network formation via pairwise peering agreements, we turn to the sequential process, modelled by a game in extensive form.

6.3.4 Network Formation in Extensive Form

The network formation games Γ and $\widetilde{\Gamma}$, which were applied to the peering game v^g in the previous sections, are noncooperative games in normal form; players decide simultaneously, or without previous knowledge of each others' choices, which strategy to play. We now want to take a quick look at network formation in extensive form, where the players choose in some specified order whether they want to form links among them or not. For details and specific models, we refer the reader to Sect. 3.6.2, p. 142. In any case, these extensive form games are still based on payoff functions f or $\widetilde{f}$, specified in (6.9) and (6.11) above, and on the value function v^g.

As we have noted before, both the value function v^g, and the allocation procedure γ display no externalities. The additional value created by the formation of a link g_{ij} is independent of the remaining network $g - g_{ij}$. This also applies to the allocation under γ, as it is split up entirely between players i and j according to their contributions.

On this foundation, distinguishing between simultaneous and sequential link formation will not make any difference in terms of the resulting network. All players are indifferent about the network around them and especially their already established neighborhood, when it comes to forming a(nother) link, because it does not affect their payoff from this link. In order for a sequential process to make a difference, we have to change the setup, e.g. the mode of link formation. We proceed to do so in the subsequent section, where peering is no longer established on the basis of pairwise agreements, but via a central peering institution which serves all its members equally. Then a decision to peer is no longer firm specific and has further implications on the payoffs.

6.4 Central Network Formation

6.4.1 The Central Peering Instance

In this section, we detach ourselves from peering agreements established on the basis of bilateral contracts between VoIP firms. Instead, we turn to a different concept, that of central peering: A central peering instance collects the routing tables of the participating VoIP firms and makes them available for access to all its members. Membership is voluntary, but VoIP firms will only gain access to the

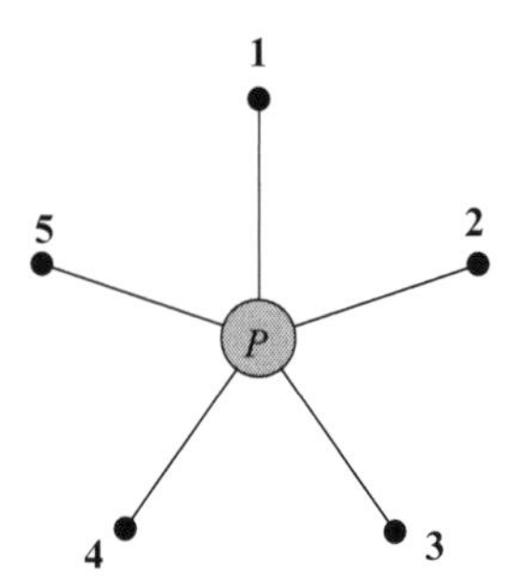

Fig. 6.1 Peering via a central instance

data hosted by the peering instance if they are a member. As such, they can access the routing tables of all other present and future members. Also, every VoIP firm can join the central peering instance, and there are no requirements other than not to engage in peering agreements on the outside. It is an all-or-nothing concept, as VoIP firms are also free to leave, whenever they deem.[6] In Fig. 6.1 we depict a stylized example where 5 firms benefit from this central peering instance, designated P. It is a mere illustration and is not intended to depict a network in the sense of graph theory, because the peering instance P is not considered to be a player (or node) as compared to the VoIP firms in our standard setting. The idea behind the concept of central peering is that all participating VoIP firms share their routing tables (the telephone numbers and respective IP-addresses of their customers) with the peering instance, to which in turn they all have unrestricted access. Such peering instances could be established by a national regulator or a neutral non-profit organisation which is collectively owned by (some of) the participating VoIP firms.

Our task now is to develop a model for this environment, with two goals in mind: For one, the technical details as to how central peering is realized would be irrelevant to the model. Secondly, for reasons of comparability, we would like to use the same framework that has been applied in the previous sections.

First, we have to realize that we are in a setting which, strictly speaking, satisfies neither unilateral nor bilateral link formation. On the one hand, players can join the peering club at will and have automatic access to the current members' routing tables, i.e. without the latters' explicit and particular consent. This reminds of unilateral link formation. On the other hand, on becoming member, a VoIP firm issues a carte blanche for all present and future members to have access to its routing data, and the severing of specific links is no longer possible. In a bilateral setting, this could be interpreted as the unrestricted or unconditional willingness to form links within the peering coalition. Further, the choice of a firm not to open itself to any peering reminds of a bilateral character.

Similar formation games, where players cannot hinder one another in joining some coalition are called *open membership games*. The decision is binary for all

[6]From this feature we make an exemption later on, but only in a technical sense.

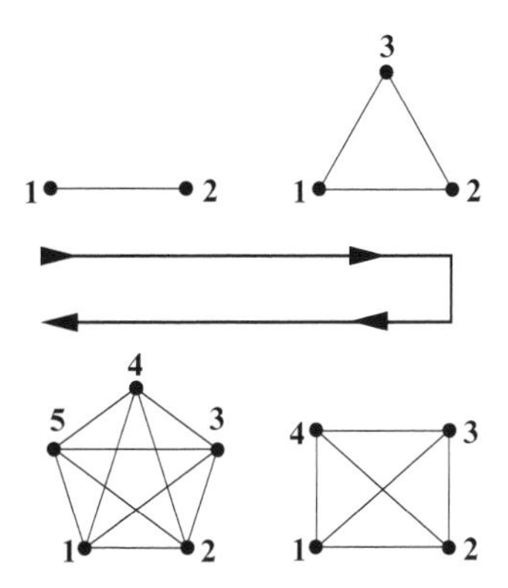

Fig. 6.2 Modelling central peering through complete networks

players, being a "yes" or a "no", and an equilibrium is reached when no one wants to deviate from his strategy: No member wants to leave the coalition and no not-member wants to join it. These conditions are referred to as *internal* and *external stability* in this context. See Bloch (2005) for a survey or Selten (1973) for an early approach of coalition formation in a Cournot-competition setting.

The next peculiarity in this setting is the peering instance itself; it cannot be regarded as a player, as it has neither strategies to choose nor any payoffs to reap. Nevertheless, all players need to be "connected" (or in technical language: linked) to it, if they want to benefit from peering. In the domain of operations research,[7] especially regarding the so-called *minimum cost spanning tree setting*, it is a standard procedure to incorporate a *root* or *source* into the network. It serves as a central hub to which all players should somehow be connected. But, in order to model central peering with the same formal toolkit as in the previous section, we have to abandon this idea, as we have not introduced the concept of a source and want to do without it.

Our approach to model the central peering instance on the basis of network theory is the following: Whenever some coalition $S \subseteq N$ peers via this central instance, we consider all members of the coalition S to be linked with one another, in a complete network restricted to S: $g|_S \overset{!}{=} g^c|_S$. More precisely,

$$N_i(g) = S \setminus \{i\} \;\text{ for all }\; i \in S,\, S \subseteq N, \tag{6.16}$$

which implies the condition that members of S cannot have any links to players who are not participating in central peering, i.e there are no pairwise peering agreements "on the side". This idea is depicted in Fig. 6.2. Beginning with an S where only two players are linked, we extend, one by one, the peering coalition to five players. The pentagon on the bottom left is the corresponding representation of the situation

[7] See, e.g. Graham and Hell (1985).

in Fig. 6.1 with the toolkit of network theory as we use it. Adding an a player i to coalition $S \subseteq N$ leads to the formation of s new links g_{ij} to all members $j \in S$.

With $g|_S = g^c|_S$ by construction, we can simplify the value function v^g given in (6.1). Because for all $i \in S$ and all $S \subseteq N$, we have $N_i(g) = S \setminus \{i\}$ and hence $N_i(g) \cap S = S \setminus \{i\}$,

$$v(S, g^c|_S) = \sum_{i \in S} \sum_{j \in N_i(g^c|_S) \cap S} \alpha_j \alpha_i l_i q(p_i) = \sum_{i \in S} \sum_{j \in S \setminus \{i\}} \alpha_j \alpha_i l_i q(p_i), \qquad (6.17)$$

which maximizes the gains from long distance fees a coalition S can realize through peering.

With the setup altered to accommodate central peering, we can now turn to the different approaches of network formation. We start with the strategic form game, followed by considerations on the extensive form. A brief treatment of stability notions of networks established through central peering concludes.

6.4.2 Network Formation in Strategic Form

As before, the formation of networks in strategic form is one in which all players pick their strategies simultaneously. But unlike before, by joining the peering instance players essentially give up their right to decline any requests for peering from other members, both present and future. This amounts to an a priori willingness to peer with all other players, if membership is selected.

Adapting the network formation game $\Gamma(N, v, \gamma)$ is readily achieved. The most significant change regarding the underlying triplet (N, S, f) is to the strategy space S, which is now binary. A player i can choose to join the peering instance, leaving him potentially open to peer with all other $j \neq i$ players, or he can decide against all peering. Thus, his strategy space reduces to $\mathsf{S}_i = \{\{\emptyset\}, N \setminus \{i\}\}$, which holds true for all $i \in N$.[8]

The resulting network $g(s)$ is one where players belong either to a completely linked component S, or its complement in $N/g(s)$ which has no links at all. The set S is the coalition joining the peering instance as a result of strategy choice $s \in \mathsf{S}$.

In the formation game $\Gamma(N, v, \gamma)$, where aggregate access payments are ignored by the payoff function f, we again obtain straightforward results as produced in the previous section. Gains from peering are nonnegative and so joining the peering instance leave any player's payoff unchanged at worse. All zero-players $j \in \mathbf{0}$ under allocation procedure γ receive payoffs of zero. All nonzero players $i \in N \setminus \mathbf{0}$, however, benefit strictly, if there is at least one other peering member. Therefore, an undominated Nash equilibrium is a strategy tuple where at least all nonzero players opt for peering:

---

[8] We stress the empty set by placing it in curly brackets, $\{\emptyset\}$, because it is an explicit strategy in addition to being an element of S_i by definition.

Proposition 6.37. *Consider the central peering network formation game $\Gamma(N,v,\gamma)$. In any undominated Nash equilibrium strategy s^*, at least all nonzero players peer: For all such s^*, we have $s_i^* = N \setminus \{i\}$ for all $i \in N \setminus \mathbf{0}$ and $s_j^* \in \mathsf{S}_j$ for all $j \in \mathbf{0}$.*

Proof. The proof is fairly basic and nontechnical. Consider first any nonzero player $i \in N \setminus \mathbf{0}$. For him $s_i^* = N \setminus \{i\}$ is a dominant strategy, because his payoff is as high as without peering (if no one else peers) or strictly higher (if peering partners exist). Hence, no incentive is given to deviate from s_i^* for all nonzero players. Next, the zero-players $j \in \mathbf{0}$ are indifferent to deviation from either strategy. Their payoffs under f are identically zero, no matter with whom they peer. Therefore, a zero-player has no incentive to deviate from either strategy he might choose. As all individual strategy spaces consist of only two different strategies, s^* is also undominated. ∎

By construction of the strategy space, any resulting network $g(s^*)$ is covered by Proposition 6.29, where

$$g^c|_{N\setminus\mathbf{0}} \subseteq g(s^*) \subseteq g^c.$$

In addition, $g(s^*)$ consists of players that are either part of a unique, fully linked component (the peering players), or completely unlinked (those not peering). In contrast to the bilateral peering environment from the previous sections, where the formation of selective links was possible, peering players here are completely linked among each other.

With the reduced strategy space $\mathsf{S}_i = \{\{\emptyset\}, N \setminus \{i\}\}$ for all $i \in N$, we omit a detailed consideration of the Coalition-Proof Nash equilibrium and the Strong Nash equilibrium (in the less restrictive sense), as they yield the same results and are based on analogue reasoning: Deviations from "peering" (with potentially everyone else) to "no peering at all" can never be strictly payoff-increasing, regardless of the size of the deviating coalition. In a Nash equilibrium, the same holds true for the other direction (i.e. deviations from "no peering at all" to "peering") and consequently, when combined, for counter-deviations of all kind.

The only significant change is incurred under the Strong Nash equilibrium in the more restrictive sense. Again, because of the reduced strategy space, the resulting efficient network is automatically the complete network.

Proposition 6.38. *Consider the central peering network formation game $\Gamma(N,v,\gamma)$. Any strategy s^* that constitutes a Strong Nash equilibrium (in the more restrictive sense) leads to the complete network: $g(s^*) = g^c$. Furthermore, the strategy s^* is unique.*

Proof. Proceeding by contradiction, let us assume that $g(s^*) \neq g^c$. By construction, this means that there exists at least one player j without any links and $s_j^* = \{\emptyset\}$.[9]

[9] Remember that with the reduced strategy space $\mathsf{S}_i = \{\{\emptyset\}, N \setminus \{i\}\}$ for all $i \in N$, components in the resulting network are either completely linked or singletons.

Suppose this single player is a nonzero player $j \in N \setminus \mathbf{0}$, who receives the payoff $f_j(s^*) = 0$. A deviation to $s_j = N \setminus \{i\}$ will strictly increase his payoff, and so, for even the smallest coalition with only j as member, s^* cannot be a Strong Nash equilibrium.

Now, let j be a zero-player, $j \in \mathbf{0}$, with empty neighborhood. If he changes his strategy to peering as member of some deviating coalition $R = \{i, j\}$ with nonzero player i, then, again, s^* cannot be a Strong Nash equilibrium, because i will profit strictly from this deviation.[10]

The proof is extended analogously to larger sets of non-peering players. That $s_i^* = N \setminus \{i\}$ for all $i \in N$ is the unique Strong Nash equilibrium strategy follows directly from the reduced strategy space $\mathsf{S}_i = \{\{\emptyset\}, N \setminus \{i\}\}$. ■

This result is quite intuitive when related to the previous finding from Proposition 6.32. The only way to achieve an efficient network in the central peering environment is via the complete network g^c: In the game Γ, even if they exist, no other efficient networks can be formed, because the strategy space S does not allow the players to peer selectively.

Now, let us again incorporate the access revenue correction, $\mathbf{A}^g$, from expression (6.8). The payoff function is then changed in the manner seen previously to

$$\widetilde{f}_i(s) = \gamma_i(\mathbf{c}^{g(s)}) - A_i^{g(s)}, \tag{6.18}$$

which consequently leads to the game $\widetilde{\Gamma}(N, v, \gamma)$. The changes in payoffs now are considerably more complicated, or at least more extensive and far-reaching. Because a player's decision to peer potentially exposes him to peering with all possible subsets of the other players, his a priori knowledge about the resulting payoff is limited. In this setting a player cannot selectively exclude, say, all those players from peering with him which yield him a negative net payoff. The only way to be shielded from such negative payoffs is by not participating in peering at all.

The incorporation of access revenues $\mathbf{A}^g$ introduces significant complications to central network formation, especially when taking into account the inability to choose or allow for individual peering partners. This is illustrated in Example 6.21, in a limited 3-player environment. There, the only equilibrium to the depicted game $\widetilde{\Gamma}$ is one in which all players decide against peering.

Example 6.21. Consider a network formation game $\widetilde{\Gamma}(N, v, \gamma)$ in a central peering situation with three players and player set $N = \{1, 2, 3\}$. The payoff function $\widetilde{f}$ is given by (6.18) and $\mathsf{S}_i = \{\{\emptyset\}, N \setminus \{i\}\}$ for all $i \in N$.

For a better understanding, we aggregate the final 3-player game from three 2-player games, because in the latter central peering and bilateral peering coincide. Beginning with players 1 and 2, we depict a normal form game in Fig. 6.3. Each

[10] In this "deviating" coalition R it is only necessary for j to actually have changed his strategy. Player i can also perform a counter-deviation, which would amount to a degenerate deviation, given the limited strategy space.

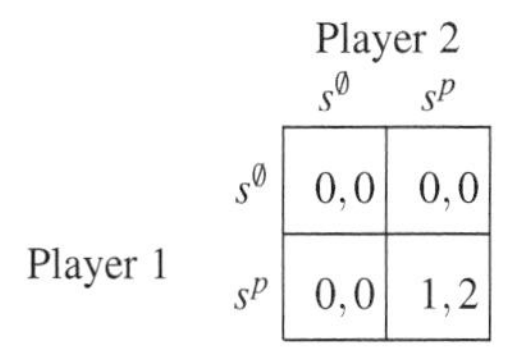

		Player 2	
		$s^{\emptyset}$	s^{p}
Player 1	$s^{\emptyset}$	$0,0$	$0,0$
	s^{p}	$0,0$	$1,2$

Fig. 6.3 Pairwise gains from peering for players 1 and 2

		Player 3	
		$s^{\emptyset}$	s^{p}
Player 1	$s^{\emptyset}$	$0,0$	$0,0$
	s^{p}	$0,0$	$-2,3$

Fig. 6.4 Pairwise gains from peering for players 1 and 3

		Player 3	
		$s^{\emptyset}$	s^{p}
Player 2	$s^{\emptyset}$	$0,0$	$0,0$
	s^{p}	$0,0$	$-1,2$

Fig. 6.5 Pairwise gains from peering for players 2 and 3

player has two strategies, s^p for peering and $s^{\emptyset}$ for remaining in the status quo, i.e. no peering. The benefits from peering will only accrue, if both players choose s^p, with payoffs $(1,2)$. Player 1 and 2 both profit strictly from peering with one another in this game. Likewise, we provide the normal form representation for players 1 and 3 in Fig. 6.4, where the payoffs for strategy tuple (s^p, s^p) are $(-2,3)$. In this situation, it is only player 3 who benefits strictly from a peering agreement, while player 1 loses out strictly.

Finally, Fig. 6.5 contains the bimatrix for players 2 and 3. Their payoffs under the peering strategy tuple are $(-1,2)$ and $(0,0)$ otherwise. Again, only player 3 benefits strictly.

Now, if we aggregate the peering payoffs into a three player game, the normal form representation of Fig. 6.6 is the result. The left table corresponds to player 3 not engaging in peering, i.e. $s_3 = s^{\emptyset}$, the right to him playing his peering strategy $s_3 = s^p$.

It is readily checked that the only two Nash equilibria in pure strategies are given by the strategy combination $(s^{\emptyset}, s^{\emptyset}, s^{\emptyset})$ and $(s^{\emptyset}, s^{\emptyset}, s^p)$.[11] In the first strategy, no individual player has an incentive to deviate to s^p, as peering, by definition, is something conducted among multiple players and will yield no gains this way. In the second, player 3 is also indifferent to changing to $s^{\emptyset}$. And for all other strategies $s \neq s^*$, there always exists a profitable deviation for some player.

[11] There also is a continuum of mixed equilibrium strategies between these two.

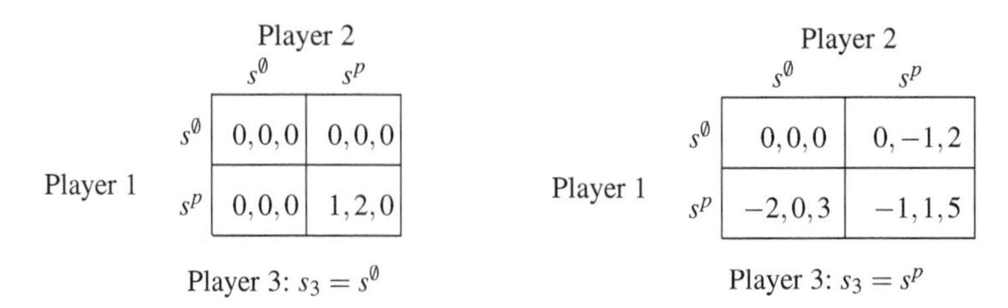

Player 3: $s_3 = s^{\emptyset}$

Player 1 \ Player 2	$s^{\emptyset}$	s^{P}
$s^{\emptyset}$	0,0,0	0,0,0
s^{P}	0,0,0	1,2,0

Player 3: $s_3 = s^{P}$

Player 1 \ Player 2	$s^{\emptyset}$	s^{P}
$s^{\emptyset}$	0,0,0	0,−1,2
s^{P}	−2,0,3	−1,1,5

Fig. 6.6 Aggregate gains from peering

Of the two pure equilibrium strategies, the first constitutes neither a Strong nor a Coalition-Proof Nash equilibrium, because for players 1 and 2 a joint deviation to s^P is strictly beneficial. The other satisfies both notions, as the reader is welcome to check.

As Example 6.21 shows, even without extraordinary circumstances (in terms of parameters of the players), a situation where no peering will take place is well conceivable in a central peering environment. With the same parameters, in the bilateral environment, an equilibrium where players 1 and 2 would peer is possible, as they could exclude player 3 from any agreement.

The only results that we could establish would rely on specifically chosen parameters, which is not desirable. Given this lack of general results, we now go on to sequential network formation in the central peering environment.

6.4.3 *Network Formation in Extensive Form*

In the central peering environment, network formation in extensive form will get an entirely new twist. To begin with, such games are now much closer to settings of unilateral link formation. If a player participates in the central peering instance, he a priori agrees to all possible requests for peering by other players. No individual selection is possible, and so it is entirely at the will of those other players to peer with him. The only way to avoid peering is by not joining the instance at all.

The standard reference on sequential link formation is the seminal work of Aumann and Myerson (1988) who introduced this type of extensive form games to the literature (see also Sect. 3.6.2, pp. 142 ff.). We now analyze their "*auxiliary linking game*" in order to establish how much of their setup can be applied to our central peering situation.

The starting point is the rule of order; it specifies the sequence in which all possible pairs of players decide one after another on the formation of a link. Naturally, this pairwise approach does not correspond to our centralistic setting anymore. The rule of order we apply has to go through the set of players one by one, instead of all possible pairs of players. At each step, the selected player has the choice to join the peering instance, thereby agreeing to peer with all previous members (if any) and all possible future ones. When the rule of order has been over its domain once, the players who have not become members get another chance to join. This process will be repeated, until in one round no new members are generated. Sticking to the original model in Aumann and Myerson (1988), we assume that memberships cannot be revoked.[12] This strengthens the unilateral character even more.

Next, there is the assumption of perfect information on the parts of the players, which is required for the existence of subgame perfect Nash equilibria. Not only need the players be informed about the rule of order and all present members of the peering instance when it is their turn to decide, but also about the distribution of gains at each possible step. The first is an institutional question, which could easily be solved. The second seems slightly more tricky but could still be plausible, as long as firms can distinguish call traffic (in both directions) in terms of VoIP firm. Then, from incoming and outgoing call volumes, each VoIP firm can calculate its partial net gains with all other firms, as access fees and long distance fees are either known or ascertained: Without specific reasons for unusual call traffic (which should be known to the firms and could be accounted for), a distribution as proposed by the BCPA is not unrealistic. From all this, the firms are able to reverse-engineer the payoffs at all conceivable configurations of central peering.

In the previous fashion, we first analyze the setting under f in which gains and hence payoffs are nonnegative. Then, we proceed to the scenario incorporating $\widetilde{f}$ in which access revenues are corrected for on the individual level.

If the payoffs f of the auxiliary linking game correspond to the allocation procedure γ, the networks resulting in equilibrium are straightforward to characterize:

Proposition 6.39. *Consider the auxiliary linking game (see Aumann and Myerson 1988) modified to a central peering environment. Then, in any subgame perfect Nash equilibrium, the resulting network g, when based on γ, is one where at least all nonzero players are members of the peering instance:*

$$N \setminus (\mathbf{0} \cup \{i\}) \subseteq N_i(g) \subseteq N \setminus \{i\} \text{ for all } i \in N \setminus \mathbf{0}.$$

Proof. For any rule of order the linking game terminates after at most n rounds. In which order the players are being called upon to join the peering instance is not important. For each nonzero player joining yields a strictly positive payoff, either immediately, or if he is the first, after another player has joined. This payoff then strictly increases with every other player who joins. The zero-players, however, are unaffected in payoffs when peering, also regardless of the round they join or of their position under the rule of order.

[12]This assumption might seem quite strong at first, but since we do not operate in a dynamic environment and only treat one– unspecified – period, firms do have a chance to redo their decisions at some point in time. One could also think of peering memberships that only last a certain period of time an would have to be renewed.

Because peering is the dominant strategy for nonzero players at every step in any round and both strategies are undominated for the zero-players, the Nash equilibrium strategy leading to the type of network described above is also subgame perfect. ■

Without access revenue correction we end up in a familiar situation when the auxiliary linking game terminates: All nonzero players peer with one another by joining the peering instance, but nothing concrete is established with respect to the zero-players. Regardless of their position under the rule of order, they are indifferent to peering, and so there are $2^{|\mathbf{0}|}$ subgame perfect Nash equilibria: One for each possible subset of zero-players, who decide to join peering. The unique efficient subgame perfect Nash equilibrium though, is the one in which all zero-players peer. Only then are all gains from peering realized as expressed through v^g.

Correcting for access revenues through $\mathbf{A}^g$ again presents us with an anything-goes scenario: Without further specification of the parameters, no general result can be established. As new members to the peering instance have an impact on the payoffs of present and future members, and for each player this impact can be positive as well as negative, any network could (a priori) be the outcome. We revisit the setting of Example 6.21 to show that regardless of the rule of order, the empty network is the only subgame perfect Nash equilibrium.

Example 6.22. Continuing with the player set and payoffs from Example 6.21, we now play the auxiliary linking game introduced by Aumann and Myerson (1988), but adapted to central peering. There are six possible rules of order, one for each permutation on the player set $N = \{1,2,3\}$. The final payoff is always identical, so we can apply backward induction to solve the game. We begin with the natural order $(1,2,3)$: It is easy to see that player 3 will always join the peering instance if there are already members, because he gains strictly with each of the other players. Player 2 will only join the instance if also player 1 has joined, because the gains with player 1 offset his losses with player 3. But player 1 would only join if player 3 was not to join, which is not the case. So player 1 will never join, regardless of the order, because his losses through player 3 are not offset by his gains with player 2. Therefore player 2 will also decline to join, and, for either strategy of player 3 not a single link will be established. If player 3 decides not to join, the game is over and the empty network $g^{\emptyset}$ is the result. If player 3 wants to join, the game continues into its next round. There, the rule of order only offers players 1 and 2 the chance to join. They both decline and the game is over. Again, the resulting network is the empty network $g^{\emptyset}$.

The reader is welcome to check this for all possible orderings. Even if player 3 turns down peering at first, players 1 and 2 will not opt for it, because they can anticipate that player 3 will join peering in the consecutive round. Player 1 declines, because for him peering in the end results in a negative payoff, which is also the case for player 2, if he peers only with player 3.

In expectation of more meaningful outcomes, we now turn to the stability notions that central peering networks may or may not exhibit.

6.4.4 Stability of Networks

The stability issues as well turn out to be more delicate, due to the far-reaching consequences involved in a decision to peer. The notions of stability will have to be altered somewhat, as we are no longer in the setting of straightforward bilateral link formation. We proceed directly to the scenario in which access fee correction $\widetilde{\mathbf{A}^g}$ is considered and all networks are the result of the formation game $\widetilde{\Gamma}(N,v,\gamma)$ with individual strategy spaces $\mathsf{S}_i = \{\{\emptyset\}, N \setminus \{i\}\}$ for all $i \in N$.

The concept of pairwise stability now reduces to plain *stability*, and is no longer defined in terms of individual links, but rather through aggregate payoffs from peering. We express its condition along the following lines:

Definition 6.106. A network $g(s)$ resulting from the network formation game $\widetilde{\Gamma}(N,v,\gamma)$ with individual strategy space $\mathsf{S}_i = \{\{\emptyset\}, N \setminus \{i\}\}$ for all $i \in N$ is stable, if

- $s_i = N \setminus \{i\}$, if $\widetilde{f}_i(s_i, s_{-i}) \geq \widetilde{f}_i(\{\emptyset\}, s_{-i})$
- $s_i = \{\emptyset\}$, if $\widetilde{f}_i(s_i, s_{-i}) > \widetilde{f}_i(N \setminus \{i\}, s_{-i})$

holds for all $i \in N$.

Under pairwise stability in a bilateral setting, any link g_{ij} leading to a Pareto improvement among the players forming it must be part of the network. Similarly, we assume that all players who are indifferent to join the central peering instance decide for participation. We remind the reader that those indifferent players now need not coincide with the set of zero-players anymore. Also, it is worthwhile to note that the consequences of an indifferent player to join the peering instance are not necessarily beneficial to all other members.

Without restricting the parameters of v^g, it is possible to end up with any kind of network as stable, even with the empty one being the unique solution. To convince oneself of the latter, we again refer the reader to Example 6.21.

Now, as opposed to the bilateral setting, the notion of strong stability does not coincide with (pairwise) stability: Because of the externalities players' decisions to join or leave the peering instance can have, coordinated actions to obtain a network g' from g via a coalition S cannot be taken apart and considered one at a time. The adapted definition is the following:

Definition 6.107. A network $g(s)$ resulting from the network formation game $\widetilde{\Gamma}(N,v,\gamma)$ with individual strategy spaces $\mathsf{S}_i = \{\{\emptyset\}, N \setminus \{i\}\}$ for all $i \in N$ is strongly stable, if for any given coalition $R \subseteq N$ and all networks $g(s'_R, s_{\overline{R}})$ obtainable from $g(s)$ via R, where $f_i(s'_R, s_{\overline{R}}) > f_i(s)$ for a player $i \in R$, there exists a player $j \in R$ for whom $f_j(s'_R, s_{\overline{R}}) < f_j(s)$.

This implies the regular stability from Definition 6.106. Also, as we will see, coordinated movements of coalitions of players only make sense in certain directions. Consider a currently peering coalition with the intention to quit. The only reason for this could be a negative payoff under peering, which for each member would increase to zero once they were out. But, for such a deviation no coordination is necessary, as this is possible individually, too. Within such a coalition, a counter-deviation also has little meaning, because it has no effect on the payoffs of the quitters, which are zero by construction.

More interesting is the case of a coalition of players planning a concerted entry into peering. Each player on its own might possibly lose through joining, but together they could secure each a nonnegative payoff. Counter-deviations are also meaningless, because no deviating player can increase his payoff above the initial value of zero. If he was expecting a negative payoff from the original deviation, he would not have agreed to deviate in the first place. In sum, we have established that only deviations to join the peering instance need to be considered in excess of regular stability of the network.

In the light of the above, let us revisit Example 6.21, where a curiosity arises:

Example 6.23. Continuing on the basis of Example 6.21, we now want to check the equilibrium network for strong stability. As we had shown, two Nash equilibria in pure strategies exist, one of them also Strong and Coalition-Proof. Interestingly, both equilibria lead to the same network, the empty network $g^{\emptyset}$. By definition, no links are present, but in one of the equilibria player 3 chooses the strategy s^p and hence indicates his willingness to peer. The empty network based on this strategy, $g(s^{\emptyset}, s^{\emptyset}, s^p) = g^{\emptyset}$, is strongly stable, because all possible deviations that increase some players' payoffs decrease some other player's payoff at the same time. In contrast, the empty network based on the other equilibrium strategy, $g(s^{\emptyset}, s^{\emptyset}, s^{\emptyset}) = g^{\emptyset}$, is only stable, but not strongly so. A deviation by players 1 and 2 yields both strictly higher payoffs, leaving player 3 unaffected.[13]

Again, with access revenue correction, no general characterization of (strongly) stable networks is possible without further assumptions on the parameters. Interestingly enough, a special feature can appear in the central peering environment: The empty network $g^{\emptyset}$ can satisfy and violate the notion of (strong) stability, depending on the strategies that gave rise to it. If only one player opted for peering, no links are created, but deviations might be profitable. If no player opted for peering, individual deviations are not profitable, but concerted might be.

We now conclude our considerations of network formation. Having inspected bilateral and central peering environments, it became clear that in both settings the introduction of access fee correction on the individual VoIP firm's level has very disturbing effects on the results. Without the correction, the formation of efficient peering networks is possible under the respective equilibrium concept. With the correction in place, these results break down if no further restrictions on the parameters of the peering game are made.

[13]The strategy $(s^p, s^p, s^{\emptyset})$ is not even a Nash equilibrium strategy, because player 3 can increase his payoff by deviating to s^p.

Based on the previous findings, the final part of this chapter contains implications for the design of an access fee regime by a national regulator with the goal to set an incentive for widespread peering among VoIP firms. We provide a simple solution that allows for efficient peering structures.

6.5 Implications for Regulators

What implications do the preceding results, or the seeming lack thereof, have for national regulators, whose only policy tool are the access fees charged for interconnection? How would such a fee have to be structured, in order to foster an environment in which possibly all VoIP firms choose to sign peering agreements with one another and realize all gains from long distance? In our case, we need to distinguish the different settings of peering, bilateral and central, and find access fee regimes for both. We maintain two assumptions: Access fees are uniform, i.e. all VoIP firms face and receive equal fees for and from terminating calls, and, for reasons of notational simplicity, the balanced calling pattern assumption.

In the bilateral setting we have shown (see (6.14)) that a VoIP firm i only engages in peering with firm j, when

$$q(p_i) \cdot l_i \geq [q(p_j) - q(p_i)] \cdot a, \tag{6.19}$$

i.e., when the gains from dropped long distance fees balance a possible loss of the pre-peering access surplus. Consequently, none of the VoIP firms with an access surplus[14] would lose from bilateral peering, if

$$a \leq \frac{q(p_i) \cdot l_i}{q(p_j) - q(p_i)} \tag{6.20}$$

is satisfied for all $i, j \in N$.

In the setting of central peering, the situation looks different. Actually, there are several situations, depending on what a regulator wants to ensure: A condition under which it is profitable for all VoIP firms to peer with one another? Or one, which is in addition resistant to deviations of (arbitrary coalitions of) players? Regarding the first case, for all firms to peer, an access fee is needed which satisfies

$$a \leq \frac{q(p_i) \cdot l_i \sum_{j \neq i} \alpha_j}{\sum_{j \neq i} \alpha_j [q(p_j) - q(p_i)]} \quad \text{for all } i \in N. \tag{6.21}$$

Interestingly, in the central setting, this also implies proofness to deviations of all kind, because the lowest payoff any player can receive if expression (6.21)

[14]Meaning $q(p_j) > q(p_i)$ for such a firm i.

is satisfied, is a payoff of zero.[15] Abandoning the peering instance then does not change the payoff at all, regardless of the deviating coalition.

However, conditions that would also satisfy subsets $S \subseteq N$ in the fashion of (6.21), could indeed have to consider the possibility of deviations, as new peering-members might make present ones worse off, even below their pre-peering payoffs. But such a condition in a general form, regardless of proofness to deviations, is likely to be too complicated to implement in the first place.

Also, implementation in general is complicated by the fact that the above conditions crucially rely on the price for long distance transit l_i, and even more problematic, on the call volumes $q(p_i)$ of the VoIP firms $i \in N$. For one, it is questionable, whether the firms want to share information about these with the regulator (in case the demand is firm-specific). Secondly, and beyond our horizon, call volumes are also subject to change over time, which in turn could require a constant readjustment of the access fees.

The solution we propose is a very simple one. Not only does it satisfy all above equations, it is also extremely easy to implement, economically justified, and need not even be monitored by the regulator for compliance: The general abolition of fees for the interconnection of genuine VoIP calls.

First of all, the IP-based networks in question are all part of the internet and as such interconnected already.[16] The real cost for this interconnection, i.e. the traffic exchanged among these networks, is already settled on the network layer through internet-based peering or transit agreements. Therefore, no additional charges should apply to the interconnection of services like VoIP on these networks. In particular not in the way it is conducted hitherto, where VoIP calls are terminated at charges based on legacy PSTN networks, the operating cost of which is much higher than that of an IP-based network.

Secondly, with the abolition of VoIP access charges, the zero-sum component vanishes from the peering game v^g: There is no longer any access revenue and the decision to peer rests only the gains from long distance transit fees. The distortion through access revenues no longer applies. With respect to the network formation approaches considered earlier in this chapter, the models incorporating access revenue correction are no longer relevant. More precisely, they coincide with the "uncorrected" ones, which yielded meaningful general results with regard to efficient peering networks. Now, already the notion of (pairwise) stability, which cannot be considered restrictive or unreasonable, leads to peering that exploits all gains from cooperation as measured by v^g.

Thirdly, the zero access fee satisfies both our scenarios of bilateral and central peering, which can be checked readily through expressions (6.20) and (6.21). Then, both concepts can even coexist, if necessary or desirable.

---

[15]Expression (6.21) is not defined for $p_i = p_j$ for all $i, j \in N$, but then access revenues are equal to zero in any case for all possible values of a.

[16]"Part of the internet" states that the telephone devices are assigned regular IP addresses, even if parts of a network are not publicly accessible.

The administrative simplicity of abolishing access fees among VoIP firms is barely worth mentioning. The implementation needs neither preparation, nor monitoring, and no readjustment of the fee structure is necessary with changing prices. Finally, even though not considered in our model, it also conforms to the European Commission Recommendation on the regulatory treatment of termination rates[17], which suggests that "*setting a common approach based on an efficient cost standard and the application of symmetrical termination rates would promote efficiency, sustainable competition and maximise consumer benefits in terms of price and service offerings.*"

[17]See *European Commission Recommendation of 7 May 2009 on the Regulatory Treatment of Fixed and Mobile Termination Rates in the EU (2009/396/EC).*

Chapter 7
Concluding Remarks, Recommendations, and Future Research

In Part I of this study selected concepts from the theory of games and from network theory are presented. It contains no new results, but some of the proofs, especially in the section on cooperative games, and the respective solution concepts have been reworked extensively. Also, some alternative definitions are provided. Doing so, we aim to give the reader a thorough understanding and intuition on the concepts at hand, as opposed to the often rigourous and overly brief exposure in the original literature. Our exposition is, hopefully, easier accessible due to the use of more basic mathematical tools and more elaborate consideration in many instances.

Nevertheless, many of the properties of communication situations or allocation rules in the network context have been reformulated to fit our purpose. This, e.g., refers to the characteristic function v^g where the players are the main ingredient, not the network, and to all allocation rules that make use of v^g. Many properties of both the characteristic function and the corresponding allocation rules have been reworked likewise. To this end, we can also mention to have unified the notation from all the sources that we draw on.

Part II turns to the application of the concepts previously introduced. First, in Chap. 4, comes a nontechnical refresher on the relevant parts of the telecommunications sector. This brief overview gives the reader a basic understanding of the situation subsequently analyzed. Building up the network architecture of the PSTN and the internet, we establish why, in conjunction with inadequate regulation, the re-routing problem even exists.

Chapter 5 introduces the main model. On the basis of a specific communication situation, which we call the *peering game*, we model the interaction of the VoIP firms in terms of their decisions to peer. The value function v^g can be decomposed not only into the sum of two games, but also over independent parts of the network. Regarding the first, we established that the peering game is based on a zero-sum game over the access charges the VoIP firms charge each other, and the savings game on the long distance transit fees saved through peering. Conveniently, the access revenue cancels out on a pairwise basis, the same basis on which the savings are created. This, and the fact that the resulting peering game turns out to be

P. Servatius, *Network Economics and the Allocation of Savings*, Lecture Notes in Economics and Mathematical Systems 653, DOI 10.1007/978-3-642-21096-9_7,

convex, makes subsequent analysis a lot easier and more tractable. Secondly, we found that the peering game is decomposable with respect to the peering network, i.e. unconnected components do not influence one another and can be analyzed separately.

We also established that any savings created can be further broken down into contributions to the marginal values of the players. From this arises the notion of a zero-player, who might exhibit no contributions but a strictly positive marginal value. In our setting such players correspond to VoIP firms with no cost for long distance transit, i.e. such that operate extensive networks. The contributions serve as cornerstone for the allocation we deem most appropriate in the peering game, called contribution allocation: Players are allocated what they save themselves in long distance fees with all their peering partners, possibly rendering obsolete complicated balancing-mechanisms needed in other allocation schemes.

The first solution concept we apply is the core, which is nonempty due to the convexity of the peering game. On the basis of the underlying network, we characterize core allocations further and can limit them from below and above: The lower bound imposing group rationality is now complemented by an upper bound incorporating the network structure.

Because the core generally is not single-valued, we proceed to the least-core in hope of achieving a unique allocation. But, as we showed for the peering game, not only can the least-core produce a set-valued solution, but the desired contribution allocation need not even be contained in it.

With this, we turn to the Shapley value, whose calculation we can simplify dramatically in the peering game. But since Shapley's operator assigns average marginal values to the players, we can not expect it to yield the contribution allocation, as zero-players with positive marginal values are strictly rewarded. The same holds true for the Myerson value. Because its restriction on the value function is based on the components of a network, nothing is gained as compared to the characteristic function of the peering game, where value arises pairwise. Hence, the solution coincides with that of the Shapley value, on which the Myerson operator is based.

We get closer to the contribution-based allocation with the Weighted Shapley value. Because the desired allocation is an element of the core, we show it to be reproducible under some weight system. Unfortunately, the weight system our approach yields has no intuitive support, neither to market shares, nor call volumes of the VoIP firms.

The last class of allocation procedures we consider are bargaining solutions. To that end, we translate the n-player peering game into the corresponding number of $\binom{n}{2}$ 2-player bargaining problems and apply two solutions, that of Nash as well as that of Kalai–Smorodinsky, and their weighted variants. Without weights, both solutions again yield the same allocation as the Shapley value, i.e. one where the surplus is split equally among pairs of players, regardless of their individual contributions to it. Under their weighted variants we can in both cases establish the contribution allocation, this time with weights that have an intuitive meaning.

In Chap. 6 we turn to the process of network formation, which in our context corresponds to the signing of peering agreements between the VoIP firms. For that purpose, we introduce the algebraic form of an allocation procedure which results exactly in the desired, contribution-based game solution. It serves for all subsequent considerations.

We first investigate a network formation environment corresponding to bilateral peering. Here, we find that the formation of efficient networks is generally possible in strategic and extensive form, and under stability considerations as well. Now that players are not only faced with the savings from transit, but also with the omission of access charges once peering is in place, we need to correct the payoffs by the access revenues on an individual level. This changes the picture dramatically, as the previous general results can no longer be maintained. All depends on the parameters of the peering game, and no meaningful characterization of the network that might be resulting (or one being stable) is possible.

The second network formation environment is central peering, where VoIP firms can join a central peering instance whose members share each others' routing tables without reservation. In order to accommodate this peering instance, we adapted the available strategies in the network formation game to yield a network that is complete among the peering participants but has no links to other players. The stability notions had to be adjusted, too, as the formation of a link does no longer require the specific content of both players. The latter fact also had an influence on the rule of order used in the sequential approach to network formation.

As before, without correction for access revenues, efficient networks are the result in equilibrium under all modes of formation. These efficient networks are under some refinements even complete. But as soon as the correction is incorporated, the results break down. In the central-peering setting the impact of the dropped access charges on payoffs is even greater: Now, if a player joins the peering instance, possibly negative effects on the payoffs of all other members might occur and are not contained to a specific link. A situation corresponding to the empty network, where no peering is conducted at all, is well conceivable in equilibrium.

Chapter 6 is concluded with a view on possible implications on the design of access charge regimes by a national regulator, based on previously gained insights. We list two separate conditions on the access charges under which comprehensive peering is made possible. The first corresponds to the bilateral scenario, the second to the centralistic approach. We then provide one simple solution that even satisfies both conditions simultaneously: To drop all access charges among VoIP firms, regardless of their peering status. This is not only efficient from an economic viewpoint, as distortions and arbitrage opportunities are disposed of. It also gives support to the reasoning that access fees should be priced to cost, which in the context of services interconnection is zero.

Naturally, with our focus on VoIP peering, this study does not cover the topic of services interconnection exhaustively, not even remotely. But it points in a direction where almost no research has been done so far. More importantly, we highlight the caveat of "double-charging" on multiple layers, as in our case applies to the interconnection of networks and that of services, independent of one another.

In part, the lack of research on services interconnection is certainly due to the fact that in many countries network operators still enjoy protection from regulators over opening their networks to competing service providers. As this protection is intended to provide incentives for initial investment, it is bound to cease over time.

For our model of the VoIP peering situation, several extensions are conceivable: As a generalization, one could drop the balanced calling pattern assumption for a more irregular flow of calls. One could also allow for more firm-specific demand functions and access fees for a less uniform setting. The first two of these changes could to some extent reflect the recent development towards flat rate billing for voice calls, where a fixed monthly fee covers all usage.

One major extension to the model is the inclusion of the call prices as decision variable with direct influence on their market shares. This would also open the door for multi-period approaches.

As far as the network formation model is concerned, extensions also come to mind readily: Allowing for unilateral peering, i.e. the availability of one firm's routing table to another, but not vice versa, or for the coexistence of bilateral and central peering.

Most generally, the above-mentioned problem of double-charging might not be unique to our setting and merits further investigation. Where else does interconnection of networks and of services overlap in a similar manner? And, is there a generalization possible as to which interconnection layer is the one that should enjoy priority over the other?

The possibility for further research is ample and we hope to have inspired some readers with our work. Also, we hope that rather sooner than later, regulators will indeed drop access fees for the interconnection of services. This should not only allow for the savings as described above, but also enable the deployment and spread of innovative services directly built on VoIP telephony, which are not possible when calls are rerouted through the PSTN. Such changes would serve equally as incentive to speed up the transition to next generation networks.

Appendix A
Selected Mathematical Concepts

Overview

This mathematical appendix is not supposed to be a comprehensive treatment, not even on the subjects it contains. It is included for mainly two reasons: First, so the reader can use it to refresh his knowledge without having to consult separate literature. Secondly, as we have brought together concepts from many different sources in this work, this chapter unifies the notation and clears up ambiguities that might arise from those original sources. Our orientation is Alós-Ferrer (2005).

The reader is assumed to be familiar with the some basic notions, for example: The natural numbers $\mathbb{N}$, the real numbers $\mathbb{R}$, the standard mathematical operations on these, such as addition, multiplication, as well as exponents. Furthermore, the reader should be accustomed to work with finite-dimensional real vector spaces[1]: What is a basis and how is it spanned, and what are the standard operations performed with vectors.[2] We also assume that the reader has gone over the notation we list on p. 287. Some of the topics treated below contain slightly more than what is required for the main part of this work. This is due to the fact that the author sometimes felt the need to include some additional aspects, in order to provide a well rounded picture or make the relevant information more intuitively available. We will only touch each subject briefly and occasionally provide references to the literature for the interested reader.

Basic Set Theory

We begin our treatise with some elementary notation and a *naïve* approach to set theory. For us, a *set* is any collection of distinct *elements*. We usually denote

[1]We particularly refer to the Euclidean space $\mathbb{R}^n$.

[2]Vectors are defined as column vectors.

P. Servatius, *Network Economics and the Allocation of Savings*, Lecture Notes in Economics and Mathematical Systems 653, DOI 10.1007/978-3-642-21096-9_8,

sets by upper case letters and elements in lower case. A set can be defined either exclusively,

$$S = \{1,2,3,4,5\} = \{3,2,4,5,1\}$$

by listing all its elements, or, through some assigned property,

$$S = \{x \in \mathbb{N} | 1 \leq x \leq 5\}.$$

In both cases, the set S has the natural numbers 1 to 5 as elements. In the first, we listed them, in the second, we gave as the required property for membership "all natural numbers between one and five." Membership to a set is denoted $s \in S$, if s is an element of the set S, and $s \notin S$ if not. Note: There is no ordering within a set and the same element cannot appear more than once! A set without any elements is called the *empty set* and is denoted by $\emptyset$. We generally write $|S|$ to refer to the cardinality of the set S, i.e. to the number of its elements. Sometimes we also use the corresponding lowercase letters s instead of $|S|$, when no confusion can arise. We say S is a *subset* of T if every element of S is also an element of T. Also, the set S is a *strict subset* of T, if there also exist elements in T that are not members of S. We denote both relations by $S \subseteq T$, and $S \subsetneq T$, respectively. Here, T would also be referred to as (*strict*) *superset* of S, where we use the notation $T \supseteq S$ and $T \supsetneq S$. If S cannot be related to T as either subset or superset (or possibly both), this is expressed by $S \nsubseteq T$ and $S \nsupseteq T$.

One very important set is the *power set* or *set of all subsets* of some finite set S. It is denoted 2^S and defined by

$$2^S := \{A | A \subseteq S\},$$

containing all possible subsets of S. Its cardinality is $2^{|S|}$ because $\emptyset \in 2^S$ and $S \in 2^S$. Also, if a set contains only one element, i.e. $S = \{s\}$, we call it a *singleton* (set).

Now, let us turn to some very basic operations on the sets we defined:

The *intersection* of two sets, $A, B \subseteq S$, consists of those elements that are members of both sets:

$$A \cap B := \{s \in S | s \in A \wedge s \in B\}.$$

If the resulting set is empty, the sets A and B are said to be *disjoint*.

The union of two sets, $A, B \subseteq S$, consists of those elements that are contained in either A or B, or both:

$$A \cup B := \{s \in S | s \in A \vee s \in B\}.$$

The notions of intersection and union generalize straightforwardly to a finite number of sets. The *difference* of two sets $A, B \subseteq S$, or *relative complement*, are all elements of A which do not at the same time belong to B:

$$A \setminus B := \{s \in A | s \notin B\}.$$

Because $A \subseteq S$, the set $S \setminus A$ is also called the complement of A in S and alternatively denoted by $\overline{A}$. Its intersection with A is empty.

A *partition* of a set S is a set of subsets of S, denoted $\mathsf{P} = \{P_1, P_2, \ldots, P_m\}$ with the following properties:

$$P_i \cap P_j = \emptyset \text{ for all } i \neq j \text{ and } \bigcup_{i=1}^{m} P_i = S.$$

Hence, all elements of partition P are to be pairwise disjoint and their union must exhaust S.

The *Cartesian product* of a finite number of, say n, sets is given by

$$S_1 \times S_2 \times S_3 \times \ldots \times S_n.$$

Its elements are *ordered n-tuples* or *vectors* denoted $(s_1, s_2, \ldots, s_3)$, where $s_i \in S_n$ for all $i = 1, 2, \ldots, n$. Here, unlike in the case of plain sets, the order of the elements is crucial. At this point we want to inform the reader that some of the expressions used in the main body occasionally produce redundant expressions such as $S \setminus \{i\}$ when element i might not even be a member of S in the first place, or even an $\emptyset \setminus \{i\}$. As the removal of an element from a set in which it is not contained does not change the set, the above expressions are equal to S and $\emptyset$, respectively. We next turn to the concept of a mapping, albeit with limited coverage on an abstract level.

On Functions and Correspondences

Given two (nonempty) sets, A and B, a *function* f is a rule which assigns to each element $a \in A$ a **unique** element $f(a) \in B$. The latter is called the *image* of a under f. The expression

$$f : A \to B$$

denotes a function from the set A to the set B, where the former is called the *domain* and the latter the *codomain* of function f. The *image* of some set $T \subseteq A$ under f is given by all elements $b \in B$ whose values are assumed for any value $a \in T$:

$$f(T) = \{b \in B \mid \exists\, a \in T \text{ with } b = f(a)\}.$$

We call $f(A)$ the *range* of f, as it includes all possible values f can assume. The *inverse image* of some set $R \subseteq B$ under f is defined as

$$f^{-1}(R) = \{a \in A \mid f(a) \in R\}.$$

It consists of all elements $a \in A$ for whom f assumes a value $b \in B$. This inverse image can well be empty!

We call a function f *onto* (or *surjective*), if

$$f(A) = B,$$

and *one-to-one* (or *injective*), if

$$f(a_1) = f(a_2) \implies a_1 = a_2$$

holds. The first property says that on its domain A, the function f will assume every value possible within codomain B, or shorter that range and codomain coincide. The second states that no two distinct elements in the domain of f yield the same value under f. Any function that is both onto and one-to-one is called a *bijection*, or *bijective*.

The graph of a function f is the set of all pairs of values from A and B each, who are associated by f:

$$\mathrm{Graph}(f) = \{(a,b) \in A \times B \mid b = f(a)\}.$$

As an extension to functions we can consider *correspondences*: A correspondence F from A to B assigns each element $a \in A$ a subset $F(a) \subseteq B$. Generally, we write

$$F : A \rightrightarrows B,$$

but we can also express such a correspondence in terms of a function,

$$F : A \to 2^B,$$

as it is nothing else than a set-valued mapping that assigns to each element $a \in A$ in the domain an arbitrary number of images. The second expression is confined to correspondences with finite codomain B.

We can now turn to a function of particular interest, a *permutation*: A permutation on some set N is a one-to-one function from N onto itself, $\pi : N \to N$. It defines a certain ordering on the $n := |N|$ elements in N, as each element is assigned a new slot in the ordered n-tuple. We use the expression $\pi(i)$ to designate the slot which element i (in the unordered set N) is assigned under permutation π. By Π_N we denote the space of permutations on the set N, but we will drop the subscript when no confusion can arise. Note, there are $n! := 1 \cdot 2 \cdot 3 \cdots (n-1) \cdot n$ such permutations on any set with cardinality n.

A special case of a permutation is the so-called *transposition*. It specifies that $\pi(i) = j$ and $\pi(j) = i$ for a pair $i \neq j$, but $\pi(k) = k$ for all other $k \neq i, j$. Only the two elements i and j are flipped in the ordering, while all others in the n-tuple retain their position.

Another commonly used type of function is called *sequence*. In its single-dimensional form, it maps to nonnegative integers $k = 0, 1, 2, \ldots$ a real number: $f : \mathbb{N}_+ \to \mathbb{R}$. Usually, sequences are denoted by $(x_k)_{k=0}^{\infty}$ or simply (x_k). A sequence

(x_k) is said to *converge* to some $x \in \mathbb{R}$, denoted by

$$\lim_{k \to \infty} x_k = x \quad \text{or} \quad x_k \to x,$$

if for every $\varepsilon > 0$ there exists an integer k_ε for which

$$\|x_k - x\| < \varepsilon,$$

whenever $k > k_\varepsilon$. We call x the unique *limit (point)* of the sequence (x_k). Sequences are naturally generalized to higher dimensions, as the domain is always identical. They are then denoted $(\mathbf{x}_k)$ and converge to some limit $\mathbf{x}$ if and only if each dimension converges accordingly.

Binary Relations

A *binary relation* on some set S, or simply relation, denoted R, consists of ordered pairs of elements drawn from this set S. We write $\mathsf{R} \subseteq S \times S$. If some element s_1 bears the desired property when related to some other element s_2, where $s_1, s_2 \in S$, we use the notation $s_1 \mathsf{R} s_2$ instead of $(s_1, s_2) \in \mathsf{R}$. Hence,

$$s_1 \mathsf{R} s_2 \Longleftrightarrow (s_1, s_2) \in \mathsf{R}.$$

Similarly, we write $s_1 \not\mathsf{R} s_2$ or $(s_1, s_2) \notin \mathsf{R}$ if s_1 is not related to s_2 via R. The universal binary relation is simply the set of all ordered pairs on S, i.e. $\mathsf{R}^u = S \times S$. For any given two elements $s_1, s_2 \in S$ compared via a binary relation R, exactly one of the four cases listed below is true:

1. $s_1 \mathsf{R} s_2 \;\wedge\; s_2 \mathsf{R} s_1$
2. $s_1 \mathsf{R} s_2 \;\wedge\; s_2 \not\mathsf{R} s_1$
3. $s_1 \not\mathsf{R} s_2 \;\wedge\; s_2 \mathsf{R} s_1$
4. $s_1 \not\mathsf{R} s_2 \;\wedge\; s_2 \not\mathsf{R} s_1$

That these cases exclude each other follows from the contradictions that arise, whenever we suppose more than one to be true at the same time. Now, we look at a list of established properties for binary relations. We refer the reader to Rubinstein (2000, pp. 9 ff.) who provides many vivid and instructive examples for the use binary relations and to Fishburn (1970) for a more formal approach. We illustrate each property along the lines of family relationships if intuitively possible.

A binary relation R on a set S is

- *Reflexive*, if $s \mathsf{R} s$ for all $s \in S$.
- *Irreflexive*, if $s \not\mathsf{R} s$ for all $s \in S$. This corresponds to the relation "is the father of", as no person can be their own father.

- *Symmetric*, if $s_1 \mathsf{R} s_2$ implies $s_2 \mathsf{R} s_1$ for all $s_1, s_2 \in S$. The relation "is siblings with" is represented here.
- *Asymmetric*, if $s_1 \mathsf{R} s_2$ implies $s_2 \not\mathsf{R} s_1$ for all such $s_1, s_2 \in S$. Here again, the relation "is the father of" can be used.
- *Complete*, if either $s_1 \mathsf{R} s_2$ or $s_2 \mathsf{R} s_1$ for all $s_1, s_2 \in S$. Among the children of one mother, the relation "is older than" could be reflected.
- *Transitive*, if $s_1 \mathsf{R} s_2$ and $s_2 \mathsf{R} s_3$ imply $s_1 \mathsf{R} s_3$ for all such $s_1, s_2, s_3 \in S$. This again corresponds to the relation "is siblings with".

This list of properties is by no means exhaustive and, we restrict ourselves to those used in this work. Also, some of these are mutually exclusive or imply one another. But they are not necessary, i.e. one can always construct a binary relation without any of the properties listed above.

Every relation R on S defines a *dual relation*, say, R' on S through the following:

$$s_1 \mathsf{R}' s_2 \iff s_2 \mathsf{R} s_1 \text{ for all } s_1, s_2 \in S$$

An important class of relations are so-called *equivalence relations*, or simply *equivalences*: Any symmetric, reflexive, and transitive relation is an equivalence. For some element $s \in S$ we can define an *equivalence class* corresponding to a certain relation R:

$$[s] := \{\tilde{s} \in S \mid s \,\mathsf{R}\, \tilde{s}\}$$

This class includes all elements $\tilde{s} \in S$ that satisfy relation R for element s. The equivalence classes $[s]$ and $[t]$ of two elements s and t are either equal or disjoint. They are equal if and only if s and t are equivalent through R.

As basic examples of binary relations, we refer to the common order relations in numbers or sets: $<, \leq, \subsetneq, \subseteq$. They are all transitive, the second and fourth are also reflexive and antisymmetric, while the first and third are irreflexive and asymmetric. The respective dual relations are given by $>, \geq, \supsetneq, \supseteq$. We now turn to a special case of binary relations which form the basis to a large part of economic theory: Preference relations.

Preference Relations and Orders

A *preference relation* is an arbitrary binary relation exhibiting the properties of completeness and of transitivity. It is usually denoted "$\succeq$" and if defined over a set A, the expression $a_1 \succeq a_2$ for $a_1, a_2 \in A$ would be interpreted as "a_1 is (weakly) preferred to a_2" or "a_1 is at least as good as a_2". The preference relation $\succeq$ can be used to derive two further relations from it: The first is the *strict preference relation* "$\succ$" defined as

$$a_1 \succ a_2 \iff a_1 \succeq a_2 \wedge a_2 \not\succeq a_1,$$

while the second is the *indifference relation* "$\sim$", an equivalence relation given by

$$a_1 \sim a_2 \iff a_1 \succeq a_2 \wedge a_2 \succeq a_1.$$

For an approach to derive these three basic preference relations on the basis of the strict preference relation we refer the reader to Shubik (1982, pp. 417 ff.), Jehle and Reny (2001, pp. 5 f.), or Ritzberger (2003, pp. 25 ff.). If a relation R is reflexive, antisymmetric, and transitive, it is a *partial order* on the set A over which it is defined. If, in addition, it is complete, it is called a *complete order*, where all elements $a \in A$ can be compared via R.

Maximal Elements

We now take advantage of the previously defined concept of a partial order to create a notion of what can be seen as a "biggest" or "smallest" element out of some set. For this, consider a partial order R on the set X with $A \subseteq X$. The following definitions arise:

$x \in X$ is an *upper bound* for the set A if $x\mathsf{R}a$ holds for all elements $a \in A$.
Likewise, $x \in X$ is a *lower bound* for the set A $a\mathsf{R}x$ holds for all elements $a \in A$. Note that both, the upper and the lower bound can be set-valued, depending on the relationship between A and X.

$x \in X$ is a *maximum* (or largest element) in A, if it is an upper bound for A and element of A, i.e. $x \in A$ at the same time.
Analogously, $x \in X$ is a *minimum* (or smallest element) in A, if it is a lower bound for A and element of A, i.e. $x \in A$ at the same time.

$x \in X$ is the *supremum* (or least upper bound) of A, if it is the minimum in the set of upper bounds of A.
$x \in X$ is the *infimum* (or greatest lower bound) of A, if it is the maximum in the set of lower bounds of A.

For an illustration, let us consider two situations: In the first, let $X = [0,1]$ and $A = [\frac{1}{4}, \frac{3}{4}]$. The upper bound of A is the set $[\frac{3}{4}, 1]$ and its lower bound is given by $[0, \frac{1}{4}]$. The notions of maximum and supremum then coincide at $\frac{3}{4}$, as do the minimum and the infimum at $\frac{1}{4}$.

In the second case, X will remain as before, but now A is an open interval, i.e. $A = (\frac{1}{4}, \frac{3}{4})$. Upper and lower bounds remain unchanged, as do the supremum and the infimum, even though they now are not elements of A anymore. But maximum and minimum do not exist in such a situation, because upper and lower bounds are both disjoint from the set A.

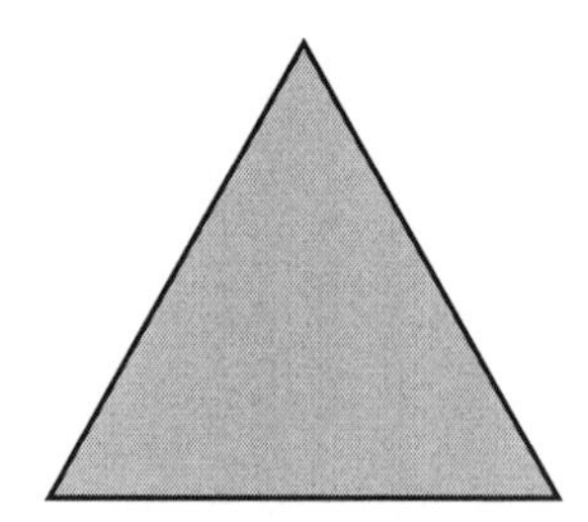
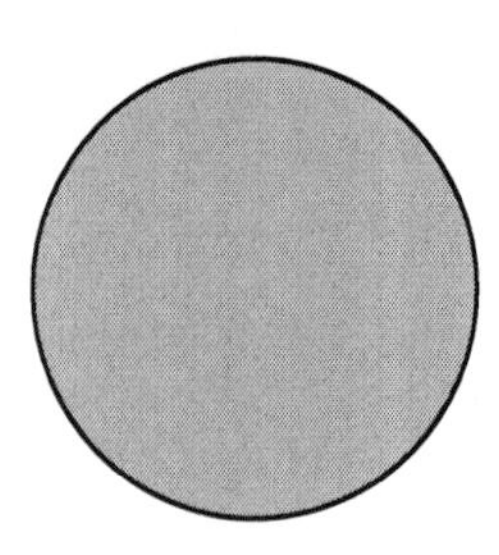

Fig. A.1 Examples for convex sets

Convexity

Convex Sets

As *convexity* arises in the context of sets and functions, we will treat both concepts in turn. We begin with the convexity of sets.

Denote by V a real vector space, say the n dimensional Euclidean space $\mathbb{R}^n$. Then, a *convex combination* of two vectors $\mathbf{x}, \mathbf{y} \in V$ is another vector given by

$$\lambda \cdot \mathbf{x} + (1 - \lambda) \cdot \mathbf{y},$$

where $\lambda \in [0,1]$. More general, a convex combination of m vectors $\mathbf{x}_1, \ldots, \mathbf{x}_m \in V$ yields a vector given by

$$\sum_{i=1}^{m} \lambda_i \cdot \mathbf{x}_i,$$

where $\lambda_i \in [0,1]$ for all $i = 1, \ldots, m$ and $\sum_{i=1}^{m} \lambda_i = 1$. Based on the first expression the following definition arises:

Definition A.108. A set $C \subseteq V$ is *convex*, if for all pairs of vectors $\mathbf{x}, \mathbf{y} \in C$ and all $\lambda \in [0,1]$,

$$\lambda \cdot \mathbf{x} + (1 - \lambda) \cdot \mathbf{y} \in C.$$

In other words, a set $C \subseteq V$ is convex whenever it contains all line segments that can be spanned by possible pairs of its elements. One can also say that a convex set contains all possible convex combinations of any finite number m of its elements.[3] As examples for convex sets, we can list circles, or more general spheres, rectangles, squares and cubes, as well as triangles. Some are depicted in Fig. A.1. A very important fact is given by the following proposition:

[3]This latter statement follows from *Carathéodory's Theorem*, if we pick the m elements to correspond to the extreme points (see Definition A.110 below) of C.

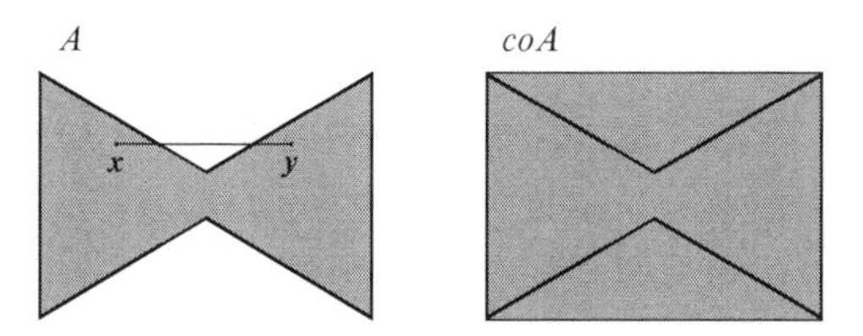

Fig. A.2 Non-convex set A and its convex hull coA

Proposition A.40. *The intersection of arbitrarily many convex subsets of a real vector space V is convex.*

Proof. Denote by C_i convex subsets of the real vector space V, indexed by $i \in N$. Given their intersection,

$$C := \bigcap_{i \in N} C_i,$$

note that $C \subseteq C_i$ for all $i \in N$. Now, for any two vectors $\mathbf{x}, \mathbf{y} \in C$ and $\lambda \in [0,1]$, it holds by assumed convexity of C_i that $\lambda \cdot \mathbf{x} + (1-\lambda) \cdot \mathbf{y} \in C_i$ for all $i \in N$. Hence, the convex combination of $\mathbf{x}$ and $\mathbf{y}$ is an element of C and so C is convex. ■

Another important concept is the *convex hull* of a set:

Definition A.109. Given a set $A \subseteq V$, where V is a real vector space, the *convex hull* of A is the intersection of all convex sets $C \subseteq V$ which contain A:

$$coA := \bigcap_{\substack{A \subseteq C \subseteq V \\ \text{with } C \text{ convex}}} C$$

Figure A.2 depicts a non-convex set A and its convex hull coA:

What follows from Definition A.109 is that if A is convex itself, then $coA = A$. Note also that a line segment between two points $\mathbf{x}, \mathbf{y} \in A$, often denoted $[\mathbf{x}, \mathbf{y}]$, is identical to the convex hull of $\mathbf{x}$ and $\mathbf{y}$, given by $co\{\mathbf{x}, \mathbf{y}\}$.

Definition A.110. Given a convex set $A \subseteq V$, a vector $\mathbf{x} \in A$ is an *extreme point*, if it cannot be expressed as a convex combination of elements of A which are all distinct from $\mathbf{x}$.

For a triangle, square, or cube, the corner points are such extreme points. For a sphere, it is the entire surface which cannot be represented by nontrivial convex combinations. Figure A.1 contains square, triangle, and circle, the latter being a "flattened" sphere whose perimeter consists exclusively of extreme points.

Convex Functions

Now, let us turn to what is referred to as convexity of a function. Whereas a set is either convex or non-convex, a function can be convex or concave and non-convex or non-concave. In addition, convexity or concavity can also be strict, a notion not applicable in set theory.

Fix a domain $D \subseteq \mathbb{R}^n$, where D is a convex set. Given a function $f : D \to \mathbb{R}$, define the following:

- The function f is *convex*, if for all $\mathbf{x}, \mathbf{y} \in D$ and all $\lambda \in [0,1]$,

$$f(\lambda \cdot \mathbf{x} + (1-\lambda) \cdot \mathbf{y}) \leq \lambda \cdot f(\mathbf{x}) + (1-\lambda) \cdot f(\mathbf{y}).$$

- The function f is *concave*, if for all $\mathbf{x}, \mathbf{y} \in D$ and all $\lambda \in [0,1]$,

$$f(\lambda \cdot \mathbf{x} + (1-\lambda) \cdot \mathbf{y}) \geq \lambda \cdot f(\mathbf{x}) + (1-\lambda) \cdot f(\mathbf{y}).$$

Strictly convex or *strictly* concave functions are the case if the inequality is strict for nontrivial convex combinations of $\mathbf{x}$ and $\mathbf{y}$, i.e. whenever $\mathbf{x} \neq \mathbf{y}$ and $\lambda \in (0,1)$.

It is also worthwhile to note that a function f is convex if and only if $-f$ is concave.

A notion related to that of convexity of a function is *supermodularity*. In principle, both notions coincide, but the latter requires less assumptions on the characteristics of underlying sets of domain and codomain.

Definition A.111. A function $f : A \to \mathbb{R}$ is *supermodular*, if

$$f(x \wedge y) + f(x \vee y) \geq f(x) + f(y), \quad \text{for all } x, y \in A.$$

Here, A is any set endowed with a partial order denoted by $\overset{P}{\geq}$, i.e. a binary relation that is reflexive, transitive, and antisymmetric. We speak of $(A, \overset{P}{\geq})$ as a partially ordered set or poset. Expressions $x \vee y$ and $x \wedge y$, respectively, denote the supremum and infimum of the set $\{x,y\}$ in A.[4] In terms of modularity, a concave function is called *submodular*.

[4] The reader is advised to skip this footnote until after having read Sect. 2.3.4, unless he is familiar with cooperative game theory. In the context of characteristic functions for coalitional games, these definitions translate as follows: The set A is the power set of the set of players, 2^N, and hence x, y are coalitions of players, $S \subseteq N$. Consequently, the partial order applied is that of set inclusion, $\subseteq$. Then, the characteristic function v of a coalitional game (N, v) is supermodular, if

$$v(S \cap T) + v(S \cup T) \geq v(S) + v(T) \quad \forall S, T \in 2^N.$$

Because $v : 2^N \to \mathbb{R}$ and $S, T \in 2^N$, the supremum, $S \vee T$, and infimum, $S \wedge T$, of the set $\{S, T\}$ in 2^N coincide with the union and the intersection of the sets S and T, which belong to 2^N as well. An upper bound for the set $\{S, T\}$ is any set $X \in 2^N$ for which it holds that $X \supseteq R$ for all $R \in \{S, T\}$. The smallest such set, i.e. the supremum of $\{S, T\}$ in 2^N, is the union of S and T, $S \cup T$.
Analogously, a lower bound for the set $\{S, T\}$ is any set $X \in 2^N$ for which it holds that $X \subseteq R$ for all $R \in \{S, T\}$. Here, the largest such set, i.e. the infimum of $\{S, T\}$ in 2^N, coincides with the intersection of S and T, $S \cap T$.

Some Topology

As we operate in the Euclidean space, we refrain from the most general definitions and confine us to topology in the vector space $\mathbb{R}^n$ endowed with the Euclidean metric, i.e.

$$\|(x_1,x_2,\ldots,x_n)\| = \sqrt{x_1^2+x_2^2+\cdots+x_n^2} \ .$$

Before we proceed, we introduce the concept of a *ball*, being essential for the distinction we want to make thereafter.

Definition A.112. A *ball* $B[\mathbf{x},\varepsilon]$ with center $\mathbf{x}$ and radius ε is the set of all elements within (Euclidean) distance ε of the point $\mathbf{x}$:

$$B[\mathbf{x},\varepsilon] := \left\{ \mathbf{y}\in\mathbb{R}^n \;\middle|\; \sqrt{\sum_{i=1}^{n}(x_i-y_i)^2} \le \varepsilon \right\}.$$

In case of an *open* ball, denoted by $B(\mathbf{x},\varepsilon)$, the inequality is strict.

A set $A \subseteq X$ is called *open*, if for every element $\mathbf{x}\in A$ there exists an $\varepsilon > 0$ for which the (open) Euclidean ball $B(\mathbf{x},\varepsilon)$ around $\mathbf{x}$ is strictly contained in A, i.e. $B(\mathbf{x},\varepsilon) \subsetneq A$. The set $A \subseteq X$ is called *closed*, if for every converging sequence $\mathbf{x}_k \to \mathbf{x} \in X$, where $\mathbf{x}_k \in A$ for all $k = 0,1,2,\ldots$, also its limit $\mathbf{x}\in A$.

Note that in Euclidean space $\mathbb{R}^n$ all singleton sets are closed. The empty set $\emptyset$, as well as $\mathbb{R}^n$ itself, are both open and closed.

The union of an arbitrary number of open sets is open, and so is the intersection of a finite number of open sets. The union of a finite number of closed sets is closed, and the intersection of an arbitrary number of closed sets is closed.

The *interior* of a set A is the largest open set which is contained in A. We denote it by $intA$. The *closure* of a set A is the smallest closed set containing A. It is denoted clA. The arising "difference", is called the *boundary* of the set A, and it is given by $bdA = clA \setminus intA$. These conjectures follow from the definitions:

- $intA \subseteq A \subseteq clA$.
- $intA = A$ if and only if A is open.
- $clA = A$ if and only if A is closed.

A set A is called *bounded*, if for all its elements $\mathbf{x}\in A$ there exists some $\varepsilon > 0$, such that A is contained in the ball $B(\mathbf{x},\varepsilon)$, i.e. $A \subseteq B(\mathbf{x},\varepsilon)$. If a set is closed and bounded, it is called *compact*.

Hyperplanes and Corresponding Half-Spaces

Using the elements introduced above, we now present the concepts of hyperplanes and half-spaces. For some vector $\mathbf{p}\in\mathbb{R}^n$ with $\mathbf{p}\neq\mathbf{0}$, i.e. not all entries equal to zero, and a scalar $\alpha \in \mathbb{R}$, we define the notion of a *hyperplane*. It is a set of points in $\mathbb{R}^n$ satisfying a linear equation:

$$H(\mathbf{p},\alpha) = \left\{\mathbf{x} \in \mathbb{R}^n | \mathbf{p}' \cdot \mathbf{x} = \alpha\right\}.$$

This hyperplane determines two (closed) *half-spaces*, which are defined as follows:

$$\begin{aligned} H_+(\mathbf{p},\alpha) &= \left\{\mathbf{x} \in \mathbb{R}^n | \mathbf{p}' \cdot \mathbf{x} \geq \alpha\right\}, \text{ and} \\ H_-(\mathbf{p},\alpha) &= \left\{\mathbf{x} \in \mathbb{R}^n | \mathbf{p}' \cdot \mathbf{x} \leq \alpha\right\}. \end{aligned}$$

Even though the hyperplane is situated in an n-dimensional space, we can express any element of it with only $n-1$ coordinates, as the "last" of the n summands on the right-hand side of the equation given by $\mathbf{p}' \cdot \mathbf{x} = \alpha$ necessarily needs to add up to α. It is straightforward that

$$H_+(\mathbf{p},\alpha) \cap H_-(\mathbf{p},\alpha) = H(\mathbf{p},\alpha),$$

i.e. the half-spaces intersect only through their corresponding hyperplane, which also happens to be the boundary for both. So-called open half-spaces are denoted as $intH_+(\mathbf{p},\alpha)$ and $intH_-(\mathbf{p},\alpha)$, and are defined by a strict inequality. Hyperplanes as well as half-spaces are convex sets.

A *polyhedron* in the real vector space $\mathbb{R}^n$ is a set of points that can be expressed as the intersection of a finite number of closed half-spaces. As each half-space is convex, so is the polyhedron. A special case of a polyhedron is the *simplex*, which is, among other things, confined to the positive orthant of $\mathbb{R}^n$. We denote the $(n-1)$-dimensional simplex by Δ^n, where n is the cardinality of the index set N, matching the dimension of $\mathbb{R}^n$. The definition of the simplex is given by

$$\Delta^n := H(\mathbf{1},\alpha) \cap \left(\cap_{i \in N} H_+(\mathbf{e}_i, 0)\right),$$

where $\mathbf{e}_i$ is the unit vector corresponding to dimension i. Also, remember from above that the hyperplane $H(\mathbf{1},\alpha)$ is already the intersection of its two corresponding half-spaces.

Two other common characterizations of the simplex exist, the first in plain of convexity, where the simplex is given by the convex hull of all n scaled unit vectors e_i in $\mathbb{R}^n$:

$$\Delta^n := co\left\{\alpha \cdot \mathbf{e}_i\right\}_{i \in N}.$$

The last characterization is probably the most basic:

$$\Delta^n := co\left\{\mathbf{x} \in \mathbb{R}^n \,\middle|\, \sum_{i \in N} x_i = \alpha \text{ and } x_i \geq 0 \text{ for all } i \in N\right\}.$$

In all three cases, if we set $\alpha = 1$, we obtain the *unit simplex*, which is usually referred to as simplex only. It should be noted that the simplex is the restriction to the nonnegative part of a given hyperplane, i.e. in addition to lying on this hyperplane, all elements must also feature nonnegative coordinates.

Notation

General:

$\wedge, \vee$	Logical operators "and", "or", as well as supremum and infimum
$\exists, \nexists$	Existence, nonexistence
$\vert, :$	"Such that" or "with the property"
$=$	Equality
$:=$	Right-hand side defines left-hand side
$\equiv$	Equivalence
$\in, \notin$	Membership (or lack thereof) of an element in a set
$\subseteq, \subsetneq, \supseteq, \supsetneq$	Weak and strict set inclusion
$\cup, \cap, \setminus$	Union, intersection, and difference of sets
$\times$	Cartesian or cross product
$\geq, >, \leq, <$	Weak and strict inequalities, element by element w.r.t. vectors
$N = \{1, 2, \ldots, n\}$	Finite set
$\vert N\vert$	Cardinality of set N
$\emptyset$	Empty set
2^N	Set of all subsets of N
P	Partition of a set
$\pi \in \Pi$	Permutation as element of the set of permutations
$\sum_i$	Summation operator over elements indexed i
$\prod_i$	Product operator over elements indexed i
Δ^n	n-dimensional (unit) simplex
$\mathbf{b} = (1, 2, \ldots, n)$	Finite-dimensional vector
$f : A \to B$	Mapping or function from set A to set B
$F : A \rightrightarrows B$	Set-valued function or correspondence from set A to set B
$\arg\max, \arg\min$	Maximizer, minimizer
co, bd, int, cl	Convex hull, boundary, interior, and closure of a set
$\succeq, \succ, \sim$	Weak and strict preference relations, indifference relation

P. Servatius, *Network Economics and the Allocation of Savings*, Lecture Notes in Economics and Mathematical Systems 653, DOI 10.1007/978-3-642-21096-9,

Game Theory:

N	Finite set of n players
$i, j, k, l \in N$	Exemplary players
$S, R, T \subseteq N$	Coalitions of players
$\Gamma = (N, \mathsf{S}, \mathbf{u})$	Strategic form game
$\mathsf{S} = \mathsf{S}_1 \times \mathsf{S}_2 \times \cdots \times \mathsf{S}_n$	n-dimensional strategy space
$s = s_1 \times s_2 \times \cdots s_n$	n-dimensional strategy tuple
$s_i \in \mathsf{S}_i$	Individual strategy
$s_{-i} \in \mathsf{S}_{-i}$	Strategy tuple of all players but i
$\mathbf{u} = u_1 \times u_2 \times \cdots \times u_n$	Vector of individual payoff functions
$\Gamma = (N, v)$	Cooperative or coalitional game
v, u ,w	Characteristic functions of a cooperative game
$\mathbf{v}$, $\mathbf{u}$,$\mathbf{w}$	Characteristic functions in vector notation
u_S, $\mathbf{u}_S$	Unanimity game with carrying coalition S
$\mathbf{V}_N$	Space of cooperative games with player set N
$\mathbf{x}$, $\mathbf{y}$, $\mathbf{z}$	n-dimensional allocations of payoffs to the players
x_i, y_i, z_i	Individual payoffs to player i under allocations $\mathbf{x}$, $\mathbf{y}$, and $\mathbf{z}$
$x(S)$	Sum over the payoffs of players in S according to allocation $\mathbf{x}$
$\mathbf{x}^{\pi}$	Marginal vector according to permutation π
$S_{\pi,i}$	Set of all players up to and including i under ordering π
$d_i(S)$	Difference operator w.r.t. player i and coalition S
χ_S	Characteristic vector of coalition S
$\mathfrak{S}$	Stable set solution in terms of von Neumann Morgenstern
$I(N, v)$	Set of imputations of the cooperative game (N, v)
$C(N, v)$	Set of core allocations of the cooperative game (N, v)
γ, γ_i	Allocation rule and corresponding individual payoff
Φ, ϕ_i	Shapley value/operator and corresponding individual payoff
$\omega = (\lambda, \Sigma)$	Weight system for weighted Shapley value Φ^{ω}
$B = (U, \mathbf{d})$	Bargaining problem with feasible set U and disagreement point $\mathbf{d}$
$\mathbf{a}(U, \mathbf{d})$	Vector of maximal individual payoffs under $(U, \mathbf{d})$
f^N, f^{KS}, f^{N_γ}, f^{KS_γ}	Bargaining solutions of Nash and Kalai–Smorodinsky and their weighted variants

Network Theory:

$\mathscr{G}_N$	Space of networks with set of nodes N
(N, g), g	Arbitrary network or graph with set of nodes N
g^c, $g^{\emptyset}$	Complete network, empty network
$g\|_S$	Network g restricted to nodes in S
g_{ij}	Link between nodes i and j
$N_i(g)$	Neighborhood or set of neighbors of node i in network g

$\eta_i(g)$	Cardinality of neighborhood or degree of player i in network g
N/g	Set of components induced by network g on set of nodes N
(N,v,g)	Communications situation with player set N and characteristic function v on network g
CS_N	Space of communications situations on player set N
v^g	Network restricted characteristic function
$d_{g_{ij}}(S)$	Difference operator w.r.t. link g_{ij} and coalition S
$\Gamma(N,\mathsf{S},f)$	Network formation game in strategic form
f	Payoff function on the basis of an allocation rule

Model-specific:

p_i	Call price of VoIP firm i
$q(p_i)$	Customer demand at price p_i
α_i	Market share of VoIP firm i
l_i	Long distance fee incurred by VoIP firm i
a	Access charges between VoIP firms
A_i^j	Access surplus/deficit of firm i w.r.t. firm j
A_i	Aggregate access revenue of firm i
$A_i(g)$	Aggregate access revenue of firm i in network g
$\mathbf{A}^g$	Vector of aggregate access revenues in network g
$v(S,g)$, v^g, $\mathbf{v}^g$	Characteristic function in the peering game
$c_i^g(S)$	Contributions of player i to coalition S in network g
$\mathbf{0}$	Set of zero-players

References

Alexander, F. and Kisrawi, N. *A Handbook on Internet Protocol (IP)-Based Networks and Related Topics and Issues*. International Telecommunication Union (ITU), Geneva, 2005.

Alós-Ferrer, C. Mathematics for Economists, 2005. Unpublished lecture notes for a Ph.D.-course given at the Institute for Advanced Studies, Vienna.

Armstrong, M. The Theory of Access Pricing and Interconnection. In Cave, M. E., Majumdar, S. K., and Vogelsang, I. (Editors), *Handbook of Telecommunications Economics*, pp. 297–384. Elsevier, 2002.

Aumann, R. J. Acceptable Points in General Cooperative n-Person Games. In Tucker, A. and Luce, R. (Editors), *Contributions to the Theory of Games IV*, pp. 287–324. Princeton University Press, Princeton, NJ, 1959.

Aumann, R. J. "game theory". In Durlauf, S. N. and Blume, L. E. (Editors), *The New Palgrave Dictionary of Economics*, pp. 529–559. Palgrave Macmillan, 2008.

Aumann, R. J. and Myerson, R. B. Endogeneous formation of links between players and of coalitions: an application of the Shapley value. In Roth, A. E. (Editor), *The Shapley Value – Essays in Honor of Lloyd S. Shapley*, pp. 175–191. Cambridge University Press, Cambridge, 1988.

Badach, A. and Hoffmann, E. *Technik der IP-Netze – Funktionsweise, Protokolle und Dienste*. Carl Hanser Verlag, München, 2007.

Bernheim, B. D., Peleg, B., and Whinston, M. D. Coalition-Proof Nash Equilibria: 1. Concepts. *Journal of Economic Theory*, 42, pp. 1–12, 1987.

Bilbao, J. M. *Cooperative Games on Combinatorial Structures*. Game Theory, Mathematical Programming and Operations Research. Kluwer Academic Publishers, Boston, MA, 2000.

Bilbao, J. M. and Edelmann, P. The Shapley Value on Convex Geometries. *Discrete Applied Mathematics*, 103, pp. 33–40, 2000.

Bloch, F. Group and Network Formation in Industrial Organisation: A Survey. In Demange, G. and Wooders, M. (Editors), *Group Formation in Economics. Networks, Clubs, and Coalitions*, pp. 335–353. Cambridge University Press, Cambridge, 2005.

Bollobás, B. *Graph Theory – An Introductory Course*. Graduate Texts in Mathematics. Springer Verlag, New York, NY, 1990.

Bondareva, O. N. Nekotorye primeneniia metodov linejnogo programmirovaniia k teorii kooperativnykh igr (Some Applications of Linear Programming Methods in/to the Theory of Cooperative Games). *Problemy Kybernetiki*, 10, pp. 119–139, 1963. Published in Russian.

Borm, P., Owen, G., and Tijs, S. On the Position Value for Communication situations. *SIAM Journal of Discrete Mathematics*, 5, pp. 305–320, 1992.

Brown, D. J. and Housman, D. Cooperative Games on Weighted Graphs, 1988. National Science Foundation Research Experiences for Undergraduates summer program held at Worcester Polytechnic Institute.

P. Servatius, *Network Economics and the Allocation of Savings*, Lecture Notes in Economics and Mathematical Systems 653, DOI 10.1007/978-3-642-21096-9,

Buigues, P.-A. and Rey, P. *The Economics of Antitrust and Regulation in Telecommunications – Perspectives for the New European Regulatory Framework*. Edward Elgar Publishing Inc., Northampton, MA, 2004.

Cawley, R. The European Union and World Telecommunications Markets. In Madden, G. (Editor), *World Telecommunications Markets*, The International Handbook of Telecommunications Economics, pp. 153–172. Edward Elgar Publishing, Northampton, MA, 2003.

Chang, C. Note: Remarks on Theory of the Core. *Naval Research Logistics*, 47, pp. 455–458, 2000.

Chun, Y. A New Axiomatization of the Shapley Value. *Games and Economic Behavior*, 1, pp. 119–130, 1989.

Chun, Y. On the Symmetric and Weighted Shapley Values. *International Journal of Game Theory*, 19, pp. 421–430, 1991.

Delley, A., Gaillet, P., Johnsen, O., Keller, H.-J., Rast, M., and Wenk, B. *Voice over IP und Multimedia*. Hochschule für Technik und Architektur, Fribourg, 2005.

Demange, G. and Wooders, M. (Editors). *Group Formation in Economics – Networks, Clubs, and Coalitions*. Cambridge University Press, Cambridge, 2005.

Dodd, A. Z. *The Essential Guide to Telecommunications*. Prentice Hall, 1999.

Dutta, B., Goshal, S., and Ray, D. Farsighted Network Formation. *Journal of Economic Theory*, 122, pp. 142–164, 2005.

Dutta, B. and Jackson, M. *Models of the Strategic Formation of Networks and Groups*. Springer Verlag, Berlin, 2003.

Dutta, B., van den Nouweland, A., and Tijs, S. Link Formation in Cooperative Situations. *International Journal of Game Theory*, 27, pp. 245–256, 1998.

Economist, The; Data, data everywhere. Special report, The Economist, 2010.

Edgeworth, F. Y. *Mathematical Psychics – An Essay on the Application of Mathematics to the Moral Sciences*. C. Kegan Paul & Co, London, 1881.

Elixmann, D., Marcus, S., and Wernick, C. The Regulation of Voice over IP (VoIP) in Europe. Technical report, wik Consult, 2008.

Epstein, R. J. and Rubinfeld, D. L. Merger Simulation: A Simplified Approach with new Applications. *Antitrust Law Journal*, 69, pp. 883–919, 2001.

European Regulators Group. Final Report on IP Interconnection. Technical Report (07)09, European Regulators Group, 2009.

Fishburn, P. C. *Utility Theory for Decision Making*, volume 18 of *Publications in Operations Research*. John Wiley & Sons, 1970.

Forgó, F., Szép, J., and Szidarovszky, F. *Introduction to the Theory of Games: Concepts, Methods, Applications*. Kluwer Academic Publishers, Dodrecht, 1999.

Fransman, M. Evolution of the Telecommunications Industry. In Madden, G. (Editor), *World Telecommunications Markets*, The International Handbook of Telecommunications Economics, pp. 15–38. Edward Elgar Publishing, Northampton, MA, 2003.

Frieden, R. M. Regulation for Internet-mediated Communication and Commerce. In Madden, G. (Editor), *Emerging Telecommunications Networks*, pp. 107–128. Edward Elgar Publishing, Northampton, MA, 2003.

Friedman, J. W. *Game Theory with Applications to Economics*. Oxford University Press, New York, NY, 1991.

Fudenberg, D. and Maskin, E. The Folk Theorem in Repeated Games with Discounting or with Incomplete Information. *Econometrica*, 54, pp. 533–554, 1986.

Gillies, D. B. *Some Theorems on n-Person Games*. Ph.D. thesis, Department of Mathematics, Princeton University, 1953.

Gillies, D. B. Solutions to General Non-Zero-Sum Games. In Tucker, A. and Luce, R. (Editors), *Contributions to the Theory of Games IV*, pp. 47–85. Princeton University Press, Princeton, NJ, 1959.

Goyal, S. *Connections*. Princeton University Press, Princeton, NJ, 2007.

Goyal, S. and Joshi, S. Bilatralism and Free Trade. *International Economic Review*, 47, pp. 749–778, 2006.

Graber, A. *Internet Pricing – Economic Approaches to Transport Services and Infrastructure*. Ph.D. thesis, University of Fribourg (Switzerland), 2004.
Graham, R. L. and Hell, P. On the History of the Minimum Spanning Tree Problem. *Annals of the History of Computing*, 7, pp. 43–57, 1985.
Greenstein, S. M. The Economic Geography of Internet Infrastructure in the United States. In Cave, M. E., Vogelsang, I., and Majumdar, S. K. (Editors), *Handbook of Telecommunications Economics*, pp. 287–372. Elsevier, 2005.
Gupta, A., Stahl, D. O., and Whinston, A. B. Pricing Traffic on Interconnected Networks: Issues, Approaches, and Solutions. In Cave, M. E., Vogelsang, I., and Majumdar, S. K. (Editors), *Handbook of Telecommunications Economics*, pp. 413–439. Elsevier, 2005.
Hall, R. L. and Hitch, C. J. Price Theory and Business Behavior. *Oxford Economic Papers*, 2, pp. 12–45, 1939.
Harsanyi, J. and Selten, R. A Generalized Nash Solution for Two-Person Bargaining Games with Incomplete Information. *Management Science*, 18, pp. 80–106, 1972.
Harsanyi, J. C. Approaches to the Bargaining Problem befor and after the Theory of Games: A Critical Discussion of Zeuthen's, Hicks', and Nash's Theories. *Econometrica*, 24, pp. 144–157, 1956.
Harsanyi, J. C. A Bargaining Model for the Cooperative *n*-Person Game. In Tucker, A. and Luce, R. (Editors), *Contributions to the Theory of Games IV*, pp. 325–356. Princeton University Press, Princeton, NJ, 1959.
Hart, S. and Kurz, M. Endogeneous Formation of Coalitions. *Econometrica*, 51, pp. 1047–1064, 1983.
Hart, S. and Mas-Colell, A. Potential, Value, and Consistency. *Econometrica*, 57, pp. 589–614, 1989.
Hildenbrand, W. and Kirman, A. *Equilibrium Analysis: Variations on themes by Edgeworth and Walras*. Advanced Textbooks in Economics. Elsevier Science Publishers, 1988.
Hyman, L. S., Toole, R. C., and Avellis, R. M. *The New Telecommunicaitons Industry: Evolution and Organization*. Public Utilities Reports, Arlington, VA, 1987.
Ichiishi, T. Super-Modularity: Applications to Convex Games and to the Greedy Algorithm for LP1. *Journal of Economic Theory*, 25, pp. 283–286, 1981.
Ichiishi, T. *Game Theory for Economic Analysis*. Economic Theory, Econometrics, and Mathematical Economics. Academic Press, New York, NY, 1983.
ICT Regulation Toolkit. *ICT Regulation Toolkit*. International Telecommunication Union, Information for Development Program, www.ictregulationtoolkit.org, 2008.
Jackson, M. O. Allocation Rules for Network Games. *Games and Economic Behavior*, 51, pp. 128–154, 2005a.
Jackson, M. O. A Survey of Network Formation Models: Stability and Efficiency. In Demange, G. and Wooders, M. (Editors), *Group Formation in Economics. Networks, Clubs, and Coalitions*, pp. 11–57. Cambridge University Press, Cambridge, 2005b.
Jackson, M. O. and van den Nouweland, A. Strongly Stable Networks. *Games and Economic Behavior*, 51, pp. 420–444, 2005.
Jackson, M. O. and Watts, A. The Evolution of Social and Economic Networks. *Journal of Economic Theory*, 106, pp. 265–295, 2002.
Jackson, M. O. and Wolinsky, A. A Strategic Model of Social and Economic Networks. *Journal of Economic Theory*, 71, pp. 44–74, 1996.
Jehle, G. A. and Reny, P. J. *Advanced Microeconomic Theory*. Addison Wesley, Boston, MA, 2001.
Jianhua, W. *Oxford Mathematical Monographs: The Theory of Games*. Oxford Mathematical Monographs. Oxford University Press, Oxford, 1988.
Kalai, E. and Samet, D. Monotonic Solutions to General Cooperative Games. *Econometrica*, 53, pp. 307–327, 1985.
Kalai, E. and Samet, D. On Weighted Shapley Values. *International Journal of Game Theory*, 16, pp. 205–222, 1987.
Kalai, E. and Smorodinsky, M. Other Solutions to Nashs Bargaining Problem. *Econometrica*, 43, pp. 513–518, 1975.

Kannai, Y. The Core and Balancedness. In Aumann, R. J. and Hart, S. (Editors), *Handbook of Game Theory with Economic Applications – Volume I*, pp. 355–395. Elsevier North Holland, 1992.

Kelly, F. The Clifford Paterson Lecture: Modelling Communications Networks, Present and Future. *Philosophical Transactions: Mathematical, Physical and Engineering Sciences*, 354, pp. 437–463, 1996.

Kelly, T., Sharifi, H. S., and Petrazzini, B. Challenges to the Network – Telecoms and the Internet. Technical report, International Telecommunication Union (ITU), Geneva, 1997.

Kreps, D. M. *A Course in Microeconomic Theory*. Harvester Wheatsheaf, New York, 1990.

Kuhn, H. Extensive Games and the Problem of Information. In Tucker, A. and Kuhn, H. (Editors), *Contributions to the Theory of Games II*, pp. 193–216. Princeton University Press, Princeton, NJ, 1953.

Laffont, J.-J., Rey, P., and Tirole, J. Network Competition: I. Overview and nondiscriminatory pricing. *RAND Journal of Economics*, 29, pp. 1–37, 1998a.

Laffont, J.-J., Rey, P., and Tirole, J. Network Competition: II. Price discrimination. *RAND Journal of Economics*, 29, pp. 38–56, 1998b.

Laffont, J.-J. and Tirole, J. *Competition in Telecommunications*. Munich Lectures in Economics. MIT Press, Cambridge, MA, 2000.

Larouche, P. *Competition Law and Regulation in European Telecommunications*. Hart Publishing, Oxford, 2000.

Linnhoff-Popien, C. and Küpper, A. Telecommunications Systems, 2002. Unpublished lecture notes for a graduate course given at the Ludwig-Maximilians University, Munich.

Lucas, W. F. A Game with no Solution. *Bulletin of the American Mathematical Society*, 74, pp. 237–239, 1968.

Lucas, W. F. The Proof that a Game may not have a Solution. *Transactions of the American Mathematical Society*, 137, pp. 219–229, 1969.

Lucas, W. F. Some Recent Developments in n-Person Game Theory. *SIAM Review*, 13, pp. 491–523, 1971.

Lucas, W. F. Von Neumann-Morgenstern Stable Sets. In Aumann, R. J. and Hart, S. (Editors), *Handbook of Game Theory with Economic Applications – Volume I*, pp. 543–590. Elsevier North Holland, 1992.

Luce, R. D. and Raiffa, H. *Games and Decisions*. John Wiley & Sons, Inc., 1957.

Marcus, S., Elixmann, D., and Carter, K. The Future of IP Interconnection: Technical, Economic, and Public Policy Aspects. Technical report, wik Consult, 2008.

Mas-Colell, A., Whinston, M. D., and Green, J. R. *Microeconomic Theory*. Oxford University Press, Oxford, 1995.

Maschler, M. The Worth of a Cooperative Enterprise to Each Member. In Deistler, M., Fürst, E., and Schwödiauer, G. (Editors), *Games, Economic Dynamics, and Time Series analysis: A Symposium in Memoriam Oskar Morgenstern*, IHS Studies, pp. 67–73. Physica Verlag, Wien, 1982.

Maskin, E. and Tirole, J. A Theory of Dynamic Oligopoly, II: Price Competition, Kinked Demand Curves, and Edgeworth Cycles. *Econometrica*, 56, pp. 571–599, 1988.

McKinsey, J. Isomorphism of Games, and Strategic Equivalence. In Tucker, A. and Kuhn, H. (Editors), *Contributions to the Theory of Games I*, pp. 117–130. Princeton University Press, Princeton, NJ, 1950.

Monderer, D. and Samet, D. Variations on the Shapley Value. In Aumann, R. J. and Hart, S. (Editors), *Handbook of Game Theory with Economic Applications – Volume III*, pp. 2055–2076. Elsevier North Holland, 2002.

Monderer, D., Samet, D., and Shapley, L. S. Weighted Values and the Core. *International Journal of Game Theory*, 21, pp. 27–39, 1992.

Moulin, H. *Axioms of Cooperative Decision Making*. Cambridge University Press, Cambridge, 1991.

Mutuswami, S. and Winter, E. Subscription Mechanisms for Network Formation. *Journal of Economic Theory*, 106, pp. 242–264, 2002.

Myerson, R. B. Graphs and Cooperation in Games. *Mathematics of Operations Research*, 2, pp. 225–229, 1977.
Myerson, R. B. *Game Theory: Analysis of Conflict*. Harvard University Press, Cambridge, MA, 1991.
Nash, J. F. The Bargaining Problem. *Econometrica*, 28, pp. 155–162, 1950.
Nash, J. F. Non-Cooperative Games. *Annals Of Mathematics*, 54, pp. 286–295, 1951.
Noam, E. M. Interconnection Practices. In Cave, M. E., Majumdar, S. K., and Vogelsang, I. (Editors), *Handbook of Telecommunications Economics*, pp. 385–421. Elsevier, 2002.
OFCOM. The Swiss Telecommunictions Market – An International Comparison. Annual Report, Federal Office of Communications – Telecom Services, 2009.
Ok, E. A. *Real Analysis with Economic Applications*. Princeton University Press, Princeton, NJ, 2007.
Owen, G. A Note on the Shapley Value. *Management Science*, 14, pp. 731–732, 1968.
Owen, G. *Game Theory*. Academic Press Limited, London, 1995.
Owen, G. *Discrete Mathematics and Game Theory*. Game Theory, Mathematical Programming and Operations Research. Kluwer Academic Publishers, Dordrecht, 1999.
Paramesh, R. Independence of Irrelevant Alternatives. *Econometrica*, 41, pp. 987–991, 1973.
Pelcovits, M. D. and Cerf, V. G. Economics of the Internet. In Madden, G. (Editor), *Emerging Telecommunications Networks*, The International Handbook of Telecommunications, pp. 27–54. Edward Elgar Publishing, Northampton, MA, 2003.
Peleg, B. and Sudhölter, P. *Introduction to the Theory of Cooperative Games*. Game Theory, Mathematical Programming and Operations Research. Kluwer Academic Publishers, Dordrecht, 2003.
Phelps, R. R. *Lectures on Choquet's Theorem*. Springer Verlag, Berlin, 2001.
Picot, A. (Editor). *The Future of Telecommunications Industries*. Springer Verlag, Berlin, 2006.
Qin, C.-Z. Endogenous Formation of Cooperation Structures. *Journal of Economic Theory*, 69, pp. 218–226, 1996.
Rafels, C. and Tijs, S. On the Cores of Cooperative Games and the Stability of the Weber Set. *International Journal of Game Theory*, 26, pp. 491–499, 1997.
Rafels, C. and Ybern, N. Even and Odd Marginal Worth Vectors, Owen's Multilinear Extension, and Convex Games. *International Journal of Game Theory*, 24, pp. 113–126, 1995.
Ransmeier, J. S. *The Tennessee Valley Authority: A Case Study in the Economics of Multiple Purpose Stream Planning*. The Vanderbilt University Press, Nashville, TN, 1942.
Ritzberger, K. *Foundations of Non-Cooperative Game Theory*. Oxford University Press, Oxford, 2003.
Rota, G. C. The Number of Partitions of a Set. *The American Mathematical Monthly*, 71, pp. 498–504, 1964.
Rothblum, U. G. Combinatorial Representations of the Shapley Value Based on Average Relative Payoffs. In Roth, A. E. (Editor), *The Shapley Value – Essays in Honor of Lloyd S. Shapley*, pp. 121–126. Cambridge University Press, Cambridge, 1988.
Royden, H. L. *Real Analysis*. Prentice Hall, 1988.
Rubinstein, A. *Economics and Language*. Cambridge University Press, Cambridge, 2000.
Sarrocco, C. and Ypsilanti, D. Convergence and Next Generation Networks. Ministerial Background Report DSTI/ICCP/CISP(2007)FINAL, Organisation for Economic Co-Operation and Development (OECD), Directorate for Science, Technology, and Industry, Committee for Information, Computer, and Communications Policy, 2007.
Schmidt, J. H. Liberalisation of Network Industries: Economic Implications and Main Policy Issues. Technical report, European Commission, Directorate-General for Economic and Financial Affairs, 1999.
Schorr, S., Sundberg, N., and Phillips, R. Trends in Telecommunication Reform 2000-2001 : Interconnection and Regulation. Technical report, International Telecommunication Union (ITU), Geneva, 2001.
Selten, R. Spieltheoretische Behandlung eines Oligopolmodells mit Nachfrageträgheit. *Zeitschrift für die gesamte Staatswissenschaft*, 121, pp. 301–324,667–689, 1965.

Selten, R. A Simple Model of Imperfect Competition, where 4 are Few and 6 are Many. *International Journal of Game Theory*, pp. 141–201, 1973.

Shapley, L. S. *Additive and Non-Additive Set Functions*. Ph.D. thesis, Department of Mathematics, Princeton University, 1953a.

Shapley, L. S. A Value for *n*-Person Games. In Tucker, A. and Kuhn, H. (Editors), *Contributions to the Theory of Games II*, pp. 307–317. Princeton University Press, Princeton, NJ, 1953b.

Shapley, L. S. On Balanced Sets and Cores. *Naval Research Logistics Quarterly*, 14, pp. 453–460, 1967.

Shapley, L. S. Cores of Convex Games. *International Journal of Game Theory*, 1, pp. 11–26, 1971.

Shapley, L. S. Let's block "block". *Econometrica*, 41, pp. 1201–1202, 1973.

Shapley, L. S. Discussant's Comments: "Equity Considerations in Traditional Full Cost Allocation Practices: An Axiomatic Perspective". In Moriarity, S. (Editor), *Joint Cost Allocations – Proceedings of the University of Oklahoma Conference on Cost Allocations*. Center for Economic and Management Research, Norman, OK, 1981.

Shapley, L. S. and Shubik, M. Quasi-Cores in a Monetary Economy with Nonconvex Preferences. *Econometrica*, 34, pp. 805–827, 1966.

Shapley, L. S. and Shubik, M. On Market Games. *Joural of Economic Theory*, 1, pp. 9–25, 1969.

Shapley, L. S. and Shubik, M. Game Theory in Economics – Characteristic Function, Core, and Stable Set. Technical report, The RAND Corporation, Santa Monica, CA, 1973.

Shubik, M. *Game Theory in the Social Sciences – Concepts and Solutions*, volume 1. MIT Press, Cambridge, MA, 1982.

Singh, R. and Raja, S. Convergence in ICT services: Emerging regulatory responses to multiple play. Technical report, World Bank, Global Information and Communication Technologies Departement, 2008.

Slikker, M. and Norde, H. Incomplete Stable Structures in Symmetric Convex Games. *Working Paper 2000-97, Center for Economic Research, Tilburg University*, 2000.

Slikker, M. and van den Nouweland, A. A one-stage Model of Link Formation and Payoff Division. *Games and Economic Bahavior*, 34, pp. 153–175, 2001a.

Slikker, M. and van den Nouweland, A. *Social and Economic Networks in Cooperative Game Theory*. Kluwer Academic Publishers, 2001b.

Straffin, P. and Heaney, J. Game Theory and the Tennessee Valley Authority. *International Journal of Game Theory*, 10, pp. 35–43, 1981.

Sweezy, P. Demand under Conditions of Oligopoly. *Journal of Political Economy*, 47, pp. 568–573, 1939.

Thompson, F. Equivalence of Games in Extensive Form. In Kuhn, H. W. (Editor), *Classics in Game Theory*, pp. 36–45. Princeton University Press, Princeton, NJ, 1997.

Thomson, W. Cooperative Models of Bargaining. In Aumann, R. J. and Hart, S. (Editors), *Handbook of Game Theory with Economic Applications – Volume II*, pp. 1237–1284. Elsevier North Holland, 1994.

Tirole, J. *The Theory of Industrial Organization*. MIT Press, Cambridge, MA, 1997.

Tirole, J. Telecommunications and Competition. In Buigues, P.-A. and Rey, P. (Editors), *The Economics of Antitrust and Regulation in Telecommunications – Perspectives for the New European Regulatory Framework*, pp. 260–265. Edward Elgar Publishing Inc., Northampton, MA, 2004.

van Damme, E. Strategic Equilibrium. In Aumann, R. J. and Hart, S. (Editors), *Handbook of Game Theory with Economic Applications – Volume III*, pp. 1521–1596. Elsevier North Holland, 2002.

van den Nouweland, A. *Games and Graphs in Economic Situations*. Ph.D. thesis, Tilburg University, 1993.

van den Nouweland, A. Models of Network Formation in Cooperative Games. In Demange, G. and Wooders, M. (Editors), *Group Formation in Economics. Networks, Clubs, and Coalitions*, pp. 58–88. Cambridge University Press, Cambridge, 2005.

van den Nouweland, A., Borm, P., van Golstein Brouwers, W., Bruinderink, R. G., and Tijs, S. A Game Theoretic Approach to Problems in Telecommunication. *Management Science*, 42, pp. 294–303, 1996.

van Velzen, B., Hamers, H., and Norde, H. Convexity and Marginal Vectors. *International Journal of Game Theory*, 31, pp. 323–330, 2002.

Varian, H. R. *Intermediate Microeconomics – A Modern Approach*. W. W. Norton & Company, New York, NY, 1999.

Vogelsang, I. Price Regulation of Access to Telecommunications Networks. *Journal of Economic Literature*, 41, pp. 830–862, 2003.

von Neumann, J. Zur Theorie der Gesellschaftsspiele. *Mathematische Annalen*, 100, pp. 295–320, 1928.

von Neumann, J. and Morgenstern, O. *Theory of Games and Economic Behavior*. Princeton University Press, Princeton, NJ, 1947, 2nd ed.

Weber, R. J. Probabilistic Values for Games. In Roth, A. E. (Editor), *The Shapley Value – Essays in Honor of Lloyd S. Shapley*, pp. 101–120. Cambridge University Press, Cambridge, 1988.

Weber, R. J. Games in Coalitional Form. In Aumann, R. J. and Hart, S. (Editors), *Handbook of Game Theory with Economic Applications – Volume II*, pp. 1285–1303. Elsevier North Holland, 1994.

Wheatley, J. Price Elasticities for Telecommunications Services with Reference to Developing Countries, 2006. London School of Economics, Department of Media and Communications.

Wilson, R. J. *Introduction to Graph Theory*. Academic Press, New York, NY, 1972.

Winter, E. The Shapley Value. In Aumann, R. J. and Hart, S. (Editors), *Handbook of Game Theory with Economic Applications – Volume III*, pp. 2026–2054. Elsevier North Holland, 2002.

Young, H. Individual Contribution and just Compensation. In Roth, A. E. (Editor), *The Shapley Value – Essays in Honor of Lloyd S. Shapley*, pp. 267–278. Cambridge University Press, Cambridge, 1988.

Young, H. Cost Allocation. In Aumann, R. J. and Hart, S. (Editors), *Handbook of Game Theory with Economic Applications – Volume II*, pp. 1193–1235. Elsevier North Holland, 1994.

Young, H. P. Monotonic Solutions of Cooperative Games. *International Journal of Game Theory*, 14, pp. 65–72, 1985.

Ypsilanti, D. The OECD and the Internationalization of Telecommunications. In Madden, G. (Editor), *World Telecommunications Markets*, The International Handbook of Telecommunications Economics, pp. 226–240. Edward Elgar Publishing, Northampton, MA, 2003.

Ypsilanti, D. and Sarrocco, C. OECD Communications Outlook 2009. Technical report, Organisation for Economic Co-Operation and Development (OECD), Directorate for Science, Technology, and Industry, Committee for Information, Computer, and Communications Policy, 2009.